Ray methods for nonlinear waves in fluids and plasmas

π Pitman Monographs and
Surveys in Pure and Applied Mathematics 57

Ray methods for nonlinear waves in fluids and plasmas

A M Anile
Università di Catania

J K Hunter
University of California

P Pantano
Università della Calabria

and

G Russo
Università dell'Aquila

CRC Press
Taylor & Francis Group
Boca Raton London New York

CRC Press is an imprint of the
Taylor & Francis Group, an **informa** business

A CHAPMAN & HALL BOOK

First published 1993 by Longman Scientific & Technical

Published 2019 by Chapman & Hall
Taylor & Francis Group
6000 Broken Sound Parkway NW, Suite 300
Boca Raton, FL 33487-2742

© 1993 by Taylor & Francis Group, LLC
CRC Press is an imprint of Taylor & Francis Group, an Informa business

First issued in paperback 2019

No claim to original U.S. Government works

ISBN-13: 978-0-367-44994-0

Visit the Taylor & Francis Web site at
http://www.taylorandfrancis.com

and the CRC Press Web site at
http://www.crcpress.com

British Library Cataloguing in Publication Data

A catalogue record for this book is
available from the British Library

Library of Congress Cataloging-in-Publication Data

Ray methods for nonlinear waves in fluids and plasmas / A.M. Anile ...
[et al.].
    p. cm. -- (Pitman monographs and surveys in pure and applied
mathematics, ISSN 0269-3666 ; 57)
    Includes bibliographical references.
    1. Wave motion, Theory of. 2. Nonlinear waves. I. Anile, Angelo
Marcello. II. Series.
QA927.R38 1991
532'.0593--dc20   91-3866
                                                      CIP

CONTENTS

PREFACE

The scope of this book is to present in a systematic and unified manner the ray method (in its various forms) for studying nonlinear wave propagation in situations of physical interest (essentially fluid dynamics and plasma physics). The book could be used for an advanced graduate course on nonlinear waves. It should also be of interest to applied mathematicians, physicists and engineers, working in areas related to nonlinear waves.

We acknowledge helpful suggestions from many collegues, in particular Professor A. Jeffrey and Professor Y. Coquet-Bruhat. One of the authors (John Hunter) would like to thank the National Science Foundation for partial support during the period this book was written. Finally we are very grateful to the typists of EditEl that with their patient and constant effort have contributed to the editing of the book.

INTRODUCTION

Wave propagation is a salient feature of many phenomena occurring in continuum physics. In most cases (short wavelength limit) an adeguate description can be obtained in terms of rays and propagation laws (for amplitude and polarization) along rays. The paradigm of this approach is *Geometrical Optics* and such a method has been extended to other linear wave theories (linearized acoustics elasticity, etc). The essence of the method lies in the derivation of ray equations and propagation laws from the field equations utilizing appropriate asymptotic expansions (in terms of the wavelength).

This method is in a mature form and systematic accounts can be found in several books (see for example Whitham, 1974 or Kline and Kay, 1965). Over the past twenty years a considerable effort has been put into extending these methods in order to deal with nonlinear waves. Many papers have appeared extending the asymptotic method first in the one-dimensional case and then in several dimensions. This book aims at presenting in a systematic way thorough treatment of the ray methods employed in order to analyze nonlinear wave propagation in several dimensions.

Ideally this book is a complement to Jeffrey and Kawahara's book *Asymptotic methods in nonlinear wave theory* where part of the ideas here expounded are already introduced (although applied mainly to 1-D propagation). The fundamental problems in nonlinear wave propagation which will be tackled by the ray method in this book are:

i)   shock wave propagation,

ii)   derivation of model equations in several dimensions for dissipative and dispersive systems,

iii)   interaction of waves, both in the hyperbolic and dispersive case.

Throughout this book particular care is devoted to the physical justification and interpretation of the results in various areas of appication in fluid dynamics and plasma physics.

The plan of the book is the following. In Chapter 1 we present the basic physical frameworks in which we shall consider nonlinear waves. In particular we introduce the fundamental equations of fluid dynamics with dissipation, magnetofluid dynamics and (two-component) plasma physics.

In Chapter 2 we introduce the basic concepts of nonlinear wave propagation starting from the simple *kinematic wave equation*. We also review the basic properties of asymptotic expansions which will be used in the sequel, and the key ideas underlying the *perturbation-reduction* methods. In Chapter 3 we recall the basic ray methods for linear wave propagation. In Chapter 4 we treat the ray methods for hyperbolic waves and discuss some applications. In Chapter 5 we study the propagation of weak discontinuities and characteristic shocks. In Chapter 6 we present an appropriate extension of the ray method, called *Generalized Wavefront Expansion*, in order to treat the propagation of weak shocks. Chapter 7 is essentially devoted to shock stability analysis using ray methods. In Chapters 8 and 9 we present the systematic development of the *perturbation-reduction* method in several dimensions for nonlinear dispersive and dissipative systems. Several applications are given to fluids and plasmas. Various model equations are obtained in several dimensions (Burgers, nonlinear Schrödinger, Kadometsev-Petviashvilii). Physical interpretation of the results are also discussed. In Chapter 10 the theory of dispersive wave interaction is treated in detail. Applications are discussed in fluid dynamics and the passage through resonance is also investigated. Chapter 11 is entirely devoted to the interaction of hyperbolic waves. Several applications to gas dynamics and elasticity are discussed in depth.

REFERENCES

- M. Kline and I.W. Kay, *Electromagnetic Theory and Geometrical Optics*, Pure and Applied Mathematics, **XII**, Interscience (1965).
- G.B. Witham, *Linear and nonlinear waves*, Wiley, New York (1974).

PHYSICAL FRAMEWORK

Introduction

Ray methods constitute a fundamental tool for studying wave propagation. The extension of these method in order to treat nonlinear waves is the subject of this book.

In this chapter we set up the basic physical frameworks in which nonlinear wave propagation will take place. The methods expounded in this book will be applied to nonlinear wave propagation in various physical contexts.

The main areas of applications will be fluid dynamics (both ideal and dissipative), magnetofluid dynamics and plasma physics. In this chapter we will recall the basic fundamentals of fluid dynamics (including dissipation), ideal magnetofluid dynamics, and plasma physics (limited to the two component fluid model).

1 Introductory concepts

In order to introduce the main physical areas of applications which will be considered later, we shall first recall some basic concepts of continuum mechanics.

The balance equations for a simple continuum are (Gurtin, 1981):

$$(1.1) \qquad \frac{\partial \rho}{\partial t} + \frac{\partial}{\partial x^i}\left(\rho u^i\right) = 0\,,$$

$$(1.2) \qquad \frac{\partial}{\partial t}\left(\rho u^i\right) + \frac{\partial}{\partial x^j}\left(\rho u^i u^j - t^{ij}\right) = f^i\,,$$

$$(1.3) \qquad \frac{\partial}{\partial t}\left(\frac{1}{2}\rho \mathbf{u}^2 + \rho E\right) + \frac{\partial}{\partial x^j}\left\{\left(\frac{1}{2}\rho \mathbf{u}^2 + \rho E\right)u^j - t^{ij}u_i + q^j\right\} = f^i u_i\,.$$

Here ρ is the mass density, u^i the velocity, E the specific internal energy, t^{ij} the stress tensor, q_j the heat flux, f^i the external force, $f^i u_i$ is the work per unit volume per unit time produced by the external forces. These equations are written in an inertial frame with t the time and $x^i (i = 1, 2, 3)$ cartesian coordinates. Repeatead indices are summed.

Equation (1.1) represents conservation of mass, (1.2) is the balance of linear momentum and Eq. (1.3) the energy balance. Conservation of angular momentum implies that the stress tensor is symmetric. By introducing the convective derivative

$$\dot{f} \equiv \frac{\partial f}{\partial t} + u^i \frac{\partial f}{\partial x^i}$$

(for any function f), Eqs. (1.1–2–3) can be rewritten as

(1.4)
$$\dot{\rho} + \rho \frac{\partial u^i}{\partial x^i} = 0,$$

(1.5)
$$\rho \dot{u}^i - \frac{\partial t^{ij}}{\partial x^j} = f^i,$$

(1.6)
$$\rho \dot{E} + \frac{\partial q^i}{\partial x^i} - t^{ij} \frac{\partial u^i}{\partial x^j} = 0.$$

Now we assume that the medium is in a state of local thermodynamic equilibrium. The thermodynamical properties of the fluid are described by the specific entropy density s. The dependence of s on ρ and E is expressed by the "fundamental relationship" $s = s(\rho, E)$. s is differentiable and monotonically increasing with E (Landau and Lifshitz, 1960 a).

Then the temperature T and the thermodynamic pressure P at equilibrium are defined by

(1.7)
$$T^{-1} = \frac{\partial s}{\partial E},$$

(1.8)
$$P = -\rho^2 T \frac{\partial s}{\partial \rho}.$$

These coincide with the usual definitions of temperature and pressure.

We define a *perfect fluid* as a continuum for which there is no heat flux,

$$\vec{q} = 0$$

and the stress tensor reduces to the thermodynamic pressure

(1.9)
$$t_{ij} = -P\delta_{ij}.$$

Equation (1.8) define P as a function of ρ and E:

(1.10)
$$P = P(\rho, E).$$

This relation is usually called *equation of state*.

Because of the monotonicity, $s(\rho, E)$ can be inverted and the internal energy can be expressed as a function of ρ and s. Legendre transforms allow us to use pressure and/or temperature as independent thermodynamical variables.

A simple example of a fluid is an ideal gas, for which the internal energy is a monotone function of temperature alone, $E = E(T)$, and the equation of state is

$$P = \overline{R}\rho T$$

where $\overline{R}$ is the gas constant.

In the case of a monotonic gas or diatomic gas at low temperatures the relation between internal energy and temperature is linear:

$$E = C_V T$$

and the equation of state can be written as

$$(1.11) \qquad\qquad P = (\gamma - 1)\rho E$$

where the adiabatic index γ is the ratio of specific heats at constant pressure and volume: $\gamma = C_P/C_V$.

However the behaviour of a real gas may be somewhat different from that of an ideal gas. In the Van der Waals model of a gas two additional effects are taken into account. First of all the proper volume of the molecules of the gas is considered and therefore the available volume (per unit mole) is reduced by an amount B. Secondly the pressure of the gas is reduced with respect to the ideal pressure, due to the attractive interaction of the molecules. The resulting equation of state is (Landau and Lifshitz, 1960 a)

$$(1.12) \qquad\qquad \left(P + \frac{A}{\tau^2}\right)(\tau - B) = \overline{R}T$$

where τ is the specific volume, $\tau = 1/\rho$, and the constants A and B depend on the particular gas.

This equation of state can be used as a simple model describing a phase transition between liquid and gaseous phase s.

The expression for the internal energy may be obtained from the first law of thermodynamics

$$(1.13) \qquad\qquad Tds = dE + Pd\tau$$

by imposing that s is a perfect differential.

Let us consider the equation of state in the form:

$$(1.14) \qquad\qquad \begin{cases} E = E(T,\tau) \\ P = P(T,\tau) \end{cases}$$

Then this requirement becomes, by rewriting equation (1.13) in terms of the independent variables T, τ

$$(1.15) \qquad\qquad \frac{\partial}{\partial \tau}\left(\frac{1}{T}\frac{\partial E}{\partial T}\right) = \frac{\partial}{\partial T}\left(\frac{1}{T}\frac{\partial E}{\partial \tau} + \frac{P}{T}\right)$$

and therefore:

$$(1.16) \qquad\qquad \frac{\partial E}{\partial \tau} = T\frac{\partial P}{\partial T} - P.$$

For a Van der Waals gas, making use of (1.12) in (1.16), we get

$$(1.17) \qquad\qquad \frac{\partial E}{\partial \tau} = \frac{A}{\tau^2}$$

and, integrating,

(1.18)
$$E = -\frac{A}{\tau} + F(T)$$

where F is an arbitrary function of temperature. The term $F(T)$ represents the kinetic energy of the gas (the internal energy of an ideal gas), while the negative term represents the potential energy of the gas due to the attraction of the molecules.

If the specific heat at constant volume is constant then

(1.19)
$$F(T) = C_V T.$$

For many materials a good approximation for C_V within a small range of temperatures is given by

(1.20)
$$C_V(T) = C_{V_0}(T/T_0)^n.$$

By plotting the isotherms in the $\tau - P$ plane we see that P is a monotone decreasing function of τ for temperatures $T \geq T_C$, while for $T < T_C$ the monotonicity is lost, and the isotherms have a maximum and a minimum (see Fig. (1.1)).

As T approaches T_C the two extrema become closer and they coincide, when $T = T_C$, at the point $P_C = (\tau_C, T_C)$. T_C is called the *critical temperature* and the point P_C is called the *critical point*.

For the generic isotherm $ABCDEF$ below the critical temperature (see fig. (1.1)) the region between C and D has $(\partial P/\partial \tau)_T > 0$. This would violate the second law of thermodynamics and imply that the material is mechanically unstable. Hence the region of the $\tau - P$ plane below the locus of the extrema of the isotherms cannot be described by the equation of state (1.12).

The monotonicity of the function $P(\tau)$ may be restored by connecting the points on the left and right branch of the curve by a horizontal line. This line corresponds to a phase transition connecting two different states: liquid (B) and gas (E). The position of the line is determined by the requirements that the chemical potentials of the two states are the same. This is equivalent to the condition that the two dashed areas BCO and ODE are equal (the so called "Maxwell construction") (Landau and Lifshitz, 1960 a, p. 261).

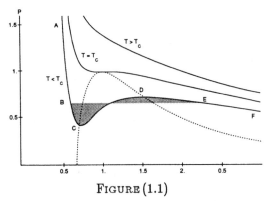

FIGURE (1.1)

The Van der Waals gas in the τ–P plane. Continuous
lines: isotherms at different temperatures. The dashed
line is the locus of extrema of the isothermal lines

The points in the branches BC and DE are metastable: they can be reached
under particular conditions and persist until thermal fluctuations restore the thermodynamical equilibrium represented by a point in the segment BE. The critical
point is found from the conditions

(1.21)
$$\left(\frac{\partial P}{\partial \tau}\right)_T = 0, \quad \left(\frac{\partial^2 P}{\partial \tau^2}\right)_T = 0.$$

Using (1.12) gives

(1.22)
$$\begin{cases} \tau_c &= 3B\,, \\[2mm] P_C &= \dfrac{A}{27B^2}\,, \\[2mm] T_C &= \dfrac{8A}{27\overline{R}B} \end{cases}$$

It is useful to express the equation of state in a dimensionless form: let

(1.23) $q \equiv P/P_C, \quad V \equiv \tau/\tau_C, \quad \Theta \equiv T/T_C, \quad \hat{E} \equiv E/(\tau_C P_C).$

Then equations (1.12) and (1.18) become

(1.24)
$$\left(q + \frac{3}{V^2}\right)(3V - 1) = 8\Theta\,,$$

(1.25)
$$\hat{E} = -\frac{3}{V} + \hat{F}(\Theta)\,.$$

In the case of constant specific heat we have

(1.26)
$$\hat{F}(\Theta) = \frac{8}{3}\frac{C_V}{\overline{R}}\,\Theta\,.$$

The structure of liquids is much more compact than that of gases. Consequently
liquids are almost incompressible. The pressure depends strongly on the density

and in some cases the dependence of pressure on entropy is negligible. In this case the fluid has a barotropic equation of state:

$$P = f(\rho)$$

and, from the first law of thermodynamics, the internal energy is separable:

$$E = E^{(1)}(\rho) + E^{(2)}(s).$$

Furthermore, the temperature is function of s only

$$T = T(s)$$

(These three statements are equivalent).

Water can be described with a good approximation by the barotropic equation of state:

$$P = A \left(\frac{\rho}{\rho_0} \right)^{\gamma} - B$$

where ρ_0, A and B are constants and $\gamma = 7$.

A more accurate equation of state which takes into account the real dependence on temperature is the Tait equation of state:

$$P = B(T)\{\exp[(\tau_0 - \tau)/\bar{\tau}] - \exp[(\tau_0 - \tau_T)/\bar{\tau}]\}$$

where $\bar{\tau}$ is a constant, τ_T is the volume at temperature T and zero pressure and τ_0 is a constant standard volume. $B(T)$ is generally an increasing function of temperature.

For water $B \cong 3000$ atm at normal conditions.

2 Irreversible thermodynamics

For a fluid which is not in local thermodynamical equilibrium, but sufficiently close to it, we proceed as follows (Landau and Lifshitz, 1960 a, Jon, Casas-Vàzguez and Lebon, 1988). We assume the existence of a non-equilibrium specific entropy density, $s = s(\rho, E)$, such that it coincides functionally with the equilibrium specific entropy density $s_{eq}(\rho, E)$,

(2.1)
$$s(\rho, E) = s_{eq}(\rho, E).$$

We define the off-equilibrium temperature T and thermodynamic pressure P analogously as with Eqs. (1.7–8), i.e.

(2.2)
$$T^{-1} = \frac{\partial s}{\partial E},$$

(2.3)
$$P = -\rho^2 T \frac{\partial s}{\partial \rho} = P(\rho, E).$$

With these definitions the Gibbs relation holds,

$$(2.4) \qquad\qquad T ds = -\frac{P}{\rho^2} d\rho + dE .$$

It is always possible to write t_{ij} in the form

$$(2.5) \qquad\qquad t_{ij} = -P\delta_{ij} - \tilde{\pi}_{ij}$$

where $\tilde{\pi}_{ij}$ is called the viscous stress. $\tilde{\pi}_{ij}$ can be decomposed into a trace-free part (shear stress) and an isotropic part

$$(2.6) \qquad\qquad \tilde{\pi}_{ij} = \pi_{ij} + \frac{\pi}{3}\,\delta_{ij} .$$

Then, by using the dynamical equations (1.4–6) it is easy to see that

$$(2.7) \qquad\qquad \rho\dot{s} = -\frac{1}{T}\,\pi_{ij}e_{ij} - \frac{\pi}{T}\,\theta - \frac{1}{T}\frac{\partial q_i}{\partial x^i}$$

where

$$e_{ij} = \frac{1}{2}\left(\frac{\partial u^i}{\partial x^j} + \frac{\partial u^j}{\partial x^i}\right) - \frac{1}{3}\delta_{ij}\frac{\partial u^k}{\partial x^k} ,$$

is the rate of strain tensor and

$$\theta = \frac{1}{3}\frac{\partial u^i}{\partial x^i}$$

is the expansion scalar of the fluid motion.

Equation (2.7) can also be rewritten as

$$(2.8) \qquad\qquad \rho\dot{s} + \frac{\partial}{\partial x^i}\left(\frac{q_i}{T}\right) = -\frac{1}{T}\left\{\pi_{ij}e_{ij} + \pi\theta + \frac{q_i}{T}\frac{\partial T}{\partial x^i}\right\} .$$

One can identify q_i/T with the entropy flux and interpret equation (2.8) as the entropy balance law.

Then

$$\sigma = \rho\dot{s} + \frac{\partial}{\partial x^i}\left(\frac{q_i}{T}\right)$$

is the entropy production rate. It is a basic tenet of irreversible thermodynamics that

$$(2.9) \qquad\qquad \sigma \geq 0 .$$

This is called the Clausius-Duhem inequality. It can also be written as

$$(2.10) \qquad\qquad \pi_{ij}e_{ij} + \pi\theta + \frac{q_i}{T}\frac{\partial T}{\partial x^i} \leq 0 ,$$

with equality holding only at equilibrium. In the language of irreversible thermodynamics the quantities $e_{ij}, \theta, \partial T/\partial x^i$ are called *thermodynamic forces*, while the quantities π_{ij}, π, q_i are called *thermodynamic fluxes*. In linear irreversible thermodynamics, we assume that the thermodynamic forces and fluxes are linearly related. Then the isotropy of the fluid and the symmetries of π_{ij} imply that

(2.11)
$$q_i = -k\frac{\partial T}{\partial x_i}$$

(2.12)
$$\pi_{ij} = -2\eta e_{ij}$$

(2.13)
$$\pi = -3\varsigma\theta$$

where k, η, ς are three phenomenological coefficients (functions, in general, of E and ρ) called the coefficients of heat conduction, shear and bulk viscosity respectively. Inequality (2.10) takes the form

(2.14)
$$2\eta e_{ij}e_{ij} + 3\varsigma\theta^2 + \frac{k}{T^2}\frac{\partial T}{\partial x_i} \cdot \frac{\partial T}{\partial x_i} \geq 0,$$

which obviously implies that k, η, ς are nonnegative.

Eqs. (2.11–13) are the constitutive laws of Fourier and of Navier-Stokes and are at the basis of the usual description of a thermo-viscous fluid.

3   The equations of magnetofluid dynamics

When a fluid is electrically conducting and is placed in an external magnetic field there is an interplay between electromagnetic and hydrodynamic effects. Therefore the flow must be described by the equations for the electromagnetic field coupled to the fluid motions. A particularly simple description arises in the magnetofluid dynamic limit. In an inertial reference frame Maxwell's equations are (Landau and Lifshitz, 1960 b)

(3.1)
$$\nabla \cdot \vec{B} = 0$$

(3.2)
$$\nabla \cdot \vec{E} = \frac{q}{\epsilon_0}$$

(3.3)
$$\nabla \times \vec{E} = -\frac{\partial \vec{B}}{\partial t}$$

(3.4)
$$\nabla \times \vec{B} = \mu_0 \left(\epsilon_0 \frac{\partial \vec{E}}{\partial t} + \vec{j} \right)$$

where $\vec{B}$ and $\vec{E}$ are the magnetic induction and electric field respectively, q is the electric charge density and $\vec{j}$ the current density, ϵ_0 and μ_0 the dielectric constant and magnetic permeability of free space (here the material medium is assumed to have the electromagnetic properties of free space).

To these equations we must add Ohm's law in a moving conductor, which takes the form

(3.5)
$$\vec{j} = \sigma(\vec{E} + \vec{v} \times \vec{B}) + q\vec{v}$$

where $\vec{v}$ is the fluid's velocity, and σ the electrical conductivity.

The approximations leading to the MHD limit are the following:

Step 1.

We assume the conductivity σ to be very large. Then, in order for $\vec{j}$ to remain finite, we must have from (3.5)

$$(3.6) \qquad\qquad \vec{E} + \vec{v} \times \vec{B} \simeq 0 .$$

Let L and τ be a length and a time characteristic of the scale of variation of the field quantities. Then, from (3.2) and (3.6),

$$q \simeq \epsilon_0 \frac{|E|}{L} \simeq \epsilon_0 \frac{v|B|}{L}$$

hence

$$(3.7) \qquad\qquad qv \simeq \epsilon_0 v^2 \frac{|B|}{L} .$$

Also, for the displacement current,

$$\left| \mu_0 \epsilon_0 \frac{\partial \vec{E}}{\partial t} \right| \simeq \mu_0 \epsilon_0 \frac{|E|}{\tau} \simeq \mu_0 \epsilon_0 \frac{v|B|}{\tau}$$

and using $\mu_0 \epsilon_0 = 1/c^2$ (where c is the light speed)

$$(3.8) \qquad\qquad \left| \mu_0 \epsilon_0 \frac{\partial \vec{E}}{\partial t} \right| \simeq \frac{v}{c} \frac{|B|}{c\tau} .$$

But

$$|\nabla \times \vec{B}| \simeq \frac{|B|}{L}$$

hence the displacement current will be negligible in (3.4) provided that

$$\frac{|B|}{L} \gg \frac{v}{c} \frac{|B|}{c\tau} , \qquad \text{i.e.}$$

$$(3.9) \qquad\qquad \tau \gg \frac{v}{c} \frac{L}{c} .$$

In the following we shall assume that the fluid motion is non relativistic, $v/c \ll 1$, and that the electromagnetic field is quasi-stationary, $\tau \gg L/c$. Therefore the inequality (3.9) will be amply satisfied. Equation (3.4) can then be simplified to

$$(3.10) \qquad\qquad \frac{1}{\mu_0} \nabla \times \vec{B} = \vec{j} .$$

We have

$$|\vec{j}| \simeq \frac{|B|}{L\mu_0} ,$$

hence

$$\frac{qv}{|\vec{j}|} \simeq \frac{v^2}{c^2} \ll 1$$

and the convection current $q\vec{v}$ is negligible compared to the conduction current $\vec{j}$.

Furthermore, let e be the electric charge, n the electron density and v_d the electron drift speed. Then

$$\frac{q}{ne} \simeq \epsilon_0 \frac{v|B|}{L} \cdot \frac{1}{ne} = \frac{\epsilon_0 \mu_0 v}{ne} \frac{|B|}{L\mu_0} \simeq \frac{v}{c} \frac{|\vec{j}|}{nec} \simeq \frac{vv_d}{c^2} \, .$$

In actual situations $v_d \ll v$, hence $q/ne \ll 1$ and the charge separation is negligible. Therefore in this approximation the fluid can be assumed to be neutral. Combining Eqs. (3.3) and (3.6) gives the evolution law for the magnetic field

(3.11)
$$\frac{\partial \vec{B}}{\partial t} = \nabla \times (\vec{v} \times \vec{B})$$

which implies that the magnetic field is frozen in the fluid (Ferraro and Plumpton, 1966).

Finally the equations of *perfect magnetofluid dynamics* can be obtained from the balance equations (1.1–3) by adding the electromagnetic stress tensor and energy flux to those of the fluid. More precisely one takes

(3.12)
$$t_{ij} = -P\delta_{ij} - \frac{1}{\mu_0} \left[\frac{1}{2} |B|^2 \delta_{ij} - B_i B_j \right]$$

and

(3.13)
$$\vec{f} = \frac{1}{\mu_0} \vec{B} \times (\vec{v} \times \vec{B}) \, .$$

To the resulting balance equations, in order to close the system, one must add the equations for the magnetic field (3.1) and (3.11).

4   The plasma model

We start with the kinetic equation

(4.1)
$$\frac{\partial f}{\partial t} + (\vec{v} \cdot \nabla) f + (\vec{F} \cdot \nabla_v) f = \left(\frac{\partial f}{\partial t} \right)_{coll}$$

for the distribution function $f(\vec{v}, \vec{r}, t)$, where $\vec{r}$ is the position vector and $\vec{v}$ the velocity of a particle. Here ∇_v is the gradient vector operator in the velocity space, $\vec{F}$ is the external force per unit mass acting on a particle and

$$\left(\frac{\partial f}{\partial t} \right)_{coll}$$

is equal to the rate of change, by collisions, of the number of particles in the range $\vec{v}, \vec{v} + d\vec{v}$, per unit volume, in the fixed space range $\vec{r}, \vec{r} + d\vec{r}$.

In the case of a ionized gas, $\vec{F}$ is given by the Lorentz force

(4.2)
$$\vec{F} = \frac{e}{m}(\vec{E} + \vec{v} \times \vec{B})$$

where $\vec{E}$ and $\vec{B}$ are the electric and magnetic field respectively, and e the particle electric charge.

We model a plasma as a mixture of two different gases, the positive ion gas (referred to by the subscript 1) and the electron gas (referred to by the subscript 2). The distribution functions for the ion and electron gases are denoted by

$$f_1(\vec{v}_1, \vec{r}, t) \quad \text{and} \quad f_2(\vec{v}_2, \vec{r}, t).$$

The kinetic equations for the two functions are

(4.3)
$$\frac{\partial f_\alpha}{\partial t} + (v_\alpha \cdot \nabla)f_\alpha + (\vec{F}_\alpha \cdot \nabla_{v_\alpha})f_\alpha = \left(\frac{\partial f_\alpha}{\partial t}\right)_{coll}, \alpha = 1, 2$$

where

$$\vec{F}_\alpha = \frac{e_\alpha}{m_\alpha}(\vec{E} + \vec{v}_\alpha \times \vec{B}).$$

Let $\Phi_\alpha(\vec{v}_\alpha)$ be a function of $\vec{v}_\alpha$. Multiplying Eq. (4.3) by $\Phi_\alpha(\vec{v}_\alpha)$, integrating over velocity space, and assuming that

$$\lim_{|\vec{v}_\alpha| \to \infty} f_\alpha(\vec{v}_\alpha, \vec{r}, t) = 0$$

we obtain the transport equation

(4.4)
$$\frac{\partial(n_\alpha < \Phi_\alpha >)}{\partial t} + \nabla \cdot (n_\alpha < \Phi_\alpha \vec{v}_\alpha >) - n_\alpha \vec{F}_\alpha \cdot < \nabla_{v_\alpha} \Phi_\alpha >=$$
$$= \int \Phi_\alpha \left(\frac{\partial f_\alpha}{\partial t}\right)_{coll} d\vec{v}_\alpha$$

where

$$n_\alpha = \int f_\alpha d\vec{v}_\alpha$$

are the particle number densities and

$$< \Phi_\alpha >= \frac{1}{n_\alpha} \int f_\alpha \Phi_\alpha d\vec{v}_\alpha.$$

Choosing $\Phi_\alpha = 1$ in Eq. (4.4) we get the particle number conservation equations

(4.5)
$$\frac{\partial}{\partial t} n_\alpha + \nabla \cdot (n_\alpha < \vec{v}_\alpha >) = 0$$

because we are considering collisions that do not change the particle number.

If we write $\vec{w}_\alpha =< \vec{v}_\alpha >$ for the mean velocity and $\rho_\alpha = m_\alpha n_\alpha$ for the densities, (4.5) becomes

(4.6)
$$\frac{\partial \rho_\alpha}{\partial t} + \nabla \cdot (\rho_\alpha \vec{w}_\alpha) = 0.$$

By choosing
$$\Phi_\alpha = m_\alpha \vec{v}_\alpha,$$

from (4.4) we obtain

(4.7)
$$\frac{\partial}{\partial t}(\rho_\alpha \vec{w}_\alpha) + \nabla \cdot (\rho_\alpha \vec{w}_\alpha \otimes \vec{w}_\alpha) + \nabla \cdot \overleftrightarrow{P}_\alpha - \rho_\alpha \vec{F}_\alpha =$$
$$= \int m_\alpha \vec{v}_\alpha \left(\frac{\partial f}{\partial t}\right)_{coll} d\vec{v}_\alpha$$

where $\overleftrightarrow{P}_\alpha$ is the pressure tensor defined by

(4.8)
$$\overleftrightarrow{P}_\alpha = \rho_\alpha < \vec{v}_\alpha \otimes \vec{v}_\alpha > -\rho_\alpha \vec{w}_\alpha \otimes \vec{w}_\alpha$$

Eq. (4.7) can also be rewritten as

(4.9)
$$\rho_\alpha \left(\frac{\partial \vec{w}_\alpha}{\partial t} + (\vec{w}_\alpha \cdot \nabla)\vec{w}_\alpha\right) + \nabla \cdot \overleftrightarrow{P}_\alpha - \rho_\alpha \vec{F}_\alpha =$$
$$= \int m_\alpha \vec{v}_\alpha \left(\frac{\partial f}{\partial t}\right)_{coll} d\vec{v}_\alpha .$$

These equations are not sufficient, by themselves, to describe the plasma, because the pressure tensor $\overleftrightarrow{P}_\alpha$ and the collision term are not known. The model which we shall adopt in the sequel (and which is a reasonable approximation in most cases) is based on some assumptions on these terms. First of all we shall assume that the pressure tensor $\overleftrightarrow{P}_\alpha$ is isotropic

(4.10)
$$\overleftrightarrow{P}_\alpha = P_\alpha \overleftrightarrow{I}$$

which is a good approximation if the plasma is unmagnetized. In this case we can define kinetic temperatures T_α for the two species by

(4.11)
$$P_\alpha = R n_\alpha T_\alpha$$

where R is the gas constant.

The collision term can be estimated in an approximate way by using the concept of relaxation time τ, i.e. by assuming that

$$\left(\frac{\partial f_\alpha}{\partial t}\right)_{coll} = \frac{f_\alpha^{(0)} - f_\alpha}{\tau}$$

where $f_\alpha^{(0)}$ corresponds to the Maxwellian distribution. A simple argument shows that, in this case,

$$\int m_1 \vec{v}_1 \left(\frac{\partial f_1}{\partial t} \right)_{coll} d\vec{v}_1 = -\frac{\rho_1 \rho_2}{\rho_0 \tau} (\vec{w}_1 - \vec{w}_2)$$

where $\rho_0 = \rho_1 + \rho_2$ (Ferraro and Plumpton, 1966).

Finally, the equations describing the unmagnetized plasma are the continuity equations (4.6) together with the momentun conservation equations

$$(4.12) \quad \begin{cases} \rho_1 \frac{\partial \vec{w}_1}{\partial t} + \rho_1 (\vec{w}_1 \cdot \nabla)\vec{w}_1 + \nabla P_1 - \rho_1 \vec{F}_1 = -\frac{\rho_1 \rho_2}{\rho_0 \tau} (\vec{w}_1 - \vec{w}_2) \\ \\ \rho_2 \frac{\partial \vec{w}_2}{\partial t} + \rho_2 (\vec{w}_2 \cdot \nabla)\vec{w}_2 + \nabla P_2 - \rho_2 \vec{F}_2 = \frac{\rho_1 \rho_2}{\rho_0 \tau} (\vec{w}_1 - \vec{w}_2) \,. \end{cases}$$

We remark that the effects of the collisions between ions and electrons is described by the RHS of Eqs. (4.12) (collisions among particles of the same species are neglected in this scheme).

For a collisionless plasma $(\tau \to \infty)$ we can neglect the RHS of Eq. (4.12).

To Eqs. (4.12) we must add Maxwell's equations

$$(4.13) \quad \begin{cases} \nabla \cdot \vec{B} \quad = 0 \\ \\ \epsilon_0 \nabla \cdot \vec{E} \quad = \dfrac{e_1 \rho_1}{m_1} + \dfrac{e_2 \rho_2}{m_2} \\ \\ \nabla \times \vec{E} \quad = -\dfrac{\partial \vec{B}}{\partial t} \\ \\ \nabla \times \vec{B} \quad = \mu_0 \left(\dfrac{\rho_1 e_1}{m_1} \vec{w}_1 + \dfrac{\rho_2 e_2}{m_2} \vec{w}_2 \right) + \mu_0 \epsilon_0 \dfrac{\partial \vec{E}}{\partial t} \,. \end{cases}$$

Here ϵ_0 is the dielectric constant in vacuo and μ_0 the magnetic permeability. For an unmagnetized plasma we may set $\vec{B} = 0$ from the beginning, and Eqs. (4.13) reduce to

$$(4.14) \quad \begin{cases} \epsilon_0 \nabla \cdot \vec{E} = \dfrac{\rho_1 e_1}{m_1} + \dfrac{\rho_2 e_2}{m_2} \\ \\ \nabla \times \vec{E} = 0 \\ \\ \epsilon_0 \dfrac{\partial \vec{E}}{\partial t} + \dfrac{\rho_1 e_1}{m_1} \vec{w}_1 + \dfrac{\rho_2 e_2}{m_2} \vec{w}_2 = 0 \,. \end{cases}$$

The plasma ion and electron frequencies Ω_1, Ω_2 are given by

$$\Omega_1^2 = \frac{n_1 e_1}{m_1 \epsilon_0}, \qquad \Omega_2^2 = \frac{n_2 e_2}{m_2 \epsilon_0} \,.$$

Neglecting the ion pressure (cold ion fluid) Eqs. (4.12) yield, with $\vec{B} = 0$,

$$\frac{\partial \vec{w}_1}{\partial t} + (\vec{w}_1 \cdot \nabla)\vec{w}_1 - \frac{\Omega_1^2}{n_1}\epsilon_0 \vec{E} = -\frac{\rho_2}{\rho_0 \tau}(\vec{w}_1 - \vec{w}_2),$$

and, since $\rho_0 \simeq \rho_1$, one obtains

(4.15) $$\frac{\partial \vec{w}_1}{\partial t} + (\vec{w}_1 \cdot \nabla)\vec{w}_1 - \frac{e_1}{m_1}\vec{E} = -\frac{\rho_2}{\rho_1 \tau}(\vec{w}_1 - \vec{w}_2).$$

Similary, Eq. $(4.12)_2$ yields

$$\rho_2 \frac{\partial \vec{w}_2}{\partial t} + \rho_2(\vec{w}_2 \cdot \nabla)\vec{w}_2 + (a_2)^2 \nabla \rho_2 - \frac{\rho_2 e_2}{m_2}\vec{E} = \frac{\rho_2}{\tau}(\vec{w}_1 - \vec{w}_2),$$

where $a_2^2 = (\partial P / \partial \rho)_s$ is the sound speed in the electron gas (assumed to be in an adiabatic state).

For a proton-electron plasma one has

$$e_2 = -e, \qquad e_1 = e$$

and the above equation becomes

(4.16) $$\frac{m_2}{m_1}\left\{ \frac{\partial \vec{w}_2}{\partial t} + (\vec{w}_2 \cdot \nabla)\vec{w}_2 \right\} + \frac{m_2}{m_1}\frac{(a_2)^2}{\rho_2}\nabla \rho_2 + \frac{e}{m_1}\vec{E}$$
$$= \frac{1}{\tau m_1}(\vec{w}_1 - \vec{w}_2).$$

REFERENCES

- V.C.A. Ferraro and C. Plumpton, *An Introduction to Magnetofluid Mechanics*, Clarendon Press, Oxford (1966).
- M.E. Gurtin, *An Introduction to continuum Mechanics*, Academic Press, New York (1981).
- D. Jon, J. Casas-Vàzquez and G. Lebon *Extended irreversible thermodynamics*, Rep. Prog. Phys. **51**, 1105 (1988).
- L.D. Landau and E.M. Lifshitz, *Statistical Physics*, Pergamon, New York (1960 a).
- L.D. Landau and E.M. Lifshitz, *Electrodynamics of Continuous Media*, Pergamon, New York (1960 b).

INTRODUCTORY CONCEPTS FOR WAVE MOTION

chapter 9

Introduction

In this chapter we introduce the fundamental concepts for studying nonlinear wave motion. In Sec. 1 we treat the prototype of hyperbolic nonlinear equation, i.e. the kinematic wave equation (Whitham, 1974) and we review the main methods emploiyed in order to solve this equation. In particular we present a detailed discussion of the shock-fitting method which holds under very general conditions (more general than what can be found in customary treatments). In Sec. 2 we recall the basic concepts pertaining to asymptotic developments. These are essential concepts in order to introduce ray methods.

In Sec. 3, as an introduction to ray methods we recall the basic perturbation-reduction methods in one space dimension (Gardner and Morikawa, 1960, Taniuti and Wei, 1968, Taniuti, 1974). In particular we treat the concepts of far fields for nonlinear hyperbolic systems, and for systems with weak dissipation and dispersion. Finally we treat also the case of modulated waves for strongly dispersive systems (Asano, 1974).

1 The kinematic wave equation and the shock-fitting

Many problems of wave propagation are described in terms of a *density*, ρ, of a certain physical quantity, a *flux*, q, and a source *density* f, of the same quantity.

In the case of one dimensional propagation the balance law describing the flow may be written, in an integral form

$$(1.1) \qquad \frac{d}{dt} \int_{x_1}^{x_2} \rho(x,t)\,dx = q(x_1,t) - q(x_2,t) + \int_{x_1}^{x_2} f(x,t)\,dx\,.$$

This formula states that the rate of growth of the quantity in a certain region is equal to the net flux plus the source term.

For smooth ρ and q the law may be written in the differential form

$$(1.2) \qquad \frac{\partial \rho}{\partial t} + \frac{\partial q}{\partial x} = f\,.$$

In many cases the flux q and the source density f are given functions of ρ, x and t

$$(1.3) \qquad \begin{aligned} q &= q(\rho,x,t) \\ f &= f(\rho,x,t) \end{aligned}$$

and

$$\frac{\partial q}{\partial x} = \left(\frac{\partial q}{\partial x}\right)_{(t,\rho)=const} + \left(\frac{\partial q}{\partial \rho}\right)_{(t,x)=const} \frac{\partial \rho}{\partial x}.$$

In the sequel we shall make use of the notation

$$\left(\frac{\partial f(x,y)}{\partial x}\right)_y$$

to mean that y is held constant during the differentiation. Equation (1.2) becomes:

(1.4)
$$\frac{\partial \rho}{\partial t} + C(\rho, x, t)\frac{\partial \rho}{\partial x} = h(\rho, x, t),$$

where $C \equiv \partial q/\partial \rho$ and $h = f - (\partial q/\partial x)_{t,\rho}$. Equation (1.4) is a scalar first order quasilinear wave equation. It can be solved by the method of characteristics
(Courant and Hilbert, 1953, Whitham, 1974).

Let us consider the initial value problem which consists of equation (1.4) together with the initial condition:

(1.5)
$$\rho(x,0) = g(x).$$

The LHS of equation (1.4) may be interpreted as a total derivative along a curve
in the $x - t$ plane:

(1.6)
$$\begin{cases} \dfrac{d\rho}{dt} = h(\rho, x, t) \quad \rho(0) = g(\xi), \\[2mm] \dfrac{dx}{dt} = C(\rho, x, t), \quad x(0) = \xi. \end{cases}$$

Note that this is a closed system of ODE's which determines the ξ-th characteristic
and the evolution of ρ along it. This system may be solved, in principle, for any
ξ, yielding a parametric solution of the form:

(1.7)
$$\begin{cases} \rho = \rho(\xi, t), \\ x = x(\xi, t). \end{cases}$$

If C is a constant the characteristics are straight parallel lines with equations
$x = \xi + Ct$. In this case equation (1.4) is *semilinear* and the integration of the
system (1.7) yields a single valued ρ as a function of x and t. On the other hand,
if C is not a constant, then two characteristics starting from different points may
eventually intersect say at a point $(\overline{x},\overline{t})$. Then the value of ρ at this point is not
uniquely determined. The function ρ becomes a multivalued function of x and t.
Single valuedness of ρ in the profile may be restored by introducing a discontinuity
(a *shock*) into the solution. The discontinuity must move with the correct speed
in order to conserve ρ.

Suppose that the solution has a jump discontinuity at $x = s(t)$. Using the
balance law (1.1), with $x_1 < s < x_2$, and letting $x_2 \to s_+ , x_1 \to s_-$, we get the
jump condition:

(1.8)
$$V[\![\rho]\!] = [\![q]\!]$$

where, for any quantity h,

$$[\![h]\!] \equiv h_- - h_+ \,,$$

h_- and h_+ being the limiting values of h just behind and in front of the discontinuity, and $V = \dot{s}$ is the velocity of the moving discontinuity.

First let us treat the general case (1.6) with parametric solution (1.7). We want to compute the time at which smooth solutions break down.

There are two ways of attacking the problem. One is to consider the intersection of neighboring characteristics. This can happen when the Jacobian of the transformation $(\xi, t) \to (x, t)$ is zero, i.e. when:

$$(1.9) \qquad \left(\frac{\partial x}{\partial \xi}\right)_t = 0 \,.$$

The other way is to consider the profile of the density in space at a fixed time. The smooth solution breaks down when the slope of the profile diverges:

$$(1.10) \qquad \left(\frac{\partial \rho}{\partial x}\right)_t = \infty \,.$$

Now

$$(1.11) \qquad \left(\frac{\partial \rho}{\partial x}\right)_t = \frac{(\partial \rho / \partial \xi)_t}{(\partial x / \partial \xi)_t}$$

and hence the two conditions are equivalent provided that

$$(\partial \rho / \partial \xi)_t \neq 0 \,.$$

Equation (1.10) yields the locus of points in the (ξ, t) plane at which the envelope of the characteristics forms. The time at which it forms first is the minimum time of this locus. It may be found, for example, by applying the method of Lagrange multipliers. The pair of equations which determine the minimum time is thus:

$$(1.12) \qquad \begin{cases} \dfrac{\partial x}{\partial \xi}(\xi, t) = 0 \\[2mm] \dfrac{\partial^2 x}{\partial \xi^2}(\xi, t) = 0 \,. \end{cases}$$

After this point a shock is fitted in order to have a single valued solution. The position in time of the shock may be found as follows.

Let us define a "derivative along the shock":

$$(1.13) \qquad \frac{\delta}{\delta t} \equiv \partial_t + V \partial_x \,,$$

then, making use of this expression in (1.4), the transport equation for ρ, on the shock, may be written:

$$(1.14) \qquad \frac{\delta \rho}{\delta t} + (C - V)\frac{\partial \rho}{\partial x} = h$$

both ahead and behind the discontinuity. Let $\rho(\xi, t)$ be the solution of system (1.7) (on both sides of the shock). Then a transport equation for the value of ξ which corresponds to the shock (see Fig. (2.1)) is obtained:

$$(1.15) \qquad \left(\frac{\partial \rho}{\partial \xi}\right)_t \frac{\delta \xi_i}{\delta t} = (V - C)\left(\frac{\partial \rho}{\partial x}\right)_t \,, \qquad i = 1, 2 \,.$$

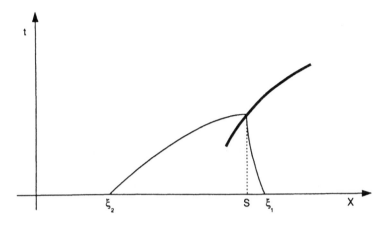

FIGURE (2.1)

Confluence of characteristics and formation of the shock.

Suppose that, at a certain time t, we know the position of the discontinuity, s, and the corresponding values of ξ on both sides, ξ_1, ξ_2. If we know the solution of (1.4) in the form (1.7) on both sides of the shock, then we can compute q_1 and q_2 from Eq. (1.3) and V from Eq. (1.8). Equation (1.15) together with the equation of motion for the shock

$$(1.16) \qquad \frac{\delta s}{\delta t} = V$$

and (1.8), constitute a closed set of ODE's which describe the propagation of the shock.

If the solution of (1.4) is not known then the evolution of the shock must be computed together with the field on both sides. The following discretized scheme can be used.

Given ξ_i, s, $(\partial x/\partial \xi)$ at time t:

1) Solve system (1.6) along the two characteristics originating in $\xi_i(t)$ up to time t and determine the value of $\rho_i(t)$.

2) Compute q_i from (1.3), $C_i = C(\rho_i, s, t)$ and V from (1.8).

3) Solve the linear system of ODE's for $(\partial x/\partial \xi)$, $(\partial \rho/\partial \xi)$ obtained by differentiating system (1.6) with respect to ξ:

$$(1.17) \qquad \frac{d}{dt}\frac{\partial \rho}{\partial \xi} = \frac{\partial h}{\partial \rho}\frac{\partial \rho}{\partial \xi} + \frac{\partial h}{\partial x}\frac{\partial x}{\partial \xi} + \frac{\partial h}{\partial t} , \qquad \frac{\partial \rho}{\partial \xi}(O) = g'(\xi)$$

$$(1.18) \qquad \frac{d}{dt}\frac{\partial x}{\partial \xi} = \frac{\partial C}{\partial \rho}\frac{\partial \rho}{\partial \xi} + \frac{\partial C}{\partial x}\frac{\partial x}{\partial \xi} + \frac{\partial C}{\partial t} , \qquad \frac{\partial x}{\partial \xi}(O) = 1$$

and determine $(\partial x/\partial \xi_i)_t$ at time t.

4) Update s and ξ_i:

(1.19) $s(t + \Delta t) = s(t) + V(t)\Delta t$

(1.20) $\xi_i(t + \Delta t) = \xi_i(t) + \Delta t(V(t) - C_i(t))/(\partial x/\partial \xi_i)(t)$.

The procedure can be iterated and the evolution of the shock can be computed up to the desidered time, unless the shock collide with another shock, as will be shown later.

At this point we make some remarks about the existence and uniqueness of shock solutions.

From a mathematical point of view a *shock* is a *generalized solution* of (1.2) which is a piecewise smooth function with a finite number of discontinuity lines in the $(x - t)$ plane. The existence of such solution is proved in (Lax, 1973) under very general conditions. As an example let us consider the problem

(1.21)
$$\frac{\partial \rho}{\partial t} + \rho \frac{\partial \rho}{\partial x} = 0$$

(1.22)
$$\rho(x,0) = \begin{cases} 0 & x \leq 0 \\ 1 & x > 0 \end{cases} .$$

Two possible solutions are in the half plane $t \geq 0$ (see Fig. (2.2))

a)
$$\rho(x,t) = \begin{cases} 0 & x \leq \frac{1}{2}t \\ 1 & x > \frac{1}{2}t \end{cases}$$

b)
$$\rho(x,t) = \begin{cases} 0 & x \leq 0 \\ \frac{x}{t} & 0 < x < t \\ 1 & x \geq t \end{cases} .$$

Both solutions satisfy equation (1.21) in the region where they are smooth and they satisfy the jump condition (1.9) across the discontinuity. The first solution is however unstable under small perturbations in the initial data. In order to rule out "a priori" such a solution an *entropy principle* has been introduced which states that the characteristics cannot originate on the shock (Lax, 1973). Mathematically this can be stated as:

(1.23)
$$\left(\frac{dx}{dt}\right)_- > V > \left(\frac{dx}{dt}\right)_+ .$$

This principle is directly related to the second law of thermodynamics when applied to a shock in gas dynamics (Whitham, 1974). The introduction of this principle restores the uniqueness of the generalized solution for the problem (1.2), (1.5).

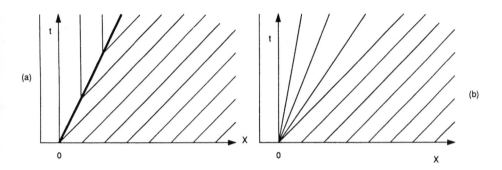

FIGURE (2.2)

Solutions of the inviscid Burgers equation that satisfy
(b) and do not satisfy (a) the entropy condition.

It is possible that two shocks form at different positions, move at different speeds and eventually collide. At this point the two shocks merge in one shock. The entropy principle ensures us that this configuration is the only possible one (see Fig. (2.3)).

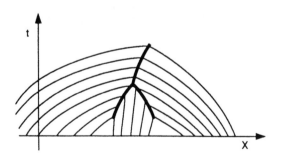

FIGURE (2.3)

Merging of two shocks.

The entropy principle introduces a strong irreversibility in the system. A certain shock configuration can be generated by different initial conditions, therefore it is impossible to go backward in time in a unique way.

In order to treat the merging of shocks the previous scheme based on the method of characteristics has to be suitably modified.

Now we discuss another method for treating shocks, the *shock-fitting* method. We consider an equation of the form (1.2) where f does not depend explicitily on ρ. Let x_a and x_b denote two points respectively to the left and to the right of the shock and let us denote by $\rho(\xi, t)$ the multivalued continuous solution and by $\tilde{\rho}(x, t)$ the discontinuous solution. Let A and $\tilde{A}$ be the areas under the two curves between points x_a and x_b at time t:

(1.24)
$$A = \int_{\xi_a}^{\xi_b} \rho(\xi, t) \left(\frac{\partial x}{\partial \xi}\right)_t d\xi$$

$$\tilde{A} = \int_{x_a}^{x_b} \tilde{\rho}(x, t) \, dx$$

where $x_a = x(\xi_a, t)$ and $x_b = x(\xi_b, t)$.

The difference in the areas is given by

$$A - \tilde{A} = \int_{\xi_1}^{\xi_2} \rho(\xi, t) \left(\frac{\partial x}{\partial \xi}\right)_t d\xi$$

where ξ_1 and ξ_2 are the values of ξ corresponding to the two sides of the shock.

The time derivative of the difference is:

(1.25)
$$\Delta \dot{A} = \frac{d}{dt}(A - \tilde{A})$$

$$= \int_{\xi_1}^{\xi_2} \left[\left(\frac{\partial \rho}{\partial t}\right)_\xi \frac{\partial x}{\partial \xi} + \rho \frac{\partial^2 x}{\partial t \partial \xi}\right] d\xi + \rho_2 \frac{\partial x}{\partial \xi_2} \dot{\xi}_2 - \rho_1 \frac{\partial x}{\partial \xi_1} \dot{\xi}_1$$

$$= \int_{\xi_1}^{\xi_2} f \frac{\partial x}{\partial \xi} d\xi - \int_{\xi_1}^{\xi_2} \left[\frac{\partial q}{\partial x} \frac{\partial x}{\partial \xi} - \rho \left(\frac{\partial C}{\partial \xi}\right)_t\right] d\xi + \rho_2 \frac{\partial x}{\partial \xi_2} \dot{\xi}_2$$

$$- \rho_1 \frac{\partial x}{\partial \xi_1} \dot{\xi}_1$$

where

$$\partial x / \partial \xi_i \equiv (\partial x / \partial \xi)(\xi_i, t), \qquad i = 1, 2$$

and (1.6) have been used. Because f does not depend explicitly on ρ the first integral vanishes. Integrating by part the second yields:

(1.26)
$$\Delta \dot{A} = -\int_{\xi_1}^{\xi_2} \left[\frac{\partial q}{\partial \rho} \left(\frac{\partial \rho}{\partial \xi}\right)_t + \frac{\partial q}{\partial x} \frac{\partial x}{\partial \xi}\right] d\xi + \rho_2 \left(C_2 + \frac{\partial x}{\partial \xi_2} \dot{\xi}_2\right)$$

$$- \rho_1 \left(C_1 + \frac{\partial x}{\partial \xi_1} \dot{\xi}_1\right) = -\int_{\xi_1}^{\xi_2} \left(\frac{\partial q}{\partial \xi}\right)_t d\xi + (\rho_2 - \rho_1)\dot{s}$$

$$= -(q_2 - q_1) + (\rho_2 - \rho_1)\dot{s}.$$

This expression is zero from the jump condition (1.9), therefore the two areas A and $\tilde{A}$ are equal (because they are equal before the formation of the shock). The shock-fitting integral condition is therefore:

$$(1.27) \qquad \int_{\xi_1}^{\xi_2} \rho(\xi,t)\frac{\partial x}{\partial \xi}\, d\xi = 0\,.$$

This condition states that the shock enters the multivalued profile so that it separates two closed regions of equal area (see Fig. (2.4)).

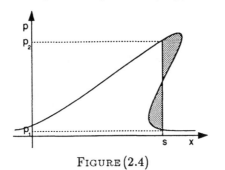

FIGURE (2.4)

Equal area rule and shock-fitting.

This is one equation relating ξ_1 and ξ_2. The other relation between these quantities is

$$(1.28) \qquad x(\xi_1,t) = x(\xi_2,t)\,.$$

This relation together with (1.8) provides a description of the propagation of the shock wave.

Now we consider some particular cases.

EXAMPLE (1.1).

$$f(x) \equiv 0, \qquad q = q(\rho)\,.$$

Here the characteristics are straight lines, and the solution of (1.2), in parametric form, is

$$(1.29) \qquad \begin{cases} \rho = g(\xi)\,, \\ x = \xi + F(\xi)t\,, \end{cases}$$

where $F(\xi) = C(\rho(\xi))$.

The family of characteristics forms an envelope, whose equation is obtained from the system

$$x = \xi + F(\xi)t\,,$$
$$0 = 1 + F'(\xi)t\,.$$

The time t^* at which the characteristics first intersect is given by

$$(1.30) \qquad t^* = \min\left(-\frac{1}{F'(\xi)}\right)\,.$$

After a shock is formed condition (1.28) yields

(1.31)
$$t = -\frac{\xi_1 - \xi_2}{F(\xi_1) - F(\xi_2)},$$

and condition (1.27) gives

(1.32)
$$\int_{\xi_1}^{\xi_2} g(\xi)\,d\xi + t \int_{\xi_1}^{\xi_2} g(\xi) F'(\xi)\,d\xi = 0\,.$$

Integrating by parts we have

$$\int_{\xi_1}^{\xi_2} g(\xi) F'(\xi)\,d\xi = F(\xi)g(\xi)\Big|_{\xi_1}^{\xi_2} - \int_{\xi_1}^{\xi_2} F(\xi) g'(\xi)\,d\xi$$

$$= C_2\rho_2 - C_1\rho_1 - \int_{\xi_1}^{\xi_2} \frac{dq}{d\xi}\,d\xi$$

$$= C_2\rho_2 - C_1\rho_1 - (q_2 - q_1)\,.$$

Using this result and (1.31) in (1.32) we obtain

(1.33)
$$\int_{\xi_1}^{\xi_2} g(\xi)\,d\xi = [C_2\rho_2 - C_1\rho_1 - (q_2 - q_1)]\frac{\xi_2 - \xi_1}{C_2 - C_1}\,.$$

Equations (1.31) and (1.33) determine ξ_1 and ξ_2.

The densities on either side of the shock and the shock position are then found by (1.29).

EXAMPLE (1.2).
$$f = f(\rho), \qquad q = q(\rho)\,.$$

In this case system (1.6) becomes

(1.34)
$$\begin{cases} \dfrac{d\rho}{dt} = f(\rho), & \rho(\xi,0) = g(\xi) \\ \dfrac{dx}{dt} = C(\rho), & x(\xi,0) = \xi \end{cases}.$$

The solution is given by

(1.35)
$$\begin{cases} t = t(\xi,\rho) = G(\rho) - G(g(\xi)) \\ x = x(\xi,\rho) = \xi + F(\rho) - F(g(\xi)) \end{cases},$$

with

(1.36)
$$\begin{cases} G(\rho) \equiv \displaystyle\int \dfrac{1}{f(\rho)}\,d\rho \\ F(\rho) \equiv \displaystyle\int \dfrac{C(\rho)}{f(\rho)}\,d\rho \end{cases}.$$

The equation of the envelope of characteristics is

$$(1.37) \qquad \left(\frac{\partial x}{\partial \xi}\right)_t = 1 + \frac{C - C_0}{f_0} g' = 0,$$

where $C = C(\rho)$, $C_0 = C(g(\xi))$, $f_0 = f(g(\xi))$ and the first equation of (1.35) has been used. The critical time at which the envelope forms is obtained by the usual method of Lagrange multipliers applied to the function

$$\tau = G(\rho) - G(g(\xi)) + \lambda\{f_0 + (C - C_0)g'\}.$$

This leads to:

$$(1.38) \qquad (fC' - f_0 C_0')g' + f_0 f_0' - \left(\frac{f_0}{g'}\right)^2 g'' = 0,$$

where

$$C' = C'(\rho), \quad C_0' = C'(g(\xi)), \quad f_0' = f'(g(\xi)).$$

Conditions (1.37–38) constitute a nonlinear system of equations which yields the time of shock formation. After this time a shock must be fitted in.

At each time the shock is defined by five unknowns, which are: the shock position s; the densities behind and ahead: ρ_1, ρ_2; the initial values of x of the two characteristics joining at the shock at time t, ξ_1 and ξ_2. For the shock position we have:

$$(1.39) \qquad \begin{cases} s = x(\xi_1, \rho_1) = \xi_1 + F(\rho_1) - F(g(\xi_1)) \\ s = x(\xi_2, \rho_2) = \xi_2 + F(\rho_2) - F(g(\xi_2)) \end{cases}.$$

The quantities ρ_1, ρ_2 on the second shock satisfy the transport equations

$$(1.40) \qquad \frac{\delta \rho_i}{\delta t} + (C_i - V)\left(\frac{\partial \rho_i}{\partial x}\right)_t = f_i, \qquad i = 1, 2.$$

From (1.35) we have

$$(1.41) \qquad \left(\frac{\partial \rho}{\partial x}\right)_t = \left(\frac{\partial \rho}{\partial \xi}\right)\left(\frac{\partial \xi}{\partial x}\right)_t = \frac{f g'}{f_0 + (C - C_0)g'}.$$

Substituting in (1.40) we get:

$$(1.42) \qquad \frac{\delta \rho_i}{\delta t} + \frac{(C_i - V)f_i g_i'}{f_{0i} + (C_i - C_{0i})g_i'} - f_i = 0.$$

Finally for the shock position we have:

$$(1.43) \qquad \frac{\delta s}{\delta t} = V = \frac{q_1 - q_2}{\rho_1 - \rho_2}.$$

The solution of the shock fitting problem is obtained in the following way: from equation (1.39) one gets ξ_1 and ξ_2 as function of s, ρ_1, ρ_2. Using this result in (1.42) all the quantities are functions of ρ_1, ρ_2 and s. Thus (1.42–43) constitute a closed system of ODE's.

EXAMPLE (1.3).

$$f = f(x), \qquad q = q(\rho).$$

The characteristic system is:

(1.44) $$\frac{d\rho}{dt} = f(x) \qquad \rho(\xi, 0) = g(\xi)$$

(1.45) $$\frac{dx}{dt} = C(\rho) \qquad x(\xi, 0) = \xi.$$

The trajectories are obtained by eliminating the time:

$$\frac{d\rho}{dx} = \frac{f(x)}{C(\rho)}$$

where d/dx is a derivative with respect to x along the characteristics. Integration yields

(1.46) $$q(\rho) = q(g(\xi)) + \int_\xi^x f(\tilde{x}) d\tilde{x}.$$

Solving for ρ gives, in a region where $C(\rho) \neq 0$,

(1.47) $$\rho = \rho(\xi, x).$$

Then the equation for $t(\xi, x)$ is determined from (1.45):

(1.48) $$t = \int_\xi^x \frac{d\tilde{x}}{C(\rho(\xi, \tilde{x}))}.$$

The derivatives of ρ and t with respect to ξ and x are given by

(1.49) $$\begin{cases} \dfrac{\partial \rho}{\partial \xi} = [C(g(\xi))g'(\xi) - f(\xi)]/C(\rho) \\[2mm] \dfrac{\partial \rho}{\partial x} = f(x)/C(\rho) \\[2mm] \dfrac{\partial t}{\partial \xi} = -\dfrac{1}{C(g(\xi))} - [C(g(\xi))g'(\xi) - f(\xi)] \int_\xi^x \dfrac{C'(\rho(\xi, \tilde{x}))}{C^3(\rho(\xi, \tilde{x}))} d\tilde{x} \\[2mm] \dfrac{\partial t}{\partial x} = \dfrac{1}{C(\rho)} \end{cases}$$

From (1.49) we get

(1.50) $$\left(\frac{\partial x}{\partial \xi} \right)_t = -C(\rho) \frac{\partial t}{\partial \xi}(\xi, x) \equiv \mathcal{F}(\xi, x).$$

The position of shock formation is determined by shock fitting. Relation (1.27) is not convenient in this case because the solution is expressed as a function of ξ and x. It is easier to solve the transport equations along the shock $(1.15\text{--}16)$:

(1.51)
$$\begin{cases} \dfrac{\delta s}{\delta t} = V \\[2mm] \dfrac{\delta \xi_1}{\delta t} = \dfrac{(V - C_1)}{\mathcal{F}(\xi_1, s)} \\[2mm] \dfrac{\delta \xi_2}{\delta t} = \dfrac{(V - C_2)}{\mathcal{F}(\xi_2, s)} \end{cases}$$

where
$$C_i = C\left(\rho(\xi_i, s)\right), \qquad i = 1, 2,$$

and V is determined by (1.8). This is a closed system of ODE's which describes the propagation of the shock.

2 Basic properties of asymptotic expansions

In order to introduce asymptotic series we start with a simple example.
Let us consider the differential equation

(2.1)
$$\frac{dy}{dx} + y = \frac{1}{x}, \qquad x \in \mathbb{R} - \{0\}.$$

The formal series
$$S(x) = \sum_{m=1}^{\infty} a_m x^{-m}$$

is called a *formal asymptotic solution* of (2.1) if when substituted into (2.1) it is a solution as a formal series i.e. by operating on $S(x)$ formally with the usual algebraic manipulations and with differentiation (for the properties of formal series, see Erdélyi, 1956, Nayfeh, 1973, O' Malley, 1974).
It is easy to see that if $S(x)$ is a formal asymptotic solution of (2.1), we must have:
$$a_1 = 1, \quad a_{m+1} = m a_m, \quad m \geq 1$$

i.e.

(2.2)
$$a_m = (m - 1)!$$

We remark that with these a_m the series $S(x)$ is divergent. In fact
$$\frac{a_{m+1} x^{-m-1}}{a_m x^{-m}} = \frac{m}{x} \to \infty \qquad \text{as} \qquad m \to \infty$$

for a fixed non vanishing x.

It is immediate to check that a particular integral of (2.1) is, for $x < 0$,

$$(2.3) \qquad \hat{y}(x) = e^{-x} \int_{-\infty}^{x} \frac{e^t}{t} dt \,.$$

An integration by parts gives

$$\hat{y}(x) = e^{-x} \left\{ \frac{1}{x} e^x + \int_{-\infty}^{x} t^{-2} e^t dt \right\}$$

and iterating

$$(2.4) \qquad \hat{y}(x) = \frac{1}{x} + \frac{1!}{x^2} + \cdots + \frac{(m-1)!}{x^m} + R_m \,,$$

with

$$R_m = m! \, e^{-x} \int_{-\infty}^{x} t^{-m-1} e^t dt \,.$$

Now one has

$$|R_m| \leq m! |x^{-m-1}| e^{-x} \int_{-\infty}^{x} e^t dt = \frac{m!}{|x^{m+1}|}$$

hence

$$\lim_{|x| \to \infty} R_m = 0 \,.$$

Therefore the formal series $S(x)$ which, as we have seen, is divergent, approximates the particular integral of (2.1), $\hat{y}(x)$, as $|x| \to \infty$ and not as $m \to \infty$ as would be the case for convergent series. It is an example of asymptotic expansion of $\hat{y}(x)$, a concept which shall be defined in a more precise way in the following.

Now, following Erdélyi (1956), we proceed by defining order relations and asymptotic expansions.

Let X denote $\mathbb{R}$ or $\mathbb{C}$ and $\mathcal{F}$ the set of real or complex functions.

First of all we define the symbols O and o as follows.

DEFINITION (2.1). Let $\varphi, \psi \in \mathcal{F}, \mathcal{A} \subset X$, then $\varphi = O(\psi)$ $\forall x \in \mathcal{A} \subset X$ if $\exists A \in \mathbb{R}^+$ such that $|\varphi(x)| \leq A|\psi(x)|$, $\forall x \in \mathcal{A}$.

DEFINITION (2.2). Let $\varphi, \psi \in \mathcal{F}$, then $\varphi = o(\psi)$ as $x \to x^0 \in X$ if $\forall \epsilon > 0 \, \exists \, U_\epsilon(x^0)$ neighborhood of $x^0 \in X$ such that

$$|\varphi(x)| \leq \epsilon|\psi(x)| \quad \forall x \in U_\epsilon(x^0) \,.$$

In general φ, ψ will depend on other parameters besides x. Hence also A and $U_\epsilon(x^0)$ will depend on these parameters. If this is not the case the order relations will be called *uniform* with respect to those parameters.

EXAMPLES:

1) $e^{-xt} - 1 = o(x)$ non-uniformly as $x \to 0$

ii) $e^{-xt} - 1 = o(x^{\frac{1}{2}})$ non-uniformly as $x \to 0$.

DEFINITION (2.3). The sequence $\{\varphi_n(x)\}, \varphi_n \in \mathcal{F}$ is called an *asymptotic sequence* as $x \to x_0$ if

$$\varphi_{n+1} = o(\varphi_n) \quad \text{as} \quad x \to x_0 \, .$$

EXAMPLES:

i) $\{(x - x_0)^n\}$ as $x \to x_0$

ii) $\{x^{-\lambda_n}\}$,   as $x \to \infty$,   with $\lambda_{n+1} > \lambda_n$

iii) $\{e^{-x} x^{-\lambda_n}\}$, as $x \to \infty$, with $\lambda_{n+1} > \lambda_n$.

Now we are able to define asymptotic expansions.

DEFINITION (2.4). Let $\{\varphi_n(x)\}$ be an asymptotic sequence as $x \to x_0$.
The formal series

$$\sum a_n \varphi_n(x)$$

is called an *asymptotic expansion of order N* of $f(x)$ as $x \to x_0$, $f \in \mathcal{F}$ if

$$f(x) = \sum_{n=1}^{N} a_n \varphi_n(x) + o(\varphi_N)$$

as $x \to x_0$.
For example if $\varphi_n = (x - x_0)^n$ then the asymptotic expansion is called an asymptotic power series. If the previous relationship holds for all N, $1 \leq N < \infty$ the above series is called an *asymptotic expansion* of $f(x)$ and is written as

$$f(x) \sim \sum_{n=1}^{\infty} a_n \varphi_n(x), \quad x \to x_0 \, .$$

From the definition one sees immediately that the asymptotic expansion of $f(x)$ with respect to the asymptotic sequence $\{\varphi_n(x)\}$ is unique. In fact

$$a_m = \lim_{x \to x_0} \frac{f(x) - \sum_{i=1}^{m-1} a_i \varphi_i(x)}{\varphi_m(x)}$$

is determined uniquely ($\varphi_m(x) \neq 0$ in a neighborhood of x_0).
The same function can have, in general, different asymptotic expansions, with respect to different asymptotic sequences.
For instance one has

$$\frac{1}{1+x} \sim \sum_{n=1}^{\infty} (-1)^{n-1} x^{-n} \quad \text{as} \quad x \to \infty, \quad \{x^{-n}\}$$

$$\frac{1}{1+x} \sim \sum_{n=1}^{\infty} (x-1) x^{-2n} \quad \text{as} \quad x \to \infty, \quad \{(x-1)x^{-2n}\} \, .$$

Also we remark that different functions can have the same asymptotic expansions with respect to the same asymptotic sequence.

For instance one has

$$\frac{1}{1+x} \sim \sum_{n=1}^{\infty} (-1)^{n-1} x^{-n} \quad \text{for} \quad x \to \infty,$$

$$\frac{1+e^{-x}}{1+x} \sim \sum_{n=1}^{\infty} (-1)^{n-1} x^{-n} \quad \text{for} \quad x \to \infty.$$

If the asymptotic sequence $\{\varphi_n\}$ depends upon some parameters, the asymptotic expansion of $f(x)$ is called *uniform* with respect to those parameters, if the relationship

$$f(x) = \sum_{n=1}^{N} a_n \varphi_n(x) + o(\varphi_N), \quad \text{for} \quad x \to x_0$$

holds uniformly. For instance let

$$f(x,t) = \sqrt{x+t}$$

and consider the asymptotic sequence

$$\{x^n\}, \quad n = 0, 1, \ldots$$

One has

$$f(x,t) \sim \sqrt{t}\left(1 + \frac{x}{2t} - \frac{x^2}{8t^2} + \frac{x^3}{16t^3} + \cdots\right)$$

non-uniformly with respect to the parameter t.

The following results concern the operations which can be performed on asymptotic power series. For more general asymptotic expansions we refer to Erdélyi (1956), Nayfeh (1973), O' Malley (1974).

PROPOSITION (2.1). *If*

$$f(x) \sim \sum_0^{\infty} a_n x^n, \quad g(x) \sim \sum_0^{\infty} b_n x^n, \quad \text{as} \quad x \to 0$$

then

$$c_1 f(x) + c_2 g(x) \sim \sum_0^{\infty} (c_1 a_n + c_2 b_n) x^n, \quad \text{as} \quad x \to 0, \quad \forall c_1, c_2.$$

PROOF.

$$f(x) = \sum_{n=0}^{N} a_n x^n + O(x^{N+1}),$$

$$g(x) = \sum_{n=0}^{N} b_n x^n + O(x^{N+1})$$

$$c_1 f(x) + c_2 g(x) = \sum_{n=0}^{N} (c_1 a_n + c_2 b_n) x^n + O(x^{N+1})$$

and the statement follows because N is arbitrary.   □

PROPOSITION (2.2). *If*

$$f(x) \sim \sum_{n=0}^{\infty} a_n x^n, \quad \text{as} \quad x \to 0,$$

$$g(x) \sim \sum_{n=0}^{\infty} b_n x^n \quad \text{as} \quad x \to 0$$

then

$$f(x)g(x) \sim \sum_{0}^{\infty} c_n x^n, \quad \text{as} \quad x \to 0,$$

with

$$c_n = a_0 b_n + a_1 b_{n-1} + \cdots + a_{n-1} b_1 + a_n b_0 .$$

PROOF .

$$f(x) = \sum_{n=0}^{N} a_n x^n + f_N x^N, \qquad g(x) = \sum_{0}^{N} b_n x^n + g_N x^N ,$$

with

$$f_N = O(x), \qquad g_N = O(x) .$$

One has

$$f(x)g(x) = \sum_{n=0}^{N} c_n x^n + d_N x^N ,$$

$$d_N = a_0 g_N + b_0 f_N + x|a_1(b_N + g_N) + b_1(a_n + f_n)| + \\ + \cdots + x^N |(a_N + f_N)(b_N + g_N)|$$

and finally $d_N = O(x)$. Similar results hold for the quotient of two asymptotic expansions. □

PROPOSITION (2.3). *Let*

$$f \sim \sum_{n=1}^{\infty} a_n x^n, \quad \text{as} \quad x \to 0, \quad F(f) = \sum_{n=0}^{\infty} A_n f^n$$

be a power series with radius of convergence $R > 0$. Then

$$F(f) \sim \sum_{n=0}^{\infty} \alpha_n x^n,$$

$$\alpha_0 = A_0, \quad \alpha_1 = a_1 A_1, \quad \alpha_2 = a_2 A_1 + a_1^2 A_2, \ldots$$

PROOF. One can take x sufficiently small such that

$$|f| < R - \epsilon, \quad \epsilon > 0$$

hence

$$\sum_{n=0}^{\infty} A_n f^n$$

converges uniformly in $[0, R - \epsilon]$ to an analytic function $F(f)$ and let

$$\psi(x) = F(f(x)).$$

Let

$$\Sigma_N = \sum_{n=0}^{N} \alpha_n x^n, \quad S_N = \sum_{n=1}^{N} a_n x^n$$

and

$$\Sigma'_N = A_0 + A_1 S_N + \cdots + A_N S_N^N.$$

Since Σ_N and Σ'_N are two polynomials whose coefficients coincide up to the degree N, it is

$$\lim_{x \to 0} \frac{\Sigma_N - \Sigma'_N}{x^N} = 0.$$

Now one has

$$f(x) = S_N + x^{N+1}(a_N + \epsilon_{N+1}), \quad \epsilon_{N+1} = O(x),$$

and therefore

$$\lim_{x \to 0} \frac{A_k f^k - A_k S_N^k}{x^N} = 0, \quad k = 0, 1, \ldots N.$$

Let $Q_N = A_0 + A_1 f + \cdots + A_N f^N$. One has

$$\lim_{x \to 0} \frac{Q_N - \Sigma'_N}{x^N} = 0$$

and furthermore

$$\left| \frac{\psi(x) - Q_N}{x^N} \right| = \frac{1}{x^N} |A_{N+1} f^{N+1} + \cdots|$$

$$\leq x \left| \frac{f}{x} \right|^{N+1} \cdot \{ |A_{N+1}| + |A_{N+2}| |f| + \cdots \}.$$

Since $f \sim a_1 x + a_2 x^2 + \cdots$, f/x is bounded as $x \to 0$, and being $|f| < R - \epsilon$, also

$$|A_{N+1}| + |A_{N+2}| |f| + \cdots,$$

is bounded; it follows

$$\lim_{x \to 0} \frac{\psi(x) - Q_N}{x^N} = 0.$$

Finally,

$$\frac{\psi(x) - \Sigma_N}{x^N} = \frac{\psi(x) - Q_N}{x^N} + \frac{Q_N - \Sigma'_N}{x^N} + \frac{\Sigma'_N - \Sigma_N}{x^N}$$

and from the previous results one obtains

$$\lim_{x \to 0} \frac{\psi(x) - \Sigma_N}{x^N} = 0.$$

□

It is possible to prove that under appropriate conditions, asymptotic power series can be integrated and differentiated termwise (Erdélyi, 1956).

This latter result does not hold for arbitrary asymptotic expansions, as can be seen from the following example:

$$\{\varphi_n(x)\}, \quad \varphi_n(x) = x^{-n}(a + \cos x^n)$$

is an asymptotic sequence for $x \to \infty$, but $\{\varphi'_n(x)\}$ is not an asymptotic sequence for $x \to \infty$.

Now we define asymptotic series.

DEFINITION (2.5). Let $\{\varphi_n(x)\}$ be an asymptotic sequence as $x \to x_0$. Then the formal series $\Sigma a_n \varphi_n(x)$ is called an asymptotic series.

In order to discuss the *sum* of an asymptotic series it is necessary of introduce first the concept of asymptotic equality of two functions.

DEFINITION (2.6). Let $\{\varphi_n(x)\}$ be an asymptotic sequence as $x \to x_0$. The functions $f(x)$ and $g(x)$ are said to be asymptotically equal with respect to the asymptotic sequence $\{\varphi_n\}$ if

$$f(x) - g(x) = o(\varphi_n), \quad x \to x_0, \quad \forall n.$$

DEFINITION (2.7). Let $\{\varphi_n(x)\}$ be an asymptotic sequence as $x \to x_0$. Furthermore let $\Sigma a_n \varphi_n(x)$ be an asymptotic series which is the asymptotic expansion of a function $f(x)$:

$$f(x) \sim \Sigma a_n \varphi_n(x), \quad x \to x_0.$$

We shall call the *sum* of the asymptotic series $\Sigma a_n \varphi_n(x)$ the class of functions asymptotically equal to f.

A theorem, due to Van Corput (Erdélyi, 1956), states that any asymptotic series can be summed.

The asymptotic series previously defined are called *regular* because, by definition, the coefficients a_m of the series $\Sigma a_m \varphi_m(x)$ are constant.

For many applications it is important to introduce generalized asymptotic series as follows.

DEFINITION (2.8). Let $\{\varphi_n(x)\}$ be an asymptotic sequence as $x \to x_0$. The formal series $\Sigma a_n(x)\varphi_n(x)$ is called a generalized asymptotic series if

$$a_n(x) = 0(1) \quad \forall n, \quad \text{for} \quad x \to x_0.$$

Similar definitions can be given for generalized asymptotic expansions.

3   The perturbation-reduction method

In Sec. 1 we have investigated the properties of the kinematic wave equation, which is the prototype for hyperbolic nonlinear waves. We have studied the phenomenon of shock formation from an initially smooth profile due to the steepening caused by nonlinearity (or equivalently the crossing of characteristics).

For nonlinear wave equations which also include dispersive and dissipative terms (thus giving rise to non-hyperbolic waves) the situation can change drastically. This case will be examined in this section. Because, in general, it is extremely difficult to find exact analytical solutions, we have to resort to asymptotic methods and thus we shall draw heavily on the concepts developed in Sec. 2.

The most interesting situation is that in which the dissipative or dispersive terms are sufficiently important in order to counteract the effects of nonlinearity. Therefore we examine the case in which dispersion or dissipation balances the nonlinearity. We shall assume the existence of an appropriate asymptotic expansion and derive an asymptotic equation (model equation) which is easier to study analytically and displays the qualitative effect of dissipation or dispersion balancing nonlinearity. This method is known as perturbation-reduction method. It was first introduced by Gardner and Morikawa (1960), and generalized by Taniuti and Wei subsequently (1968) in order to investigate general nonlinear dispersive or dissipative systems.

In the cases which will be considered this method treats the far field of a system in the approximation of weak nonlinearity and weak dispersion.

Far Field Equations

Let us consider a linear hyperbolic system (Courant and Hilbert, 1953)

$$(3.1) \qquad\qquad \mathbf{U}_t + A\mathbf{U}_x = 0,$$

where A is a constant $N \times N$ matrix and $\mathbf{U}$ the field vector of dependent variables. $\mathbf{U}$ can be written as a superposition of N progressive waves

$$(3.2) \qquad\qquad \mathbf{U}(x,t) = \sum_{j=1}^{N} f_j(x - \lambda_j t)\mathbf{R}_j,$$

where λ_j are the eigenvalues of the matrix A, $\mathbf{R}_j$ the corresponding eigenvectors and $f_j(\xi)$ arbitrary functions to be determined from the initial conditions.

At $t = 0$ (3.2) gives

$$\mathbf{U}(x,0) = \sum_{j=1}^{N} f_j(x)\mathbf{R}_j .$$

A progressive wave solution of (3.1) is a wave of arbitrary profile which propagates at constant velocity without changing its shape. Hence, the expression (3.2) gives a superposition of progressive waves. If the initial disturbance is localized, the functions $f_j(x)$ have compact support and, after a sufficiently long time, the solution consists of a superposition of progressive decoupled (spatially separated) waves.

Therefore, if we consider localized initial values, an observer, after a long time, observes separated waves. In this sense the progressive waves are the *far field* of the system (3.1).

When we consider a nonlinear system, *simple waves* (Whitham, 1974, chapter 6) take the place of progressive waves as elementary waves but, in this case, the shape of the waves changes as they propagate. In the remaining section we restrict our considerations to times before shocks form. Let us consider, for example, a 2×2 nonlinear hyperbolic system of PDE's

$$(3.3) \qquad\qquad \mathbf{U}_t + A(\mathbf{U})\mathbf{U}_x = 0 .$$

The general solution of this system, can be written as (Taniuti, 1974)

$$(3.4) \qquad\qquad \mathbf{U} = \sum_{j=1}^{2} \int \Pi^j(\varphi_j)\mathbf{R}_j(\varphi_j)\,d\varphi_j ,$$

with $\varphi_j(x,t)$ the phases, related to the eigenvalues of A by

$$(3.5) \qquad\qquad \lambda_j = -\frac{\varphi_{jt}}{\varphi_{jx}}$$

where $\lambda_j(\mathbf{U})$ are the eigenvalues of A and Π^1 and Π^2 are arbitrary functions and

$$\varphi_{jt} \equiv \frac{\partial \varphi_j}{\partial t} , \qquad \varphi_{jx} \equiv \frac{\partial \varphi_j}{\partial x} .$$

The term

$$\int \Pi^j(\varphi_j)\mathbf{R}_j(\varphi_j)\,d\varphi_j$$

is known as a simple wave. As in the linear case, if we consider a time sufficiently long but smaller than the critical time, the decoupled simple waves can be considered as the far field of the system (3.3).

For a system with more than two equations the solution cannot be represented as a superposition of simple waves. But a theorem assures that the wave with largest velocity has the structure of a simple wave (Taniuti, 1974). In this case the simple wave with the largest velocity can be considered as the far field of the nonlinear system. Therefore the problem of describing the far field of a nonlinear system is equivalent to deducing an asymptotic equation which describes the fastest simple wave.

We suppose that the field vector $\mathbf{U}$ is expanded in terms of a small parameter ϵ in the following way

$$(3.6) \qquad\qquad \mathbf{U} = \mathbf{U}_0 + \epsilon \mathbf{U}_1 + \epsilon^2 \mathbf{U}_2 + \cdots$$

Here $\mathbf{U}_0$ is a known constant state, and $\mathbf{U}_1$ represents the simple wave

$$(3.7) \qquad\qquad \mathbf{U}_1 = \mathbf{U}_1(\varphi),$$

where $\varphi = \varphi(x,t)$.

We suppose that λ is the largest eigenvalue of $\mathcal{A}$ and we denote the characteristic variable (the *phase*) corresponding to the unperturbed state by

$$(3.8) \qquad\qquad \varphi = x - \lambda_0 t,$$

where $\lambda_0 = \lambda(\mathbf{U}_0)$. Hence,

$$(3.9) \qquad \lambda(\mathbf{U}_0 + \epsilon \mathbf{U}_1 + \cdots) = \lambda_0 + \epsilon \nabla_{\mathbf{U}} \lambda \cdot \mathbf{U}_1 + O(\epsilon^2).$$

From this expression we see that the characteristic velocity deviates from the unperturbed velocity by a quantity of order ϵ. For long times, ($t = O(1/\epsilon)$) the characteristics deviate significantly from those of the linear theory. After such times nonlinear effects must be taken into account, even though the amplitude of the perturbation is small. To study the far field, it is convenient to write $\mathbf{U}_1$ as a function of the unperturbed characteristic variable φ and the slow time $\tau = \epsilon t$.

Expanding (3.3) in ϵ, by using (3.6), we have

$$\epsilon \frac{\partial \mathbf{U}_1}{\partial \varphi} \varphi_t + \epsilon^2 \frac{\partial \mathbf{U}_1}{\partial \tau} + \epsilon^2 \frac{\partial \mathbf{U}_2}{\partial \varphi} \varphi_t + O(\epsilon^3)$$

$$+ (\mathcal{A}_0 + \epsilon \mathbf{U}_1 \cdot \nabla_{\mathbf{U}} \mathcal{A}_0 + O(\epsilon^2)) \left(\epsilon \varphi_x \frac{\partial \mathbf{U}_1}{\partial \varphi} + \epsilon^2 \varphi_x \frac{\partial \mathbf{U}_2}{\partial \varphi} + O(\epsilon^3) \right) = 0.$$

Equating coefficients of the same power of ϵ we have, to the first order,

$$(3.10) \qquad\qquad (-\lambda_0 I + \mathcal{A}_0) \frac{\partial \mathbf{U}_1}{\partial \varphi} = 0.$$

A solution of (3.10) is

$$(3.11) \qquad\qquad \mathbf{U}_1 = \Pi(\varphi, \tau) \mathbf{R}.$$

where Π is an arbitrary function which will be determined to the next order, and $\mathbf{R}$ is a right eigenvector of $\mathcal{A}_0$ corresponding to the eigenvalue λ_0 (we denote the corresponding left eigenvector by $\mathbf{L}$). To the second order, after left multiplication by $\mathbf{L}$, we obtain

$$(3.12) \qquad\qquad \frac{\partial \Pi}{\partial \tau} + \alpha \Pi \frac{\partial \Pi}{\partial \varphi} = 0,$$

where

$$\alpha = \frac{\mathbf{L}(\nabla \mathcal{A}_0 \mathbf{R})\mathbf{R}}{\mathbf{L}\mathbf{R}}.$$

and ∇ denotes the differentiation with respect to $\mathbf{U}$. Here and in the rest of the chapter we make use of the following notation for the directional derivative of a matrix $\mathcal{A}$:

$$\nabla \mathcal{A} \mathbf{V} = \sum_I \frac{\partial \mathcal{A}}{\partial U^I} V^I,$$

$$\underbrace{\nabla \cdots \nabla}_{n \text{ times}} \mathcal{A} \mathbf{V}_1 \cdots \mathbf{V}_n = \sum_{I_1, \cdots, I_n} \frac{\partial^n \mathcal{A}}{\partial U^{I_1} \cdots \partial U^{I_n}} V_1^{I_1} \cdots V_n^{I_n}.$$

Equation (3.12) is invariant under a large class of transformations, known as the Gardner-Morikawa transformations which can be written as

(3.13)
$$\begin{cases} \varsigma = \epsilon^\gamma \varphi \\ \tau = \epsilon^{\gamma+1} t \end{cases}$$

The parameter γ identifies the far field. As we observe in the next section $\gamma = 1$ $(1/2)$ for dissipative (dispersive) system when a balance between nonlinearity and dissipation (dispersion) exists. Therefore we have derived the kinematic wave equation as the asymptotic equation describing the far field of a quasilinear hyperbolic system. The results obtained for the kinematic wave equation can then be applied, in an asymptotic sense, to the far field of the system.

Far field equations with dissipation or dispersion

Now we consider a general partial differential system with dissipative or dispersive terms of the following type

(3.14)
$$\frac{\partial \mathbf{U}}{\partial t} + \mathcal{A}\frac{\partial \mathbf{U}}{\partial x} + \left\{ \sum_{\beta=1}^{S} \prod_{\alpha=1}^{P} \left(\mathcal{H}_\alpha^\beta \frac{\partial}{\partial t} + \mathcal{K}_\alpha^\beta \frac{\partial}{\partial x} \right) \right\} \mathbf{U} = 0$$

where $\mathcal{A}, \mathcal{H}_\alpha^\beta, \mathcal{K}_\alpha^\beta$ are $N \times N$ matrices functions of $\mathbf{U}$.

Proceeding as in the previous paragraph our aim is to derive an asymptotic equation describing the far field when weak dissipation (or dispersion) balance weak nonlinearity.

We look for a solution of the system (3.14) as an asymptotic series in terms of a small parameter ϵ about an unperturbed state $\mathbf{U}_0$:

(3.15)
$$\mathbf{U} = \mathbf{U}_0 + \epsilon \mathbf{U}_1 + \epsilon^2 \mathbf{U}_2 + \cdots$$

We introduce rescaled variables

(3.16)
$$\begin{cases} \varsigma = \epsilon^\gamma (x - \lambda_0 t) \\ \tau = \epsilon^{\gamma+1} t \end{cases}$$

where λ_0 is an eigenvalue of the matrix $\mathcal{A}_0$ and $\gamma = 1/(P-1)$. As in the previous section, to the first order in ϵ, we obtain

$$(-\lambda_0 I + \mathcal{A}_0)\frac{\partial \mathbf{U}_1}{\partial \varphi} = 0$$

and, consequently, by a suitable choice of initial conditions,

(3.17)
$$\mathbf{U}_1 = \Pi \mathbf{R}.$$

To the second order we obtain the solubility condition

(3.18)
$$\frac{\partial \Pi}{\partial \tau} + \alpha \Pi \frac{\partial \Pi}{\partial \varsigma} + b\frac{\partial^P \Pi}{\partial \varsigma^P} = 0,$$

where

$$\alpha = \frac{\mathbf{L}(\nabla \mathcal{A}_0 \mathbf{R})\mathbf{R}}{\mathbf{L}\mathbf{R}},$$

$$b = \frac{\mathbf{L}\left[\sum_{\beta=1}^{S}\prod_{\alpha=1}^{P}(-\lambda_0 \mathcal{H}_{\beta 0}^{\alpha} + \mathcal{K}_{\beta 0}^{\alpha})\right]\mathbf{R}}{\mathbf{L}\mathbf{R}}.$$

Equation (3.18) reduces to the Korteweg-deVries (KdV) for $P = 3$ and to Burgers, for $P = 2$ (see, for example, Whitham, 1974).

Modulated waves

In this paragraph we treat the case of strongly dispersive waves and investigate how they are modulated by weak nonlinearity.

We follow the approach of Asano (1974).

Let us consider the quasilinear system

(3.19)
$$\frac{\partial \mathbf{U}}{\partial t} + \mathcal{A}\frac{\partial \mathbf{U}}{\partial x} + \mathbf{B} = 0, \qquad \mathcal{A} = \mathcal{A}(\mathbf{U}), \ \mathbf{B} = \mathbf{B}(\mathbf{U}).$$

We introduce the slow variable

$$\eta = \epsilon^2 x.$$

Let $\mathbf{U} = \mathbf{U}^{(0)}(\eta)$ be such that $\mathbf{B}(\mathbf{U}^{(0)}) = \mathbf{0}$. $\mathbf{U}^{(0)}$ can be interpreted as unperturbed state (although it is not an exact solution of (3.19)). We also introduce the phase variable ξ given by

(3.20)
$$\xi = \epsilon\left(\int \frac{dx}{\lambda_0} - t\right)$$

with $\lambda_0 = \lambda_0(\eta)$ to be determined.

Now we look for solutions of (3.19) of the form

(3.21)
$$\mathbf{U} = \mathbf{U}^{(0)}(\eta) + \sum_{\alpha=1}^{\infty}\sum_{l=-\infty}^{+\infty}\epsilon^{\alpha}\mathbf{U}_l^{(\alpha)}(\xi, \eta)e^{il(kx-\omega t)}$$

$$\omega = \text{const}, \quad k = k(\eta), \quad \mathbf{U}_{-l}^{(\alpha)*} = \mathbf{U}_l^{(\alpha)} \quad (\text{reality}).$$

Then it follows

$$(3.22) \qquad \frac{\partial \mathbf{U}}{\partial t} = \sum_{\alpha=1}^{+\infty} \sum_{l=-\infty}^{+\infty} \epsilon^\alpha \cdot \left\{ \left(-\epsilon \frac{\partial \mathbf{U}_l^{(\alpha)}}{\partial \xi} \right) e^{il(kx-\omega t)} \right.$$

$$\left. + \mathbf{U}_l^{(\alpha)} (-i\omega l) e^{il(kx-\omega t)} \right\}$$

$$\frac{\partial \mathbf{U}}{\partial x} = \sum_{\alpha=1}^{\infty} \sum_{l=-\infty}^{+\infty} \epsilon^\alpha \cdot \left\{ \left(\frac{\epsilon}{\lambda_0} \frac{\partial \mathbf{U}_l^{(\alpha)}}{\partial \xi} + \epsilon^2 \frac{\partial \mathbf{U}_l^{(\alpha)}}{\partial \eta} \right) \right.$$

$$\left. + il \frac{d(k\eta)}{d\eta} \mathbf{U}_l^{(\alpha)} \right\} \cdot e^{il(kx-\omega t)} + \epsilon^2 \frac{d\mathbf{U}^{(0)}}{d\eta}$$

To the order ϵ it is

$$(3.23) \qquad \begin{cases} \dfrac{\partial \mathbf{U}}{\partial t} = \epsilon \displaystyle\sum_{l=-\infty}^{+\infty} (-i\omega l) \mathbf{U}_l^{(1)} e^{il(kx-\omega t)} + O(\epsilon^2) \\[4mm] \dfrac{\partial \mathbf{U}}{\partial x} = \epsilon \displaystyle\sum_{l=-\infty}^{+\infty} \left(il \dfrac{d(k\eta)}{d\eta} \right) \mathbf{U}_l^{(1)} e^{il(kx-\omega t)} + O(\epsilon^2). \end{cases}$$

By substituting these expressions into (3.19) we obtain

$$(3.24) \qquad \left[-i\omega l \mathcal{I} + \mathcal{A}_0 \, il \frac{d(k\eta)}{d\eta} + (\nabla \mathbf{B})_0 \right] \mathbf{U}_l^{(1)} = 0.$$

Let

$$(3.25) \qquad \mathcal{W}_l \equiv -i\omega l \mathcal{I} + il \frac{d(k\eta)}{d\eta} \mathcal{A}_0 + (\nabla \mathbf{B})_0.$$

We assume that

$$(3.26) \qquad \det \mathcal{W}_l \neq 0, \quad |l| \neq 1,$$

hence, in order to have non trivial solutions, Eq. (3.25) implies the dispersion relation

$$(3.27) \qquad \det \mathcal{W}_{\pm 1} = 0,$$

and furthermore

$$(3.28) \qquad \mathbf{U}_1^{(1)} = \mathbf{R}(\eta) \varphi(\xi, \eta), \qquad \mathbf{U}_l^{(1)} = 0 \quad \text{for} \quad |l| \neq 1,$$

with $\mathbf{R}(\eta)$ right nullvector of $\mathcal{W}_1$.

The phase φ is governed by the following equation (see Appendix 1).

$$(3.29)$$

$$L\mathcal{A}_0 R \frac{\partial \varphi}{\partial \eta} - \frac{\partial \mathcal{W}_1}{\partial \omega} L \frac{\partial R}{\partial \omega} \frac{\partial^2 \varphi}{\partial \xi^2} + |\varphi|^2 \varphi L \left\{ i \frac{d(k\eta)}{d\varphi} \left[2 \nabla \mathcal{A}_0 R^* R_2^{(2)} + \right. \right.$$

$$+ \nabla \mathcal{A}_0 R_0^{(2)} R - \nabla \mathcal{A}_0 R_2^{(2)} R^* + \nabla \nabla \mathcal{A}_0 R^* RR + \frac{1}{2} \nabla \nabla \mathcal{A}_0 RR^* R \right] +$$

$$+ \nabla \nabla B_0 RR_0^{(2)} + \nabla \nabla B_0 R^* R_2^{(2)} + \frac{1}{6} \nabla \nabla \nabla B_0 (2 R R^* R + RR^* R) \right\} +$$

$$+ \varphi L \left\{ \mathcal{A}_0 \frac{\partial R}{\partial \eta} + \nabla \mathcal{A}_0 R \frac{dU^{(0)}}{d\eta} + i \frac{d(k\eta)}{d\eta} [\nabla \mathcal{A}_0 VR] + \nabla \nabla B_0 RV \right\} = 0$$

where

> L is left nullvector of $\mathcal{W}_1$,

$$(3.30)$$

$$R_0^{(2)} = -\mathcal{W}_0^{-1} \left\{ \left(i \frac{d(k\eta)}{d\eta} \nabla \mathcal{A}_0 R^* R - c.c. \right) + \frac{1}{2} (\nabla \nabla B_0 RR^* + c.c.) \right\}$$

$$R_2^{(2)} = -\mathcal{W}_2^{-1} \left\{ i \frac{d(k\eta)}{d\eta} \nabla \mathcal{A}_0 RR + \frac{1}{2} \nabla \nabla B_0 RR \right\}$$

$$V(\eta) = -\mathcal{W}_0^{-1} \mathcal{A}_0 \frac{dU^{(0)}}{d\eta} \, .$$

Eq. (3.29) can be rewritten as

$$(3.31) \qquad\qquad i\frac{\partial \varphi}{\partial \eta} - \alpha \frac{\partial^2 \varphi}{\partial \xi^2} + \mu |\varphi|^2 \varphi + k\varphi = 0 \, .$$

This is the so called *nonlinear Schrödinger equation*.

It is a dispersive wave and its group velocity λ_0 is given by

$$(3.32) \qquad\qquad \frac{1}{\lambda_0} = \frac{\partial k}{\partial \omega} + \frac{\partial^2 k}{\partial \omega \partial \eta} \eta \, .$$

The generalization to propagation in several dimensions will be treated in Section 8.5.

REFERENCES

- N. Asano, *Modulation for nonlinear waves in dissipative or unstable media*, J. Phys. Soc. Japan **36**, 861 (1974).
- R. Courant and D. Hilbert, *Methods of Mathematical Physics*, Interscience, New York (1953).
- A. Erdélyi, *Asymptotic Expansions*, Dover, New York (1956).
- C. S. Gardner and G. K. Morikawa, *Similarity in the asymptotic behavior of collision-free hydromagnetic waves and water waves*, Courant Inst. Math. Sci. Rep., N. 90 - 9082, 1 (1960).
- P. D. Lax, *Hyperbolic system of conservation laws and the mathematical theory of shock waves*, Conf. Board Math. Sci. **11**. SIAM, Philadelphia (1973).
- A. Nayfeh, *Perturbation methods*, Interscience, New York (1973).
- R. E. O'Malley, *Introduction to Singular Perturbations*, Academic Press, New York (1974).
- T. Taniuti, *Reductive perturbation methods and far fields of wave equations* Progr. Theor. Phys. Suppl., **55**, 1 (1974).
- T. Taniuti and C. C. Wei, *Reductive perturbation method in nonlinear wave propagation. I*, J. Phys. Soc. Japan **24**, 941 (1968).
- G. B. Whitham, *Linear and Nonlinear Waves*, Wiley, New York (1974).

RAY METHODS FOR LINEAR WAVES

chapter 3

Introduction

Ray methods for the general study of wave motion originated from geometrical optics. Geometrical optics at the elementary level is usually presented as a closed theory based on the concepts of phase, rays and amplitude propagation along the rays. It is also usually stated that such a theory describes well short wave radiation and in this form the subject is fairly old. However it is only in the second half of this century that a systematic (and whenever possible also rigorous) derivation of the laws of geometrical optics from the fundamental equations of electromagnetism has been extensively carried out (for a comprehensive review see Kline and Kay (1965), Keller, Lewis and Seckler (1956), Lünebourg (1944).

The power of the geometrical optics method and the success of the related W.K.B. approximation in quantum mechanics prompted many people to extend the underlying ideas to wave motion in general as described by linear PDE's waves in fluids, elastic media, etc.).

Fundamental advances in this project were made by Lax (1957), Lewis (1958), Ludwig (1966), Friedlander (1975), among others.

The main results of these investigations is that, in a suitable asymptotic sense and for a suitable class of linear PDE's (e.g. hyperbolic) a theory closely resembling geometrical optics can be constructed, which approximates the true wave solution for sufficiently short wavelenghts (or high frequencies).

In this Chapter we sketch the basic ideas underlying these approaches.

In Sec. 1 we recall briefly the concepts of local phase, local wave number and frequency for wave trains solutions of PDE's with constant coefficients. In Sec. 2 these concepts are extended to linear PDE's with varying coefficients by resorting to the two-timing asymptotic method.

In order to focus on the fundamental physical and mathematical ideas, the treatment will be heuristic. In particular convergence of the asymptotic methods will not be treated. An account of the rigorous results which can be obtained can be found in the work by Majda and Di Perna (1987). Application of these methods to linear wave propagation has been very fruitful in various areas besides electromagnetism. Fundamental results have been obtained in elasticity theory (Hudson, 1980), fluid dynamics (Germain, 1977), plasma physics (Dougherty, 1970), ocean waves (Lighthill, 1980), meteorology (Bretherton, 1971), seismology (Ben-Menahem and Singh, 1981), etc.

1   Phase and group velocity of a wave train

Let us consider a system of linear, first order PDE's

$$(1.1) \qquad\qquad A_{IJ}^{\alpha}(x)\partial_{\alpha}U^{J} + B_{IJ}(x)U^{J} = 0$$

where

$$I, J = 1, 2, \ldots, N, \qquad \alpha = 0, 1, 2, 3, \qquad x^{0} = t$$

and

$$A_{IJ}^{\alpha}(x), \; B_{IJ}(x)$$

are smooth functions of $x \in \mathbb{R}^4$. Repeated indices are summed. In this section we assume that A_{IJ}^{α} and B_{IJ} are constant.

We seek solutions of (1.1) in the form of plane waves

$$(1.2) \qquad\qquad U^{I} = U_{0}^{I} \exp(ik_{\nu}x^{\nu}).$$

Then the following dispersion relation must hold, as consequence of the solvability of (1.1)

$$(1.3) \qquad\qquad \det(ik_{\alpha}A_{IJ}^{\alpha} + B_{IJ}) = 0.$$

Let $\omega = -k_0$ and $\vec{k} = (k_i)$. ω and $\vec{k}$ can be interpreted as frequency and wave number of the plane wave. We assume Eq. (1.3) admits solutions and denote a solution for ω by

$$(1.4) \qquad\qquad \omega = W(\vec{k}),$$

with $\vec{k}$ real. Eq. (1.4) is called *dispersion relation.*

We assume that $W(\vec{k})$ is real and analytic in $\vec{k}$. When $W(\vec{k})$ is complex (for k real) we have an instability for $\text{Im}\,\omega > 0$ and decay for $\text{Im}\,\omega < 0$.

Each such solution determines a mode of propagation. The quantity

$$\theta = \vec{k} \cdot \vec{x} - \omega t$$

is called the phase of the plane wave (1.2) and U_0^I is the amplitude. The phase velocity is

$$\vec{v}_p = \frac{\omega}{|\vec{k}|}\hat{k}, \qquad \hat{k} \equiv \frac{\vec{k}}{|\vec{k}|}.$$

These definitions can be extended to wave trains, which are defined as Fourier integrals

$$(1.5) \qquad\qquad U^{I}(\vec{x}, t) = \int_{\mathbb{R}^3} F^{I}(\vec{k}) e^{i(\vec{k}\cdot\vec{x} - W(\vec{k})t)} d^3\vec{k}.$$

In order for the integral in (1.5) to converge it is sufficient to assume that

$$F^{I} \in L^{1}(\mathbb{R}^3).$$

It is convenient to restrict $F^I(\vec{k})$ to the space

$$S_3 = \left\{ f \in C^\infty(\mathbb{R}^3) \, , \right.$$

such that

$$\left. \sup_{|\alpha| \le N} \sup_{x \in \mathbb{R}^3} (1 + |\vec{x}|^2)^N \, |(D_\alpha f)(x)| < \infty \right\} \, ,$$

for $N = 0, 1, 2, \ldots$ and α is a multiindex,

$$|\alpha| = \alpha_1 + \alpha_2 + \alpha_3 \, ,$$

with

$$D_\alpha = \left(\frac{\partial}{\partial x^1} \right)^{\alpha_1} \left(\frac{\partial}{\partial x^2} \right)^{\alpha_2} \left(\frac{\partial}{\partial x^3} \right)^{\alpha_3} \, .$$

This ensures that (1.5) is smooth and decays rapidly as $\vec{x} \to \infty$. Then $U^I(\vec{x}, t)$ is a solution of (1.1), provided that

$$(1.6) \qquad (ik_\alpha A^\alpha_{IJ} + B_{IJ}) F^J(\vec{k}) = 0 \, .$$

Now we study the one-dimensional case,

$$(1.7) \qquad U^I(x, t) = \int_{\mathbb{R}} F^I(k) e^{ikx - iW(k)t} \, dk \, .$$

Let

$$\chi(k) = W(k) - k \frac{x}{t} \, ,$$

so that

$$(1.8) \qquad U^I(x, t) = \int_{\mathbb{R}} F^I(k) e^{-i\chi(k)t} \, dk \, .$$

We shall study the asymptotic expansion of this integral as $t \to \infty$ keeping x/t constant.

We assume that the wave motion is dispersive, meaning that

$$(1.9) \qquad W''(k) \ne 0 \, .$$

We can then apply the method of stationary phase (Erdélyi, 1956).

The stationary points of $\chi(k)$ are given by

$$(1.10) \qquad W'(k) - \frac{x}{t} = 0 \, .$$

Let $l(x, t)$ be the (assumed unique) solution of this equation. Then the stationary phase expansion of (1.8) is

(1.11)

$$U^I(x, t) = F^I(l) \sqrt{\frac{2\pi}{t \, |W''(l)|}} \, \exp \left[ilx - iW(l)t - \frac{i\pi}{4} \operatorname{sgn} W''(l) \right] + o \left(\frac{1}{\sqrt{t}} \right) \, .$$

If we put

$$\omega(x,t) = W(l(x,t))$$

$$\theta(x,t) = xl(x,t) - t\omega(x,t)$$

$$a^I(x,t) = F^I(l)\sqrt{\frac{2\pi}{t \mid W''(l) \mid}} \exp\left[-\frac{\pi i}{4} \operatorname{sgn} W''(l)\right].$$

We can rewrite Eq. (1.11) in the form

(1.12) $$U^I(x,t) = \operatorname{Re}\left\{a^I(x,t)e^{i\theta(x,t)}\right\}$$

which has the form of a plane wave, but with a varying amplitude, wave number and frequency.

For the plane wave (1.2) one has

(1.13) $$\omega = -\theta_t, \quad k = \theta_x.$$

In the case (1.12) one has

$$\theta_x = l + x\frac{\partial l}{\partial x} - t\frac{\partial \omega}{\partial x} = l + \frac{\partial l}{\partial x}(x - tW'(l)) = l(x,t)$$

$$\theta_t = x\frac{\partial l}{\partial t} - \omega(x,t) - t\frac{\partial \omega}{\partial t} = -\omega(x,t) + \frac{\partial l}{\partial t}(x - tW') = -\omega(x,t)$$

and therefore relations (1.13) are preserved also for wave trains.

We call $l(x,t)$ the *local wave number* of the wave train and $\omega(x,t)$ the *local frequency*. They satisfy the dispersion relation (1.4) and are related to the *local phase* $\theta(x,t)$ by

(1.14) $$l(x,t) = \frac{\partial\theta}{\partial x}$$

(1.15) $$\omega(x,t) = -\frac{\partial\theta}{\partial t}.$$

It is easy to see that

$$\frac{l_x}{l} = \frac{W'(l)}{lW''(l)}\frac{1}{x}, \qquad \frac{l_t}{l} = -\frac{W'(l)}{lW''(l)}\frac{1}{t}$$

hence as $x \to \infty$, $t \to \infty$, $l(x,t)$ varies slowly compared to the phase θ. Consequently the amplitude $a^I(x,t)$ varies slowly as well. Therefore Eq. (1.12) can be interpreted as a locally plane wave with slowly varying wave number, frequency, and amplitude, which propagates with the velocity

$$\frac{x}{t} = W'(l)$$

called the *group velocity*. These asymptotic waveforms will be called *locally plane waves*.

It follows from (1.14–15) that

$$\frac{\partial l}{\partial t} + \frac{\partial \omega}{\partial x} = 0,$$

which is a conservation law for l (conservation of the number of crests per unit length). Also, since $\omega = W(l(x,t))$, one has

$$\frac{\partial l}{\partial t} + W'(l)\frac{\partial l}{\partial x} = 0,$$

which states that the local wave number $l(x,t)$, propagates with the group velocity $W'(l)$. The group velocity concept is also related to the propagation of the wave energy as shown by following example.

Let us assume that A^α in (1.1) is symmetric and that B is antisymmetric (if B is symmetric then the solution of the dispersion relation cannot be real!)

$$A^\alpha_{IJ} = A^\alpha_{JI}, \quad B_{IJ} = -B_{JI}.$$

Also, for simplicity, we assume that $A^0_{IJ} = \delta_{IJ}$ and consider the one-dimensional case.

Then from (1.1) one can obtain the following conservation law

$$(1.16) \qquad\qquad \partial_t \mathcal{E} + \partial_x \mathcal{F}^1 = 0$$

$$(1.17) \qquad\qquad \mathcal{E} = \delta_{IJ} U^I U^J, \qquad \mathcal{F}^1 = A^1_{IJ} U^I U^J.$$

Here, $\mathcal{E}$ and $\mathcal{F}^1$ can be interpreted as an energy density and flux respectively. Now, by substituting (1.12) into (1.17) we obtain

$$\mathcal{E} = \frac{1}{4}\left(a_I a_I e^{2i\theta} + 2a_I \bar{a}_I + \bar{a}_I \bar{a}_I e^{-2i\theta}\right)$$

$$\mathcal{F}^1 = \frac{1}{4}\left(A^1_{IJ} a_I a_J e^{2i\theta} + 2A^1_{IJ} a_I \bar{a}_J + A^1_{IJ} \bar{a}_I \bar{a}_J e^{-2i\theta}\right).$$

When these expressions are used in (1.16), we find

$$(1.18) \qquad\qquad \partial_t < \mathcal{E} > + \partial_x < \mathcal{F}^1 > = 0$$

where

$$< \mathcal{E} > = \frac{1}{2} a_I \bar{a}_I, \quad < \mathcal{F}^1 > = \frac{1}{2} A^1_{IJ} a_I \bar{a}_J,$$

are the energy density and flux averaged with respect to θ over a period of oscillation. From (1.6), (1.11) follows

$$(1.19) \qquad\qquad C_{IJ}(l) a^J = 0$$

where

(1.20) $$C_{IJ}(l) = -iW(l)\delta_{IJ} + il A^1_{IJ} + B_{IJ}.$$

By differentiating (1.19) with respect to l, and contracting with $\bar{a}^I$ we find

$$C'_{IJ}\bar{a}^I a^J = 0$$

because

$$C_{IJ}\bar{a}^I = 0.$$

Hence

(1.21) $$A^1_{IJ} a^I \bar{a}^J = W'(l) a^I \bar{a}^J$$

From (1.20) one has

(1.22) $$< \mathcal{F}^1 > = W'(l) < \mathcal{E} >$$

which states that energy propagates with the group velocity $W'(l)$.

2   Two-timing methods

In the previous section we have briefly recalled the concepts of phase, phase velocity and group velocity for wave trains. These concepts were defined for Fourier integral solutions of constant coefficients linear PDE's. An important concept which was also introduced was that of local phase, local wave number and frequency. Whereas for a plane harmonic wave the wave number and frequency are constant, for a general wave train in the form a Fourier integral, the concepts of wave number and frequency are only local in the sense that they depend on space and time. By introducing the local phase $\theta(x,t)$ it was possible to define the local wave number and frequency in such a way that the relationship which hold for a plane harmonic wave remain valid also for a general wave train.

One would like to extend these concepts to wave solutions of linear PDE's with nonconstant coefficients. In general this is not possible in an exact fashion but one has to resort to approximate methods. A variety of methods exists for this purpose but most of them are variants of a basic approximation tecnique, called the two-timing method, or *method of multiple scales*, which allows us to introduce the concepts of local phase, local wave number and frequency in an asymptotic sense. A deep discussion of the two-timing method can be found in the book of Whitham (1974).

The idea at the root of the two-timing method is that there are two widely different scales in the problem, the scale L of the variation of the background state into which the wave propagates and the mean wavelenght λ of the wave train, with $L \gg \lambda$. We shall investigate wave trains for which the amplitude, wave number and frequency vary slowly (on the background scale) and the phase varies rapidly. Therefore we consider wave solutions of (1.1) (with coefficients which are not, in general, constant) which depend on two characteristic scales: the scale L of variation of the amplitude, wave number and frequency; and the scale $\lambda \ll L$ over which the phase θ varies of 2π. Let $\epsilon = \lambda/L$. Then $0 < \epsilon \ll 1$ is a dimensionless parameter characteristic of the solution.

We define slow variables X^α in $\mathbb{R}^4$ by

$$X^\alpha = \epsilon x^\alpha .$$

Let $\Theta(X^\alpha)$ be a C^1 function of X^α and define $\theta(x^\alpha)$ by

(2.1) $$\theta(x^\alpha) = \frac{1}{\epsilon}\,\Theta(\epsilon x^\alpha) .$$

Then we call θ a *phase* or fast variable.

We define

$$\omega = -l_0 = -\frac{\partial\theta}{\partial t}, \quad \vec{l} = \nabla\theta$$

and we call ω the local frequency and $\vec{l}$ the local wave number. Clearly ω and $\vec{l}$ are slowly varying functions of x^α.

Expanding (2.1) as $\epsilon \to 0$ with small x^α fixed and by developing $\Theta(\epsilon x^\alpha)$ around $\epsilon = 0$ one has

$$\theta(x^\alpha) = \frac{1}{\epsilon}\{\Theta(0) + l_\alpha x^\alpha \epsilon + O(\epsilon^2)\} ,$$

whence $\theta(x^\alpha) = l_\alpha x^\alpha$ apart from an inessential phase factor. This amounts to a locally plane wavefront.

We define a locally plane wave as a formal series of the kind

(2.2) $$\hat{U}^I(X^\alpha,\theta) = e^{i\theta} \sum_{p=0}^{\infty} \epsilon^p U^I_{(p)}(X^\alpha) .$$

If $f(x) = \hat{f}(X,\theta)$, let

$$\hat{f}_{,\alpha} \equiv \frac{\partial \hat{f}}{\partial X^\alpha}, \quad \dot{\hat{f}} \equiv \frac{\partial \hat{f}}{\partial \Theta}$$

then

(2.3) $$\frac{\partial f}{\partial x^\alpha} = \epsilon \hat{f}_{,\alpha} + \dot{\hat{f}} l_\alpha .$$

We seek formal solutions of the system (1.1) which are locally plane waves. That is, we require (2.2) to satisfy (1.1) formally as an identity in ϵ.

We allow the coefficient matrices to depend in an arbitrary smooth way on the slow variables X^α.

To the zeroth order we find

(2.4) $$C_{IJ}U^J_{(0)} = 0$$

with

$$(2.5) \qquad\qquad C_{IJ} = -i\omega A^0_{IJ} + il_a A^a_{IJ} + B_{IJ},$$

where the index a is summed from 1 to 3.

Equation (2.4) has a non-trivial solution for $U^J_{(0)}$ if

$$(2.6) \qquad\qquad \det(C_{IJ}) = 0,$$

which is the local dispersion relation.

Let us assume that a solution of Eq. (2.6) is

$$(2.7) \qquad\qquad \omega = W(\vec{l}, X^\alpha),$$

with W real and analytic in $\vec{l}$. The explicit dependence of W on X^α arises from the dependence of the coefficients of (1.1) on X^α.

From

$$\omega = -\frac{\partial\theta}{\partial t}, \qquad l_a = \frac{\partial\theta}{\partial x^a}$$

one has

$$\frac{\partial l_a}{\partial t} + \frac{\partial\omega}{\partial x^a} = 0, \qquad \frac{\partial l_a}{\partial x_b} - \frac{\partial l_b}{\partial x^a} = 0$$

hence

$$(2.8) \qquad\qquad \frac{\partial l_a}{\partial t} + V^b \frac{\partial l_a}{\partial x^b} = -\frac{\partial W}{\partial x^a},$$

where

$$(2.9) \qquad\qquad V^b = \frac{\partial W(\vec{l}, X^\alpha)}{\partial l_b}.$$

By introducing the rays as the curves defined by

$$(2.10) \qquad\qquad \frac{dx^a}{dt} = \frac{\partial W}{\partial l_a} = V^a.$$

Eq. (2.8) can be written as

$$(2.11) \qquad\qquad \frac{dl_a}{dt} = -\epsilon \frac{\partial W}{\partial x^a}$$

along the rays.

Eqs. (2.10–11) are Hamilton's equations with moments $p_a = l_a$ and Hamiltonian

$$H(p_a, x_a, t) = W(\vec{l}, \vec{x}, t).$$

Similarly from $\omega = -\partial\theta/\partial t$ one obtains the Hamilton-Jacobi equation

$$\frac{\partial\theta}{\partial t} + W(\nabla\theta, \vec{x}, t) = 0.$$

Eqs. (2.10–11) allow us to determine $\vec{l}(\vec{x}, t)$ if we know $W(\vec{l}, \vec{x}, t)$ and the initial values $\vec{l}(\vec{x}, 0) = \vec{f}(\vec{x})$. For instance if W is independent of $(\vec{x}, t)$. Eq. (2.11) gives $\vec{l}$ =constant on any ray $x^a = \xi^a + V^a t$ with

$$V^a = \frac{\partial W}{\partial l_a}(\vec{f}(\vec{\xi})),$$

where $\vec{\xi}$ =constant parametrizes the ray.

Now let us assume that the matrix $C = (C_{IJ})$ corresponding to the solution $\omega = W(\vec{l}, \vec{x}, t)$ has maximum rank $N - 1$ (the method can be easily extended to the general case).

Let $L^I(\vec{l}, \vec{x}, t)$, $R^I(\vec{l}, \vec{x}, t)$ be the corresponding left and right eigenvectors, defined up to a multiplicative factor

$$L^I C_{IJ} = 0, \quad C_{IJ} R^J = 0.$$

From (2.4) we obtain

(2.12)
$$U_{(0)}^I = \Pi R^I$$

with $\Pi = \Pi(X^\alpha)$ a scalar to be determined.

By substituting Eq. (2.2) into (1.1) one has, to the higher orders,

(2.13)
$$C_{IJ} U_{(p+1)}^J + A_{IJ}^\alpha U_{(p),\alpha}^J = 0.$$

By left multiplication by L^I and putting $p = 0$ one gets

(2.14)
$$k^\alpha \Pi_{,\alpha} + \psi\Pi = 0$$

where

(2.15)
$$k^\alpha = L^I A_{IJ}^\alpha R^J$$

(2.16)
$$\psi = L^I A_{IJ}^\alpha R_{,\alpha}^J.$$

Now let us assume $A_{IJ}^0 = \delta_{IJ}$ (which is always possible if $\det A^0 \neq 0$) and normalize L^I, R^I such that $L^I R^I = 1$.

Then

$$k^0 = 1, \quad k^a = L^I A_{IJ}^a R^J.$$

From $C_{IJ} R^J = 0$, by differentiating with respect to l_a and left multiplying by L^I, one obtains

$$L^I \frac{\partial C_{IJ}}{\partial l_a} R^J = 0.$$

From (2.5), (2.7) and (2.9) it follows

$$k^a = V^a .$$

Therefore Eq. (2.14) can be put in the form

(2.17)
$$\frac{d\Pi}{dt} + \psi\Pi = 0$$

along the rays

$$\frac{d\vec{x}}{dt} = \vec{V} .$$

Therefore if $\vec{x}(0) = \vec{\xi}$ and $F(\vec{\xi})$ is the initial value of Π, one has

(2.18)
$$\Pi(X^\alpha) = F(\vec{\xi})e^{-\int_0^t \psi(s)ds} .$$

In the case where the system is such that

$$A_{IJ}^a = A_{JI}^a, \quad B_{IJ} = -B_{JI}$$

we can repeat the same considerations as in the previous section about the energy density and the energy flux.

In particular it is easy to see that

(2.19)
$$< \mathcal{F}^a >= V^a < \mathcal{E} >$$

and that

(2.20)
$$\partial_t < \mathcal{E} > + \partial_a < \mathcal{F}^a >= 0 .$$

For higher orders one proceeds as follows: for $p \geq 1$ let

(2.21)
$$U_{(p)}^I = Z_{(p)} R^I + S_{(p)}^I .$$

Then (2.13) gives, after left multiplication by L^I,

(2.22)
$$k^\alpha Z_{(p),\alpha} + \psi Z_{(p)} + L^I A_{IJ}^\alpha S_{(p),\alpha}^J = 0$$

and furthermore

(2.23)
$$C_{IJ} S_{(p)}^J + A_{IJ}^\alpha U_{(p-1),\alpha}^I = 0 .$$

The solubility conditions for (2.23) are satisfied, and therefore it is possible to obtain $S_{(p)}$ as function of $U_{(p-1)}$.

Then Eq. (2.22) is the transport equation for $Z_{(p)}$ along the rays which allows us to obtain, by recursion, $Z_{(p)}$ from the initial data.

We can also solve formally the initial value problem. Let us assume that, at $t = 0$, the following locally plane wave is given

(2.24)
$$Y^I(\vec{x}) = \hat{Y}^I(\vec{X}, \varphi) = e^{i\varphi} \sum_{p=0}^{\infty} \epsilon^p Y_{(p)}^I(\vec{X})$$

where

(2.25)
$$\varphi = \varphi(\vec{x}) = \frac{1}{\epsilon}\phi(\vec{x}).$$

Let us assume for the sake of simplicity, that the local dispersion relation (2.6) admits N real distinct propagation modes

(2.26)
$$\omega = W^{(k)}(\vec{l}, \vec{x}, t), \qquad k = 1, \ldots, N.$$

First we solve the Hamiltonian system

$$\frac{d\vec{x}}{dt} = \frac{\partial W^{(k)}}{\partial \vec{l}}, \qquad \frac{d\vec{l}}{dt} = -\frac{\partial W^{(k)}}{\partial \vec{x}}$$

with the initial conditions

$$\vec{x}(0) = \vec{\xi}, \qquad \vec{l}(0) = \nabla\varphi$$

thereby obtaining the rays of the k-th mode.

The phase $\theta^{(k)}(\vec{x}, t)$ of the k-th mode is obtained by solving the Hamilton-Jacobi equation

$$\frac{\partial \theta}{\partial t} + W^{(k)}(\nabla\theta, \vec{x}, t) = 0$$

with $\theta(0, \vec{x}) = \varphi(\vec{x})$.

Let $\mathbf{R}^{(k)}$ be the right eigenvector of C corresponding to the k-th mode. Since $\mathbf{R}^{(k)}, k = 1, \ldots, N$ span $\mathbb{R}^N$ one has

(2.27)
$$Y_{(0)} = \Pi^{(k)} R^{(k)}, \qquad \Pi^{(k)} = \Pi^{(k)}(\vec{x}).$$

By solving the transport equation (2.17) along each k-th ray one obtains $\Pi^{(k)}(X^\alpha)$. By integrating and using Eq. (2.22–23) we obtain $U_{(p)}^{(k)}(X^\alpha)$.

The formal solution is then a superposition of locally plane waves

(2.28)
$$\mathbf{U}(\vec{x}, t) = e^{i\theta^{(1)}} \sum_{p=0}^{\infty} \epsilon^p U_{(p)}^{(1)}(X^\lambda) + \cdots + e^{i\theta^{(N)}} \sum_{p=0}^{\infty} U_{(p)}^{(N)}(X^\lambda).$$

A variant of the method we have just expounded is the well known high-frequency-expansion method (Courant and Hilbert, 1962, Lighthill, 1980, Kline and Kay, 1965, Lax, 1957, Lewis, 1965, Ludwig, 1966, Whitham, 1974).

In this latter method the stretching transformation $X^\alpha = \epsilon x^\alpha$ is not performed and one works on the original coordinates (x^α). Therefore, with this method, only non-dispersive waves can be treated.

One looks for solutions depending on the parameter ϵ and also on an (unknown) phase $\phi(x^\alpha)$, of the form

$$(2.29) \qquad \mathbf{U}(x^\alpha, \varphi) = e^{i\phi/\epsilon} \sum_0^\infty \epsilon^p \mathbf{U}_{(p)}(x^\alpha).$$

By substituting the series (2.29) formally into system (1.1) and equating termwise to zero the coefficients of the various powers of ϵ one obtains, to the order ϵ^{-1},

$$(2.30) \qquad l_\alpha A^\alpha_{IJ} U^J_{(0)} = 0$$

where $l_\alpha = \dfrac{\partial \phi}{\partial x^\alpha}$.

From (2.30), for a non vanishing $\mathbf{U}_{(0)}$ we have the characteristic equation

$$(2.31) \qquad \det(l_\alpha A^\alpha_{IJ}) = 0.$$

Let $\phi(x^\alpha) = $ const. be a simple root of the characteristic Eq. (2.31), R^J and L^I the corresponding right and left eigenvectors, normalized by $L^I R^I = 1$, then from (2.30) it follows

$$(2.32) \qquad \mathbf{U}_{(0)} = \Pi(x^\alpha)\mathbf{R}$$

with $\Pi(x^\alpha)$ to be determined.

From system (2.29), to the order ϵ^0 we obtain

$$il_\alpha A^\alpha_{IJ} U^J_{(1)} + A^\alpha_{IJ} \partial_\alpha U^J_{(0)} + B_{IJ} U^J_{(0)} = 0$$

which, after multiplication by L^J, yields the transport equation for Π

$$(2.33) \qquad \frac{d\Pi}{d\sigma} + \psi\Pi = 0$$

along the rays given by

$$(2.34) \qquad \frac{dx^\alpha}{d\sigma} = k^\alpha$$

where

$$(2.35) \qquad k^\alpha = L^I A^\alpha_{IJ} R^J$$

and

$$\psi = L^I A^\alpha_{IJ} \partial_\alpha R^J + L^I B_{IJ} R^J.$$

A similar treatment can be performed in the case when $\varphi = $ constant is a multiple root of the characteristic equation.

Notice that when $B_{IJ} = 0$ this method essentially coincides with the two-timing one. In the next chapter we shall take up again, in more detail, the high-frequency method for weakly nonlinear waves. Finally we mention that, among the several variants of the two-timing methods, one of the most successful is the averaged Lagrangian technique (for systems allowing a Lagrangian description) (Dougherty, 1970, Whitham, 1974).

REFERENCES

- A. Ben Menahem, S. J. Singh, *Seismic Waves and Sources*, Springer-Verlag, New York (1981).
- F. P. Bretherton, *The generalised theory of wave propagation*, on *Mathematical Problems in the Geophysical Sciences*, Lecture Notes in Applied Mathematics, **13**, 61–102, William H. Reid Editor, American Mathematical Society, Providence RI (1971).
- R. Courant and D. Hilbert, *Methods of Mathematical Physics*, vol. II Interscience, New York (1962).
- R. J. Di Perna and A. Majda, *Oscillations and Concentrations in Weak Solutions of the Incomprensible Fluid Equations*, Comm. Math. Phys. **108**, 667 (1987).
- J. P. Dougherty, *Lagrangian methods in plasma dynamics. Part 1. General theory of the method of averaged Lagrangian*, J. Plasma Phys. **4**, 761 (1970).
- A. Erdélyi, *Asymptotic Expansions*, Dover Publications, New York (1956).
- F. G. Friedlander, *The wave equation on a curved space-time*, Cambridge University Press (1975).
- P. Germain, *Methodes Asimptotiques en Mécanique des Fluides*, on *Fluid Dynamics*, edited by Balian and J. L. Peube, Gordon and Breach Science Publishers, London (1977).
- J. A. Hudson, *The excitation and propagation of elastic waves*, Cambridge University Press (1980).
- J. B. Keller, R. M. Lewis and B. D. Seckler, *Asymptotic solution of some diffraction problem*, Comm. Pure and Appl. Math. **9**, 207 (1956).
- M. Kline and I. W. Kay, *Electromagnetic Theory and Geometrical Optics*, John Wiley & Sons, New York (1965).
- P. D. Lax, *Asymptotic Solutions of oscillatory Initial Value Problems*, Duke Math. J. **24**, 627 (1957).
- R. M. Lewis, *Asymptotic Expansion of Steady-State Solutions of Symmetric Hyperbolic Linear Differential equations*, J. Math. Phys. **7** (1958).
- R. M. Lewis, *Asymptotic theory of wave propagation*, Arch. Rat. Mech. Anal. **20**, 191 (1965).
- J. Lighthill, *Waves in Fluids*, Cambridge University Press (1980).
- D. Ludwig, *Uniform Asymptotic Expansion at a Caustic*, Comm. Pure and Appl. Math. Vol. XIX, 215 (1966).
- R. K. Lünebourg, *The Mathematical Theory of Optics*, Lecture Notes, Brown University (1944).
- G. B. Whitham, *Linear and Nonlinear waves*, Wiley, New York (1974).

RAY METHODS
FOR NONLINEAR
HYPERBOLIC WAVES

Introduction

In the previous chapter we have introduced the fundamentals of ray methods for linear waves. In particular we considered two cases. The first is that of linear dispersive waves and in this context we introduced the two-timing method. The second in that of linear non-dispersive (or hyperbolic) waves and this case is best discussed by using the high-frequency expansion method.

Now the problem naturally arises of extending the concepts developed for discussing linear waves (local phase, local wave number and frequency, rays, transport laws, etc.) to the nonlinear case. For this purpose it is convenient to simplify the problem by separating the effects of dispersion from those of nonlinearity. Therefore in this chapter we shall discuss weakly nonlinear hyperbolic waves. Hence in this chapter we will present a nonlinear extension of the high-frequency expansion method previously introduced. The combined effects of dispersion, dissipation and nonlinearity will be discussed in Chapters 8 and 9.

Let L, λ be, as in the previous chapter, the scales of variation of the background state, and perturbation respectively. We write $\epsilon = \lambda/L$ and assume that $\epsilon \ll 1$.

In a nonlinear problem we must also consider the order of magnitude of the perturbation amplitude, a, and therefore there is one parameter more than in the linear case.

In order to extend the high-frequency expansion method we must assume a definite relationship between ω and a. Drawing from the exact solutions representing simple waves for hyperbolic systems one infers that the most interesting case is $1/\omega \sim a$ (Choquet-Bruhat, 1969). Under this assumption Choquet-Bruhat constructed a nonlinear high-frequency expansion method which has proved itself to be extremely successful. In this theory the wave propagates along the bicharacteristics of the unperturbed state and the nonlinearity shows itself in a modified transport equation for the amplitude. Such a transport equation, using suitable variables, reduces to the standard kinematic wave equation and therefore the relevant results of Sec. 2.1 are readily applicable. This method has been applied successfully to several problems in fluid mechanics. (Boillat, 1965, Chin et al., 1986, Eckhoff, 1981) and relativity theory (Anile, 1989).

In this chapter in Sec. 1 we shall expound the nonlinear high-frequency method for a single wave, following the approaches of Choquet-Bruhat (1969) and Boillat (1965). In Sec. 2 we present an application to the study of an acoustic wave propagating in a stratified atmosphere. Interacting waves will be considered in Chapter 11.

It is remarkable that in recent years the convergence of the method has been rigorously established for hyperbolic waves in one space dimension (Di Perna and Majda, 1985).

1   Asymptotic waves for quasilinear systems

Let us consider the quasilinear system

$$
(1.1) \qquad \mathcal{A}^\alpha(\mathbf{U}, x^\mu) \partial_\alpha \mathbf{U} = \mathbf{f}(\mathbf{U}, x^\mu)
$$

where $\mathbf{U} = (U^1, \ldots, U^N)$ is the field vector, $\mathcal{A}^\alpha$ are $N \times N$ matrices, with $\mathcal{A}^0 = I$, and x^μ are space-time coordinates with $x^0 = t$ and x^i spatial cartesian coordinates.

Generalizing the approch expounded in Chapter 3, Sec. 2, we seek solutions of (1.1) in the form of asymptotic series (Choquet-Bruhat, 1969, Boillat, 1965)

$$
(1.2) \qquad \mathbf{U} = \mathbf{U}_0(x^\alpha) + \epsilon \mathbf{U}_1(x^\alpha, \theta) + \epsilon^2 \mathbf{U}_2(x^\alpha, \theta) + \cdots
$$

where $\mathbf{U}_0(x^\alpha)$ is a given solution of (1.1), $\epsilon \ll 1$ is a parameter related to the frequency of the wave $\theta = \Theta(x^\alpha)/\epsilon$, with the phase $\Theta(x^\alpha)$ a function to be determined.

Assuming that $\mathcal{A}^\alpha$ and $\mathbf{f}$ are analytic in $\mathbf{U}$ in a suitable neighborhood of $\mathbf{U}_0$, we obtain the following expansions

$$
(1.3) \qquad
\begin{cases}
\mathcal{A}^\alpha(\mathbf{U}, x^\mu) = \mathcal{A}^\alpha(\mathbf{U}_0, x^\mu) + \epsilon \nabla \mathcal{A}^\alpha(\mathbf{U}_0, x^\mu) \mathbf{U}_1 + 0(\epsilon^2) \\[2mm]
\partial_\alpha \mathbf{U} = \partial_\alpha \mathbf{U}_0 + \epsilon \left(\dfrac{\partial \mathbf{U}_1}{\partial x^\alpha} + \dfrac{1}{\epsilon} \dfrac{\partial \mathbf{U}_1}{\partial \theta} \Theta_\alpha \right) + \epsilon \dfrac{\partial \mathbf{U}_2}{\partial \theta} \Theta_\alpha + 0(\epsilon^2) \\[2mm]
\mathbf{f}(\mathbf{U}, x^\alpha) = \mathbf{f}(\mathbf{U}_0, x^\alpha) + \epsilon \nabla \mathbf{f}(\mathbf{U}_0, x^\alpha) \mathbf{U}_1 + 0(\epsilon^2)
\end{cases}
$$

where $\Theta_\alpha = \partial_\alpha \Theta$ and

$$
\nabla \equiv \left(\frac{\partial}{\partial U^1}, \ldots, \frac{\partial}{\partial U^N} \right).
$$

Substituting the expansions (1.3) into (1.1) and equating to zero termwise the coefficients of the powers in ϵ gives, to the order 0

$$
(1.4) \qquad \mathcal{A}_0^\alpha \Theta_\alpha \frac{\partial \mathbf{U}_1}{\partial \theta} = 0
$$

and to the order ϵ

$$
(1.5) \qquad \mathcal{A}_0^\alpha \left(\frac{\partial \mathbf{U}_1}{\partial x^\alpha} + \Theta_\alpha \frac{\partial \mathbf{U}_2}{\partial \theta} \right) + \nabla \mathcal{A}_0^\alpha \mathbf{U}_1 \left(\partial_\alpha \mathbf{U}_0 + \frac{\partial \mathbf{U}_1}{\partial \theta} \Theta_\alpha \right) = \nabla \mathbf{f}_0 \mathbf{U}_1
$$

where

$$
(1.6) \qquad \mathcal{A}_0^\alpha = \mathcal{A}^\alpha(\mathbf{U}_0, x^\mu).
$$

Let

$$n_i = \frac{\Theta_i}{|\operatorname{grad}\Theta|}, \qquad \Theta_i \equiv \partial_i\Theta$$

be a unit normal vector, and consider the eigenvalue problem

$$(\mathcal{A}^i n_i - \lambda \mathcal{A}^0)\mathbf{R} = 0.$$

We assume the system (1.1) to be hyperbolic in the t-direction (Jeffrey, 1976). This implies that $\mathcal{A}$ is a non singular, the N eigenvalues $\lambda(\mathbf{U}, x^\alpha, \vec{n})$ are real and the corresponding N eigenvectors $\mathbf{R}_I$ (where I runs from 1 to the multiplicity of the eigenvalue) span $\mathbb{R}^N$.

Whithout loss of generality we shall assume that $\mathcal{A}^0 = \mathcal{I}$. Then (1.4) implies, for a particular choice of the eingenvalue λ,

(1.7) $$\mathbf{U}_1 = u^I(x^\alpha;\theta)\mathbf{R}_I(\mathbf{U}_0, x^\alpha;\vec{n}) + \mathbf{V}_1(x^\alpha)$$

with u^I m scalar functions (where m is the multiplicity of the eigenvalue λ) and $\mathbf{V}_1$ an arbitrary vector function).

Furthermore the phase Θ satisfies by construction the "eikonal" equation

(1.8) $$\psi(\mathbf{U}, x^\alpha; \Theta_\mu) = 0$$

where

(1.9) $$\psi(\mathbf{U}, x^\alpha, \Theta_\mu) = \Theta_t + |\operatorname{grad}\Theta|\lambda(\mathbf{U}, x^\alpha;\vec{n}).$$

The eikonal equation (1.8) is solved by the method of characteristics. One introduces the rays, $x^\alpha = x^\alpha(\sigma)$, defined by the equations (Boillat, 1965)

(1.10) $$\frac{dx^\alpha}{d\sigma} = \frac{\partial\psi_0}{\partial\Theta_\alpha}, \qquad \frac{d\Theta_\alpha}{d\sigma} = -\frac{\partial\psi_0}{\partial x^\alpha}$$

where

$$\psi_0(x^\alpha, \Theta_\beta) = \psi(\mathbf{U}_0(x^\alpha), x^\alpha; \Theta_\beta).$$

From Eq. (1.10) with $\alpha = 0$, one finds $\sigma = t$.
Then one assigns the initial value for x^i,

$$(x^i)_{t=0} = x_0^i$$

and for the phase Θ,

$$(\Theta)_{t=0} = \phi(x_0^i),$$

whence one obtains also

$$(\Theta_i)_{t=0} = \phi_i(x_0^j), \qquad \phi_i \equiv \partial_i\phi$$

which defines the normal $\vec{n}_0$ at the point x_0^i.

By integrating system (1.10) with these initial values one obtains the explicit equations for the rays

(1.11) $$x^i = x^i(x_0^j, \sigma), \quad x^i(x_0^j, 0) = x_0^i, \quad t = \sigma.$$

If the Jacobian

$$J = \frac{D(x^i)}{D(x_0^i)}$$

is non-vanishing, Eqs. (1.11) can be inverted yielding

(1.12) $$x_0^i = x_0^i(x^j, t), \quad \sigma = t$$

and the phase is given by

$$\Theta(x^\alpha) = \phi(x_0^i(x^j, t)).$$

In Eq. (1.7) the arbitrary vector $\mathbf{V}_1$ can be chosen to vanish, $\mathbf{V}_1 = 0$, without loss of generality.

Let $\mathbf{L}_I(\mathbf{U}_0, x^\alpha; \vec{n})$ be the left eigenvectors of the matrix $A_0^i n_i$ corresponding to the eigenvalue $\lambda(\mathbf{U}_0, x^\alpha; \vec{n})$ (where $I = 1, \dots, m$, m being the multiplicity of λ).

By substituting (1.7) into (1.5), and left multiplying by $\mathbf{L}_J(\mathbf{U}_0, x^\alpha; \vec{n})$ we obtain

(1.13) $$\mathbf{L}_J A_0^\alpha \left(\frac{\partial u^I}{\partial x^\alpha} \mathbf{R}_I + u^I \partial_\alpha \mathbf{R}_I \right) + \mathbf{L}_J \nabla A_0^\alpha \mathbf{U}_1 \cdot \left(\partial_\alpha \mathbf{U}_0 + \phi_\alpha \frac{\partial u^I}{\partial \theta} \mathbf{R}_I \right) =$$
$$= \mathbf{L}_J \nabla \mathbf{f}_0 \mathbf{U}_1.$$

By differentiating Eq. (1.9) with respect to Θ_α one can prove that

$$\mathbf{L}_J A_0^\alpha \mathbf{R}_I = \frac{\partial \psi_0}{\partial \Theta_\alpha} \mathbf{L}_J \mathbf{R}_I.$$

Differentiating the identity

$$A^\alpha \Theta_\alpha \mathbf{R}_I = \psi \mathbf{R}_I,$$

with respect to $\mathbf{U}$ in the direction $\mathbf{U}_1$ yields

$$\nabla A^\alpha \Theta_\alpha \mathbf{U}_1 \mathbf{R}_I + (A^\alpha \Theta_\alpha - \psi I) \nabla \mathbf{R}_I \mathbf{U}_1 = (\nabla \psi \mathbf{U}_1) \mathbf{R}_I,$$

whence

$$\mathbf{L}_J \nabla A_0^\alpha \Theta_\alpha \mathbf{U}_1 \mathbf{R}_I = (\nabla \psi_0 \mathbf{U}_1) \mathbf{L}_J \mathbf{R}_I.$$

Therefore one can rewrite (1.13) as

(1.14) $$\left\{ \frac{\partial u^I}{\partial \sigma} + (\nabla \psi_0 \mathbf{U}_1) \frac{\partial u^I}{\partial \theta} \right\} \mathbf{L}_J \mathbf{R}_I + C_{JI}^0 u^I = f_{JI}^0 u^I,$$

with $\mathbf{U}_1 = u^I \mathbf{R}_I$ and where

(1.15) $$\begin{cases} C_{JI}^0 = \mathbf{L}_J (A_0^\alpha \partial_\alpha \mathbf{R}_I + \nabla A_0^\alpha \mathbf{R}_I \partial_\alpha \mathbf{U}_0) \\ f_{JI}^0 = (\mathbf{L}_J \nabla \mathbf{f} \mathbf{R}_I)_0 \\ \dfrac{\partial}{\partial \sigma} = \dfrac{\partial}{\partial t} + \lambda n^i \dfrac{\partial}{\partial x^i}. \end{cases}$$

The exceptional case occurs when the wave is exceptional (or linearly degenerate), i.e.

(1.16) $$(\nabla\lambda) \cdot \mathbf{R}_I = 0.$$

Then

$$\nabla\psi_0 \cdot \mathbf{U}_1 = |\,\text{grad}\Theta|\nabla\lambda_0 \cdot \mathbf{R}_I u^I = 0$$

and (1.14) reduces to a linear system.

Now we recall the following property of the Jacobian J. One has

(1.17)
$$\frac{1}{J}\frac{\partial J}{\partial\sigma} = \frac{D(x_0^i)}{D(x^i)}\left\{ \frac{D\left(\dfrac{\partial x^1}{\partial\sigma},\dots,x^n\right)}{D(x_0^1,\dots,x_0^n)} + \dots + \frac{D\left(x^1,\dots,\dfrac{\partial x^n}{\partial\sigma}\right)}{D(x_0^1,\dots,x_0^n)} \right\} =$$
$$= \frac{D(\Lambda_0^1,\dots,x^n)}{D(x^1,\dots,x^n)} + \dots + \frac{D(x^1,\dots,\Lambda_0^n)}{D(x^1,\dots,x^n)} = \partial_i\Lambda_0^i$$

where

(1.18) $$\Lambda_0^i = \frac{\partial\psi_0}{\partial\Theta_i}$$

are the components of the radial speed (Boillat, 1965).

It is convenient to introduce the quantity

(1.19) $$\tilde\theta(\sigma, x_0^i) = \sqrt{J}.$$

Let us consider now propagation into a constant state,

$$\mathbf{U}_0 = \text{const.}$$

We shall also assume that A^α does not depend on x^μ. Then φ_0 does not depend on x^μ and (1.10) gives

$$\Theta_i = \text{const.}$$

Along the rays one has

(1.20) $$x^i = x_0^i + \Lambda^i(\mathbf{U}_0, \vec{n}_0)\sigma.$$

Also, the coefficients in the transport equations (1.14) simplify. From (1.15), one has

$$C_{JI}^0 = \mathbf{L}_J A_0^\alpha \frac{\partial\mathbf{R}_I}{\partial\Theta_\mu}\Theta_{\alpha\mu}$$

and it follows that

(1.21) $$C_{JI}^0 = (\mathbf{L}_J\mathbf{R}_I)\frac{\partial}{\partial\sigma}\log\tilde\theta.$$

For isotropic wave motion, when λ_0 is independent of $\vec{n}$, $\tilde{\theta}$ can be expressed as a function of σ and the total and mean curvature of surface

$$\phi(x_0^i) = \text{const.}$$

When λ_0 is a simple eigenvalue, (1.14) can be written as

(1.22)
$$\frac{\partial u^1}{\partial \sigma} + (\nabla\psi \cdot \mathbf{R})_0 u^1 \frac{\partial u^1}{\partial \theta} + \frac{1}{\tilde{\theta}}\frac{\partial \tilde{\theta}}{\partial \sigma} u^1 + \nu_0 u^1 = 0$$

where

$$\nu_0 = \frac{(\mathbf{L}\nabla \mathbf{fR})_0}{\mathbf{LR}}.$$

By the change of variables

(1.23)
$$\begin{cases} u^1 = \dfrac{u}{\tilde{\theta}}e^w \\[2mm] w = \int_0^\sigma \nu_0 d\sigma \\[2mm] \tau = \int_0^\sigma (\nabla\psi \cdot \mathbf{R})_0 \dfrac{e^w}{\tilde{\theta}} d\sigma, \end{cases}$$

(1.22) reduces to

$$\frac{\partial u}{\partial \tau} + u\frac{\partial u}{\partial \theta} = 0$$

which is the canonical form for the kinematic wave equation. When $\mathbf{f}$ does not depend on x^α, $w = \nu_0\sigma$. Let the initial values for u be given by

$$u = F(x_0^i; \theta_0), \quad \theta_0 = \phi/\epsilon.$$

Then u is given by

(1.24)
$$u = F(x_0^i; \theta - u\tau).$$

The singularities of u^1 appear when $\tilde{\theta} = 0$, corresponding to a caustic, or when $\partial u/\partial \tau \to \infty$, i.e.

$$1 + \tau\frac{\partial F}{\partial \theta_0}(x_0^i; \theta - u\tau) = 0.$$

2   Acoustic waves in a gravitational atmosphere

We recall Eq. (1.22) and rewrite it as

(2.1)
$$\frac{\partial \Gamma}{\partial \sigma} + N\Gamma\frac{\partial \Gamma}{\partial \theta} + M\Gamma = 0.$$

Where

$$\Gamma \equiv u^1, \quad N = (\nabla \psi \cdot \mathbf{R})_0, \quad M = \frac{1}{\tilde{\theta}} \frac{\partial \tilde{\theta}}{\partial \sigma} \quad \text{and} \quad \nu_0 = 0.$$

We assume that, for $\sigma = 0$, the values of Γ, θ, x^α have been assigned, i.e.

(2.2) $\qquad \theta(\sigma = 0) = \mu, \quad x^\nu(\sigma = 0) = x_0^\nu, \quad \Gamma(\sigma = 0) = \Gamma_0(\mu).$

The implicit solution of (2.1), (2.2.) is

(2.3) $$\Gamma(\sigma) = \Gamma_0(\mu) e^{-Q(\sigma)},$$

(2.4) $$\theta = \mu + \Gamma_0(\mu) \int_0^\sigma N(\sigma') e^{-Q(\sigma')} d\sigma',$$

where

$$Q(\sigma) = \int_0^\sigma M(\sigma') d\sigma'.$$

In (2.3) it is understood that $\mu = \mu(\sigma)$ from (2.4). The infimum value of the σ's such that (2.4) is not invertible with respect to μ is called the critical parameter.

Obviously conditions (2.2) will be interpreted differently in the case of initial or mixed initial-boundary value problems.

In this section we use the general theory developed above to analyze the propagation of sound waves through a stratified athmosphere. For one-dimensional propagation along the z-axis the gas dynamics equations are

(2.5) $$\mathbf{U}_t + \mathcal{A} \mathbf{U}_z + \mathbf{B} = 0$$

where

$$\mathbf{U} = \begin{pmatrix} v \\ \rho \\ s \end{pmatrix}, \quad \mathcal{A} = \begin{pmatrix} v & \dfrac{1}{\rho}\dfrac{\partial P}{\partial \rho} & \dfrac{1}{\rho}\dfrac{\partial P}{\partial s} \\ \rho & v & 0 \\ 0 & 0 & v \end{pmatrix}, \quad \mathbf{B} = \begin{pmatrix} g \\ 0 \\ 0 \end{pmatrix}$$

and $P = P(\rho, s)$. Here v is the fluid velocity, ρ is the density, s is the specific entropy and g is the gravitational constant. The pressure P is given by the equation of state $P = P(\rho, s)$. We neglect thermoviscous effects.

First we consider an isothermal athmosphere. The unperturbed solution is

(2.6) $$v_0 = 0, \quad \rho_0 = \tilde{\rho}_0 e^{-z/h}, \quad s_0 = s_0(z),$$

where $h = \overline{R} T_0 / g$ is the density scale height and the fluid is assumed to be perfect gas (see Chapter 1, Sec. 1) with polytropic constant γ. Note that in this case the sound speed $c_0 = \sqrt{\gamma g h}$ is independent of z. Here, for the sake of simplicity, we will consider upward propagating acoustic waves.

Then for the phase we choose the "retarded time"

$$\Theta = t - \frac{z}{c_0} .$$

The rays are straight lines

$$z - c_0 t = \text{constant} ,$$

and we use $t = \sigma$ as a parameter along the rays. The asymptotic solution for the sound wave is

$$\begin{pmatrix} v \\ \rho \\ s \end{pmatrix} = \begin{pmatrix} 0 \\ \rho_0 \\ s_0 \end{pmatrix} + \epsilon \Gamma(t, \theta) \begin{pmatrix} 1 \\ \rho_0/c_0 \\ 0 \end{pmatrix} + O(\epsilon^2)$$

where $\Gamma(t, \xi)$ satisfies

$$\Gamma_t - \frac{\gamma+1}{2c_0} \Gamma\Gamma_\theta - \frac{c_0}{2h} \Gamma = 0 .$$

Let us consider the initial value problem:

$$\Gamma(0, \theta) = \overline{\Gamma} \sin \theta .$$

The parametric solution of (2.1) along the ray $z = c_0 t + z_0$ is

$$(2.7) \qquad \theta = \mu + \frac{\overline{\Gamma}(\gamma+1)h}{c_0^2} \{ 1 - e^{(z-z_0)/2h} \} \sin \mu$$

$$(2.8) \qquad \Gamma = \overline{\Gamma} e^{(z-z_0)/2h} \sin \mu,$$

where μ parametrizes the characteristics. When (2.7) is invertible for μ we obtain (Anile, Mulone and Pluchino 1980)

$$\mu(\theta) = \theta + \sum_{q=1}^{\infty} \frac{2J_q(qE)}{q} \sin q\theta ,$$

$$E = \frac{\overline{\Gamma}(\gamma+1)h}{c_0^2} \left\{ 1 - e^{\frac{z-z_0}{2h}} \right\}$$

where J_q is the Bessel function of order q. Also, from (2.7) one obtains the critical height z_c which occurs on the $z_0 = 0$ ray

$$(2.9) \qquad z_c = 2h \ \log \left\{ 1 + \frac{c_0^2}{(\gamma+1)h\overline{\Gamma}} \right\} .$$

Next we consider an isentropic athmosphere. The equations of motion are (2.5) with $s = \text{constant}$ and $P = P(\rho)$.

For some special equations of state it is then possible (Rozdestvenskii and Yanenko, 1983) to find exact solutions which represent nonlinear plane waves. In particular we shall consider the case of the equation of state

$$P = A\rho^3 + B,$$

where A and B are constants.

Then the change of variables

$$v = \frac{1}{2}(\alpha + \beta), \quad c = \sqrt{dP/d\rho} = \frac{1}{2}(\alpha - \beta)$$

reduces (2.5) to

(2.10)
$$\frac{\partial \alpha}{\partial t} + \alpha \frac{\partial \alpha}{\partial z} = -g,$$

(2.11)
$$\frac{\partial \beta}{\partial t} + \beta \frac{\partial \beta}{\partial z} = -g.$$

Henceforth we shall restrict ourselves to (2.10). It is easy to see that the implicit solution of the initial value problem for (2.10) with $\alpha(z,0) = f(z)$ is

(2.12)
$$\alpha = -gt + f(\eta),$$

(2.13)
$$z = -\frac{1}{2}gt^2 + f(\eta)t + \eta.$$

In order to compare the exact solution with the asymptotic solution we choose $f(\eta)$ as follows

(2.14)
$$f(\eta) = \alpha_0(\eta) + \frac{G}{\omega} \sin \frac{\omega}{g} \alpha_0(\eta),$$

where

(2.15)
$$\alpha_0(z) = \sqrt{\overline{\alpha}_0^2 - 2gz}$$

and G, ω, and $\overline{\alpha}_0$ are constants. We remark that $\alpha = \alpha_0(z)$ is the static solution of (2.10), and we regard it as the unperturbed state. This static solution is defined only in the half-plane $-\infty \leq z \leq \overline{\alpha}_0^2/2g$, which shows that the isentropic assumption breaks down in a static atmosphere at sufficiently large heights. For the sake of physical interpretation, we restrict the solution to the strip $0 \leq z \leq \overline{\alpha}_0^2/2g$. It is easy to see that the domain of existence of the solution (2.12–14) can exceed this strip for some values of ω. The critical time and critical height for the solution (2.12–14) occur on the characteristic issuing from $\eta = 0, t = 0$ and are

(2.16)
$$t_c = \overline{\alpha}_0/(g + G)$$

and

(2.17)
$$z_c = \frac{\overline{\alpha}_0^2}{2g}\left\{1 - \frac{G}{(g+G)}\right\} + \frac{\overline{\alpha}_0 G}{w(g+G)}\sin\frac{w}{g}\overline{\alpha}_0.$$

Now we seek the asymptotic solution of (2.10) corresponding to the initial value (2.14).

(2.18)
$$\alpha(z,t) = \alpha_0(z) + \epsilon\alpha_1(z,t,w) + O(\epsilon^2)$$

$$\theta = \Theta(z,t)/\epsilon.$$

For the phase Θ we choose:

$$\Theta = t + \frac{\alpha_0(z)}{g}$$

and the rays are given by

(2.19)
$$z = -\frac{1}{2}gt^2 + \sqrt{\overline{\alpha}_0^2 - 2gEt} + E - \overline{\alpha}_0.$$

From (2.14) the initial condition for α_1 is:

(2.20)
$$\alpha_1 = G\sin\theta, \quad \text{at } t = 0.$$

We note that the domain of existence of the asymptotic wave (2.18) is restricted to the strip $0 \le z \le \overline{z} = \overline{\alpha}_0^2/2g$, which need not be the case for the exact solution.

The transport equation for α_1 is solved as usual. After some algebra one obtains that, along a ray issuing from $z = E$, $t = 0$ the solution for α_1 is given implicitly by

(2.21)
$$\alpha_1(z,t,\xi) = G\alpha_0(E)/\alpha_0(z)\sin\mu,$$

(2.22)
$$\theta = \mu + G((\alpha_0(z) - \alpha_0(E))/(g\alpha_0(z))\sin\mu.$$

In the final solution θ is evaluated at $\Theta(z,t)/\epsilon$. The critical time and the critical height occur on the ray issuing from $z = 0$ and are found to be

(2.23)
$$t'_c = \overline{\alpha}_0/(g + |G|),$$

(2.24)
$$z'_c = \frac{\overline{\alpha}_0^2}{2g}\left\{1 - \frac{|G|^2}{(g+|G|)^2}\right\}.$$

We note that the critical time for the asymptotic wave coincides with that for the exact solution. However the critical heights differ by the quantity

$$\frac{\epsilon\overline{\alpha}|G|}{(g+|G|)}\sin\left(\frac{\overline{\alpha}_0}{g\epsilon}\right)$$

which vanishes as $\epsilon \to 0$.

REFERENCES

- A.M. Anile, G. Mulone and S. Pluchino, *Critical time for asymptotic acoustic waves in a gravitational atmosphere*, Wave Motion **2**, 267 (1980).
- A.M. Anile, *Relativistic Fluid and Magnetofluids*, Cambridge University Press (1989).
- G. Boillat, *La Propagation des Ondes*, Gauthier-Villars, Paris (1965).
- G. Boillat, *Caracteristiques complexes et instabilité des champs quasi lineaires*, Rendiconti del Circolo Matematico di Palermo **30**, 416 (1981).
- R.C.Y. Chin, J.C. Garrison, C.D. Levermore and J. Wong, *Weakly nonlinear acoustic instabilities*, Wave Motion **8**, 537 (1986).
- Y. Choquet-Bruhat, *Ondes Asymptotiques et approaches pour systemes d'equations aux derivees partielles nonlineares*, J. Math. Pures et Appl. **48**, 117 (1969).
- R. Di Perna and A. Majda, *The validity of geometrical optics for weak solutions of conservation laws*, Comm. Math. Phys. **98**, 313 (1985).
- K.S. Eckhoff *On stability for Symmetric Hyperbolic Systems, I*, Journal of Differential Equations **40**, 94 (1981).
- A. Jeffrey, *Quasilinear Hyperbolic Systems and Waves*, Pitman, London (1976).
- B.L. Rozdestvenskii and N.N. Yanenko, *Systems of Quasilinear Equations and their Application to Gas Dynamics*, translations of Mathematical Monograph vol. 55, American Mathematical Society (1983).

RAY METHOD FOR THE PROPAGATION OF DISCONTINUITIES

chapter 5

Introduction

In the previous chapter we have introduced ray methods for nonlinear hyperbolic waves by an asymptotic method, the nonlinear high-frequency expansion. We found that the fundamental concepts of ray theory, local phase, local wave number and frequency, rays, transport equations along the rays, can be introduced in an asymptotic sense when the wave frequency is sufficiently large. Obviously such a situation describes a continuous wave train propagating into a slowly varying background. This situation, although quite general, does not encompass the equally interesting case of impulsive waves.

In this chapter we shall concentrate on impulsive waves and show how ray methods can be developed for this class of waves.

Impulsive waves represent pulses with time and length scales negligible compared to the other scales present in the physical problem under investigation. They are represented as singular surfaces, i.e. propagating surfaces across which the field variables or their derivatives undergo a jump discontinuity. When the fields are continuous but some of their derivatives are discontinuous the surface is a weak discontinuity surface (Truesdell and Toupin, 1960, Jeffrey, 1976). Otherwise the surface represents a shock wave. One of the great advantages of the concept of singular surfaces is that, under suitable conditions, it leads to an *exact ray theory*, free of any approximation.

In fact, in the case of a weak discontinuity, it can be shown that its motion decouples from that of the smooth part of the solution. Therefore the evolution of the wavefront can be uniquely determined once the state ahead of the wavefront has been specified as well as the initial data for the wave parameters. In particular it is possible to introduce a class of curves (rays) by solving the characteristic equation. Then the wave parameters (amplitude, polarization, etc.) are determined by a set of ordinary differential equations (transport laws) along the rays.

This theory, in its essence, goes back to the work of Hadamard (1903) and has been extensively developed in the context of continuum mechanics by Thomas (1957), Boillat (1965), Truesdell and Toupin (1960), Chen (1976) and others. It has proved to be an invaluable tool for studying acceleration waves in continuous media (Chen, 1976). Sections 1–2 of this Chapter are devoted to a concise exposition of the formalism underlying this theory. We also present a simpler approach to the transport laws which reduces the problem to a pure ODE's one. In Secs. 3 and 4 the fundamental techniques employed for studying weak discontinuities are extended in order to cover the case of characteristic shocks and intermediate discontinuities.

1 General formalism

We consider a system of conservation laws in $\mathbb{R}^4$,

(1.1) $$\partial_t \mathbf{F}^0 + \partial_i \mathbf{F}^i = \mathbf{f}$$

where $\mathbf{U} = (\mathbf{U}^a), a = 1, \ldots, N$, is the field vector; $\mathbf{F}^0, \ldots, \mathbf{F}^3$ are smooth functions of $\mathbf{U}$ in an open domain $\Omega \subset \mathbb{R}^N$; $\mathbf{f} = \mathbf{f}(\mathbf{U}, x^i, t)$ smooth on $\Omega \times \mathbb{R}^4$. Let

$$\mathcal{A}^i \equiv \nabla_\mathbf{U} \mathbf{F}^i, \quad \mathcal{A}^0 \equiv \nabla_\mathbf{U} \mathbf{F}^0.$$

Then, for differentiable solutions $\mathbf{U}$, (1.1) yields:

(1.2) $$\mathcal{A}^0 \partial_t \mathbf{U} + \mathcal{A}^i \partial_i \mathbf{U} = \mathbf{f}.$$

We shall assume that (1.2) is hyperbolic in the t-direction (Jeffrey, 1976). Hence $\mathcal{A}^0$ is non-singular and, letting

$$\mathcal{M}^i \equiv (\mathcal{A}^0)^{-1} \mathcal{A}^i; \quad \mathbf{g} \equiv (\mathcal{A}^0)^{-1} \mathbf{f}$$

(1.2) becomes

(1.3) $$\partial_t \mathbf{U} + \mathcal{M}^i \partial_i \mathbf{U} = \mathbf{g}.$$

Let $\Sigma(t)$ be a moving surface of discontinuity for the system (1.1), V_Σ the normal speed of propagation of Σ and $\vec{n}$ its unit normal. Then the jump relations across Σ are

(1.4) $$-V_\Sigma [\![\mathbf{F}^0]\!] + n_i [\![\mathbf{F}^i]\!] = 0.$$

For any quantity $h(\mathbf{U})$, we denote

$$[\![h(\mathbf{U})]\!] = h_- - h_+ ; \quad h_+ \equiv h(\mathbf{U}_+), \quad h_- \equiv h(\mathbf{U}_-),$$

with $\mathbf{U}_+$ and $\mathbf{U}_-$ the states just ahead and behind $\Sigma(t)$, and the field $\mathbf{U}$ is assumed to be smooth everywhere except across $\Sigma(t)$.

Let us consider the eigenvalue problem:

(1.5) $$(\mathcal{M}^i(\mathbf{U})n_i - \lambda I)\mathbf{R} = 0$$

where λ is the eigenvalue and $\mathbf{R}$ the right eigenvector. By hyperbolicity Eq. (1.5) admits N real eigenvalues

$$\lambda^{(1)}(\mathbf{U}, \vec{n}), \ldots, \lambda^{(N)}(\mathbf{U}, \vec{n})$$

and the corresponding eigenvectors form a basis of $\mathbb{R}^N$.

We shall treat several kinds of discontinuities: weak discontinuities, characteristic shocks, intermediate discontinuities, and k-shocks (Jeffrey, 1976).

When the discontinuity is a k-shock, corresponding to the single eigenvalue $\lambda^{(k)}(\mathbf{U}, \vec{n})$, we have (among other relationships)

$$(1.6) \qquad \lambda^{(k)}(\mathbf{U}_-, \vec{n}) > V_\Sigma > \lambda^{(k)}(\mathbf{U}_+, \vec{n})$$

and the jump conditions (1.4) can be solved for $[\![\mathbf{U}]\!]$ in the form

$$(1.7) \qquad [\![\mathbf{U}]\!] = \mathbf{G}(\mathbf{U}_+, \vec{n}, V_\Sigma) \,.$$

For an intermediate discontinuity the wavefront speed V_Σ coincides with one of the eigenvalues $\lambda^{(k)}(\mathbf{U}_\pm, \vec{n})$. In this case we shall also assume that the jump $[\![\mathbf{U}]\!]$ can be expressed in terms of $\mathbf{U}_+$, and some other parameter v:

$$(1.8) \qquad [\![\mathbf{U}]\!] = \mathbf{G}(\mathbf{U}_+, \vec{n}, v) \,.$$

Weak discontinuities and characteristic shocks will be considered in the next section.

In the sequel we shall make use of the 1st and 2nd order compatibility relations across Σ (Truesdell and Toupin, 1960, Kosinski, 1986, Anile, 1982).

Let

$$(1.9) \qquad \frac{\delta}{\delta t} \equiv \frac{\partial}{\partial t} + V_\Sigma n^i \frac{\partial}{\partial x^i}$$

denote the Thomas displacement derivative.

The kinematic and geometric compatibility relations of first order are, respectively:

$$(1.10) \qquad [\![\partial_t \mathbf{U}]\!] = \frac{\delta}{\delta t}[\![\mathbf{U}]\!] - V_\Sigma \mathbf{Y}^1$$

$$(1.11) \qquad [\![\partial_i \mathbf{U}]\!] = \tilde{\partial}_i [\![\mathbf{U}]\!] + n_i \mathbf{Y}^1$$

where $\mathbf{Y}^1 \equiv [\![n^j \partial_j \mathbf{U}]\!]$ and $\tilde{\partial}_i$ denotes the transverse derivative on $\Sigma(t)$ defined by $\tilde{\partial}_i = h_i^j \partial_j$, with $h_{ij} = \delta_{ij} - n_i n_j$ the projection tensor on $\Sigma(t)$.
Note that in cartesian coordinates covariant and contravariant components of tensors have the same numerical value.

The kinematic and geometric compatibility relations of second order may be obtained by iterating the first order ones and are given by:

$$(1.12) \qquad \begin{aligned} [\![\partial_{tj}^2 \mathbf{U}]\!] = {} & n_j \frac{\delta \mathbf{Y}^1}{\delta t} + \mathbf{Y}^1 \frac{\delta n_j}{\delta t} + \frac{\delta}{\delta t}(\tilde{\partial}_j [\![\mathbf{U}]\!]) \\ & - V_\Sigma \tilde{\partial}_j \mathbf{Y}^1 - V_\Sigma n_j \mathbf{Y}^2 - V_\Sigma n^i \tilde{\partial}_j \tilde{\partial}_i [\![\mathbf{U}]\!] \end{aligned}$$

(1.13)
$$[\![\partial_{ij}^2 \mathbf{U}]\!] = \tilde{\partial}_i \tilde{\partial}_j [\![\mathbf{U}]\!] + \mathbf{Y}^1 \chi_{ij} + n_i \tilde{\partial}_j \mathbf{Y}^1$$
$$+ n_j \tilde{\partial}_i \mathbf{Y}^1 + n_i n_j \mathbf{Y}^2 + n_i n^k \tilde{\partial}_j \tilde{\partial}_k [\![\mathbf{U}]\!].$$

Here $\mathbf{Y}^2 \equiv n^i n^j [\![\partial_{ij}^2 \mathbf{U}]\!]$ and

(1.14)
$$\chi_{ij} \equiv \tilde{\partial}_i n_j$$

is the 2nd fundamental form of $\Sigma(t)$.

We shall also need the following expression for the displacement derivative of $\vec{n}$

(1.15)
$$\frac{\delta n^i}{\delta t} = -\tilde{\partial}_i V_\Sigma .$$

Let the surface Σ be represented in a parametric form

(1.16)
$$x^i = x^i(v^\Gamma, t), \quad \Gamma = 1, 2$$

and let $a_{\Gamma\Delta}$ be its first fundamental form, given by

(1.17)
$$a_{\Gamma\Delta} = \frac{\partial x^i}{\partial v^\Gamma} \frac{\partial x_i}{\partial v^\Delta} .$$

The relationship between the transverse derivative $\tilde{\partial}_i$ and the covariant derivative on the surface $\Sigma(t)$ (which we denote by a semicolon;) is

(1.18)
$$\tilde{\partial}_i \phi = a^{\Gamma\Delta} x_{i;\Gamma} \phi_{;\Delta}$$

for any quantity ϕ defined on $\Sigma(t)$. Also,

$$h^{ij} = a_{\Gamma\Delta} \frac{\partial x^i}{\partial v^\Gamma} \frac{\partial x^j}{\partial v^\Delta} .$$

2 The evolution of weak discontinuities

A surface across which the field is continuous but its derivatives are discontinuous is called a weak discontinuity (Jeffrey, 1976).

In this case relations (1.10–11) simplify.

Let us take the jump of equation (1.3) across Σ

(2.1)
$$[\![\partial_t \mathbf{U}]\!] + \mathcal{M}^i [\![\partial_i \mathbf{U}]\!] = 0$$

(note that $\mathcal{M}^i(\mathbf{U})$ and $\mathbf{g}(\mathbf{U})$ are continuous across Σ).

Using (1.10–11) we obtain

(2.2)
$$(\mathcal{M}^i n_i - V_\Sigma I) \mathbf{Y}^1 = 0 .$$

This means that V_Σ is an eigenvalue and $\mathbf{Y}^1$ an eigenvector of $\mathcal{M}^i n_i$, therefore a weak discontinuity surface is a characteristic surface.

Let $\lambda = \lambda(\mathbf{U}, \vec{n}) = V_\Sigma$ be the corresponding eigenvalue and let m be its multiplicity.

Equation (2.2) gives

$$\mathbf{Y}^1 = \sum_{I=1}^{m} \phi^I \mathbf{R}_I$$

where $\mathbf{R}_I$ are m (independent) eigenvectors corresponding to λ. Now if we take the jump in the space derivative of system (1.3) across Σ, use the compatibility relations (1.10–13) and multiply by n^j, we get

(2.3)
$$\frac{\delta \mathbf{Y}^1}{\delta t} + (\mathcal{M}^i n_i - \lambda I)\mathbf{Y}^2 + \mathcal{N}_a^i n_i Y^{1a} \mathbf{Y}^1 + \mathcal{N}_a^i n_i n^j \partial_j U_0^a \mathbf{Y}^1$$
$$+ \mathcal{N}_a^i Y^{1a} \partial_i \mathbf{U}_0 + \mathcal{M}^i \tilde{\partial}_i \mathbf{Y}^1 - \nabla_a \mathbf{g} Y^{1a} = 0$$

where

$$\mathcal{N}_a^i \equiv \frac{\partial \mathcal{M}^i}{\partial U^a}.$$

Next we multiply (2.3) by $\mathbf{L}_J$, the J-th left eigenvector of $\mathcal{M}^i n_i$, and then use the following identities (which are proved in appendix A2)

(2.4)
$$\mathbf{L}_J \mathcal{M}^i \mathbf{R}_I = \frac{\partial \lambda}{\partial n_i} \mathbf{L}_J \mathbf{R}_I$$

(2.5)
$$\mathbf{L}_J \mathcal{N}_a^i n_i \mathbf{R}_I = \nabla_a \lambda \mathbf{L}_J \mathbf{R}_I$$

the resulting equation may be written as

(2.6)
$$\mathbf{L}_J \mathbf{R}_I \frac{d\phi^I}{dt} + \phi^I \phi^K \nabla_a \lambda R_I^a \mathbf{L}_J \mathbf{R}_K + \phi^I \left(\mathbf{L}_J \nabla_a \mathbf{R}_I \frac{\delta U_0^a}{\delta t} \right.$$
$$+ \mathbf{L}_J \frac{\partial \mathbf{R}^I}{\partial n_j} \frac{\delta n^j}{\delta t} + \mathbf{L}_J \mathcal{M}^i \nabla_a \mathbf{R}_I \tilde{\partial}_i U_0^a + \frac{1}{2} \frac{\partial^2 \lambda}{\partial n_i \partial n_j} \chi_{ij} \mathbf{L}_J \mathbf{R}_I$$
$$+ \chi_{ij} \frac{\partial \lambda}{\partial n_i} \mathbf{L}_J \frac{\partial \mathbf{R}^I}{\partial n_j} + \mathbf{L}_J \mathbf{R}_I \nabla_a \lambda n^j \partial_j U_0^a + \mathbf{L}_J \mathcal{N}_a^i R_I^a \partial_i \mathbf{U}_0$$
$$\left. - \mathbf{L}_J \nabla_a \mathbf{g} R_I^a \right) = 0$$

where

$$\frac{d}{dt} = \frac{\delta}{\delta t} + \frac{\partial \lambda}{\partial n_i} \tilde{\partial}_i$$

is the total derivative along the "rays" defined by

(2.7)
$$\frac{dx^i}{dt} = \lambda n^i + h_j^i \frac{\partial \lambda}{\partial n_j}$$

and U_0 is the unperturbed field ahead of Σ and is supposed to be a known function of x^i.

Note that in (2.6) one can solve for the derivatives $d\phi^I/dt$ because

$$(2.8) \qquad\qquad \det(\mathbf{L}_J \mathbf{R}_I) \neq 0.$$

In fact (Boillat, 1965) let us denote by $\Delta^{(k)}$ each determinant of the kind (2.8) corresponding to the eigenvalue $\lambda^{(k)}$ and by $\mathcal{L}$ and $\mathcal{R}$ the $N \times N$ matrices of left and right eigenvectors. By the orthogonality of left and right eigenvectors corresponding to different eigenvalues we have:

$$\det(\mathcal{L}\mathcal{R}) = \prod_{k=1,\ldots,n} \Delta^{(k)}$$

By hyperbolicity the eigenvectors are linearly independent, therefore

$$\det(\mathcal{L}\mathcal{R}) \neq 0.$$

Hence each $\Delta^{(k)} \neq 0$.

This system must be supplemented with transport equations for n^i and χ_{ij}. The trasport equation for n^i given by (1.15) can be written:

$$(2.9) \qquad\qquad \frac{dn_i}{dt} + \nabla_a \lambda \tilde{\partial}_i U^a = 0.$$

Hereafter we suppose that the unperturbed field ahead of the discontinuity is a known function of the position. Therefore the eigenvalue λ can be considered a known function of x^i and $\vec{n}$:

$$(2.10) \qquad\qquad \lambda = \lambda(\mathbf{U}(x^i,\vec{n}),\vec{n}) = \hat{\lambda}(x^i,\vec{n}).$$

To derive the transport equations for χ_{ij} we shall make use of the following properties, which are proved in Appendix A3

(i)   symmetry of the second fundamental form:

$$(2.11) \qquad\qquad \chi_{ij} = \chi_{ji},$$

(ii)   first commutation relation:

$$(2.12) \qquad \frac{\delta}{\delta t}\tilde{\partial}_j - \tilde{\partial}_j \frac{\delta}{\delta t} = \left[\left(\frac{\partial\hat{\lambda}}{\partial x_k} + \frac{\partial\hat{\lambda}}{\partial n_s}\chi_s^k\right)n_j - \hat{\lambda}\chi_j^k\right]\tilde{\partial}_k,$$

(iii)   second commutation relation:

$$(2.13) \qquad \tilde{\partial}_j\tilde{\partial}_i - \tilde{\partial}_i\tilde{\partial}_j = (n_j\chi_i^s - n_i\chi_j^s)\tilde{\partial}_s.$$

Let us take the derivative of Eq. (1.15) with $V_\Sigma = \lambda$,

$$\tilde{\partial}_j \frac{\delta n_i}{\delta t} + (\tilde{\partial}_j h_i^k) \frac{\partial \lambda}{\partial x^k} + h_i^k \left(\tilde{\partial}_j \frac{\partial \lambda}{\partial x^k} \right) + \left(\tilde{\partial}_j \frac{\partial \lambda}{\partial n^k} \right) \chi_i^k + \frac{\partial \lambda}{\partial n^k} \tilde{\partial}_i \chi_i^k = 0 \,.$$

by developing the derivatives we obtain

$$(2.14) \quad \tilde{\partial}_j \frac{\delta n_i}{\delta t} - (\chi_{ji} n^k + n_i \chi_j^k) \frac{\partial \lambda}{\partial x^k} + h_{ik} \left(h_{js} \frac{\partial^2 \lambda}{\partial x_k \partial x_s} + \frac{\partial^2 \lambda}{\partial x_k \partial n_s} \tilde{\partial}_j n_s \right)$$

$$+ \left(h_{js} \frac{\partial^2 \lambda}{\partial x_s \partial n_k} + \frac{\partial^2 \lambda}{\partial n_k \partial n_s} \chi_{js} \right) \chi_{ki} + \frac{\partial \lambda}{\partial n_k} \tilde{\partial}_j \tilde{\partial}_i n_k = 0 \,.$$

Now, making use of (2.11–13) we finally get

$$(2.15) \quad \frac{d\chi_{ij}}{dt} + h_{jk} h_{is} \frac{\partial^2 \lambda}{\partial x_s \partial x_k} + h_{ik} \frac{\partial^2 \lambda}{\partial x_k \partial n_s} \chi_{js} + h_{jk} \frac{\partial^2 \lambda}{\partial x_k \partial n_s} \chi_{is}$$

$$+ \frac{\partial^2 \lambda}{\partial n_k \partial n_s} \chi_{js} \chi_{ik} - \frac{\partial \lambda}{\partial x_k} (\chi_{ik} n_j + \chi_{ji} n_k + \chi_{kj} n_i)$$

$$+ \left(\lambda - \frac{\partial \lambda}{\partial n_k} n_k \right) \chi_j^s \chi_{si} = 0 \,.$$

Notice that in this way we do not need to solve the eikonal equation for the phase, in order to determine the rays. In fact Eqs. (2.6), (2.7), (2.9) and (2.15) form a set of ODE's in the unknowns $\phi^I, x^i, n^i, \chi_{ij}$, which in principle can be solved given the initial data. This approach therefore avoids solving the eikonal equation (which is a PDE) and could be useful in numerical calculations (Russo, 1986).

3 Characteristic shocks

For a *characteristic shock* the shock surface Σ coincides with a characteristic surface and its velocity V_Σ coincides with an eigenvalue of the system, both ahead of and behind the shock (Jeffrey, 1976, Boillat, 1982)

$$(3.1) \qquad\qquad V_\Sigma = \lambda(\mathbf{U}_+, \vec{n}) = \lambda(\mathbf{U}, \vec{n}) \,.$$

If the corresponding eigenvalue is single then the shock is characteristic iff the wave is exceptional (also called linearly degenerate), i.e. iff

$$(\nabla_a \lambda R^a) = 0 \,.$$

A shock corresponding to an eigenvalue of constant multiplicity $m > 1$ is always exceptional (Boillat, 1982).

We shall consider the general case $m \geq 1$.

The jump conditions at the shock surface (1.4) constitute a set of N relations, of which $N - m$ are independent. By solving the system one determines the field behind the shock as a function of the field ahead of the shock and of m parameters $\{u^I, I = 1, \ldots, m\}$

(3.2)
$$[\![\mathbf{U}]\!] = \mathbf{G}(\mathbf{U}_0, \vec{n}, u^I).$$

By taking the jump of the field equations (1.3) and using the first order compatibility relations we get:

(3.3)
$$\frac{\delta \mathbf{G}}{\delta t} + (M^i_- n_i - V_\Sigma I)\mathbf{Y}^1 + M^i_- \tilde{\partial}_i \mathbf{G} + [\![M^i]\!]\partial_i \mathbf{U}_0 - [\![\mathbf{g}]\!] = 0.$$

By developing the derivatives of $\mathbf{G}$ as

(3.4)
$$\frac{\delta \mathbf{G}}{\delta t} = \nabla_a \mathbf{G}\frac{\delta U_0^a}{\delta t} + \frac{\partial \mathbf{G}}{\partial u^I}\frac{\delta u^I}{\delta t} + \frac{\delta \mathbf{G}}{\delta n^i}\frac{\delta n^i}{\delta t}$$
$$\tilde{\partial}_i \mathbf{G} = \nabla_a \mathbf{G}\tilde{\partial}_i U_0^a + \frac{\partial \mathbf{G}}{\partial u^I}\tilde{\partial}_i u^I + \frac{\partial \mathbf{G}}{\partial n_j}\tilde{\partial}_i n_j$$

and using the definition of derivative along the rays, we get:

(3.5)
$$\mathbf{L}_J\frac{\partial \mathbf{G}}{\partial u^I}\frac{du^I}{dt} + \mathbf{L}_J\nabla_a \mathbf{G}\frac{\delta U_0^a}{\delta t} + \mathbf{L}_J M^i_- \nabla_a \mathbf{G}\tilde{\partial}_i U_0^a$$
$$- \frac{\partial \mathbf{L}_J}{\partial n_i}(M^k_- n_k - \lambda I)\frac{\partial \mathbf{G}}{\partial n_j}\chi_{ij} - \mathbf{L}_J\frac{\partial \mathbf{G}}{\partial n_i}\nabla_a \lambda \tilde{\partial}_i U_0^a + \mathbf{L}_J[\![M^i]\!]\partial_i \mathbf{U}_0$$
$$- \mathbf{L}_J[\![\mathbf{g}]\!] = 0$$

where the fact that $(\partial \mathbf{G}/\partial u^I)$ is a right eigenvector of $M^i n_i$ and relation (2.5) have been used.

System (3.5), together with equations (2.9), (2.15) and (2.7) constitute a set of $m + 12$ ODE's for the unknown quantities u^I, x^i, n^i, χ_{ij}. Applications to the evolution of contact discontinuities in gas dynamics and Alfvén shocks in magnetohydrodynamics have been made (Russo, 1986).

4 Intermediate discontinuities

A discontinuity whose speed is equal to one of the eigenvalues of the state behind the wavefront is called an *intermediate discontinuity* (Jeffrey, 1976). In this case

$$V_\Sigma = \lambda(\mathbf{U}_-, \vec{n}).$$

The method described in the previous section is not directly applicable to the propagation of intermediate discontinuities in several dimensions because the spatial derivatives of the field just behind the shock are not known. But in the one dimensional case the extension is straightforward.

Let us consider such a situation, described by the system

(4.1)
$$\partial_t \mathbf{U} + M\partial_x \mathbf{U} = \mathbf{g}.$$

By proceeding as in Sec. 3 we obtain

(4.2) $$\frac{\delta[\![\mathbf{U}]\!]}{\delta t} + (\mathcal{M}_- - V_\Sigma I)\mathbf{Y}^1 + [\![\mathcal{M}]\!]\partial_x\mathbf{U}_0 - [\![\mathbf{g}]\!] = 0.$$

We solve the jump conditions in the form

(4.3) $$[\![\mathbf{U}]\!] = \mathbf{G}(\mathbf{U}_0, u).$$

where u is a parameter describing the jump of the field across Σ.

By multiplying (4.2) by the left eigenvector of $\mathcal{M}_-$ corresponding to the eigenvalue $\lambda(V_-) = V_\Sigma$ we get the transport equation

(4.4) $$\mathbf{L}\frac{\partial\mathbf{G}}{\partial u}\frac{\delta u}{\delta t} + \mathbf{L}\nabla_a\mathbf{G}\frac{\delta U_0^a}{\delta t} + \mathbf{L}[\![\mathcal{M}]\!]\partial_x\mathbf{U}_0 - \mathbf{L}[\![\mathbf{g}]\!] = 0.$$

The equation of motion of the discontinuity is

(4.5) $$\frac{dx}{dt} = \lambda(\mathbf{U}_-).$$

Here $\mathbf{L}, \mathbf{G}, [\![\mathcal{M}]\!], [\![\mathbf{g}]\!]$ and λ are determined as functions of $\mathbf{U}_0$ and u from (4.3). Thus (4.4) and (4.5) constitute a pair of ODE's for the strength and position of the discontinuity.

REFERENCES

- A. M. Anile, *A geometric characterization of the compatibility relations for regularly discontinuous tensor dields*, Le Matematiche, Vol. XXXVII, 106 (1982).
- G. Boillat, *La Propagation des Ondes*, Gautier-Villars, Paris (1965).
- G. Boillat, Article in *Wave Propagation*, G. Ferrarese Editor, International Mathematical Summer Center, Bressanone, 8–17 June 1980, Liguori, Naples (1982).
- P. J. Chen, *Selected Topics in Wave Propagation*, Nordhoff, Leyden (1976).
- J. Hadamard, *Leçons sur la Propagation des Ondes*, Hermaner, Paris (1903).
- A. Jeffrey, *Quasilinear Hyperbolic Systems and Waves*, Pitman, London (1976).
- W. Kosinski, *Field singularities and wave analysis in continuum mechanics*, PWN-Polish Scientific Publishrs, Warszawa (1986).
- G. Russo *On the evolution of ordinary discontinuities and characteristic shocks*, Le Matematiche, Vol. XLI, 123 (1986).
- T. Y. Thomas, *Extended compatibility condition for the study of surfaces of discontinuity in continuum mechanics*, Journal of Mathematics and Mechanics 6, 311 (1957).
- C. Truesdell and R. Toupin, *The Classical Field Theories*, Handbuch der Physik, Vol. III/1, Springer, Berlin (1960).

GENERALIZED WAVEFRONT EXPANSION FOR WEAK SHOCKS

chapter

6

Introduction

In the previous chapter, we have introduced ray methods for treating the propagation of weak discontinuities, characteristics shocks and intermediate discontinuities. The rays are the bicharacteristic curves of the original quasilinear system and the propagation of the discontinuity is governed by a set of ordinary differential equations along the rays. This theory is exact in the sense that no approximation is made in deriving the equations.

However, characteristic shocks and intermediate discontinuities are very special cases of shock waves. It would be very interesting to have a ray theory which applies to shock waves of general type as well. This seems very difficult because, for an arbitrary shock wave, it is not possible to separate the motion of the wavefront from the flow behind the shock. Several attempts have been made in this direction. We mention in particular geometrical shock dynamics of Whitham (1974), the theories of Prasad (1982) and Maslov (1980) among others. In general these theories are simply heuristic and, although they can lead to good results in some cases, there is no (even purely formal) justification of their validity.

In the case of plane weak shocks propagating into a uniform medium there is a mathematically valid theory based on the concept of shock-fitting which dates back to Chandrasekhar (1943) and Landau (1945). This theory has been extended to cover non planar geometry by Whitham (1974) and to more general situations (non-uniform background state) by Hunter and Keller (1983), giving rise, in the case of gas dynamics, to nonlinear geometrical acoustics. In such a theory, however, it is necessary to know the full initial profile of the wave pulse in order to determine the subsequent evolution of the shock front, and this feature leads to difficulties when applying these methods to pratical problems. Therefore it would be derivable to have a (more limited in scope) theory describing approximately the evolution of the front itself analogously to the propagation of weak discontinuities. This has been achieved by Anile (1984), Anile and Russo (1986, 1988) and the method is expounded in Sec. 1. An application to the propagation of acoustic shocks in a stratified atmosphere is presented in Sec. 2. Finally, in Sec. 3 the theory is recast in a form which makes it easy to compare with nonlinear geometrical acoustics.

1 Derivation of the basic equations

In this section we describe an asymptotic technique for treating weak shock waves.

We shall consider the propagation of a k-shock. We denote by $\mathbf{L}$ and $\mathbf{R}$ the normalized left and right eigenvectors of $M^i(\mathbf{U}_+)n_i$, corresponding to the eigenvalue $\lambda_+^{(k)}$, written below simply as λ_+. We consider a system of conservation laws in the form (1.3) of Chapter 5 i.e.

$$(1.1) \qquad \partial_t \mathbf{U} + M^i \partial_i \mathbf{U} = \mathbf{g}.$$

We define

$$(1.2) \qquad \mathbf{Y}^0 \equiv [\![\mathbf{U}]\!], \dots, \mathbf{Y}^k \equiv [\![n^{i_1} \dots n^{i_k} \partial_{i_1 \dots i_k}^k \mathbf{U}]\!]$$

where $[\![\mathbf{U}]\!]$ denotes the jump in $\mathbf{U}$ across the shock and $\vec{n}$ is the unit normal to the shock front. For a weak shock $\mathbf{Y}^0 = O(\epsilon)$, where ϵ is a small parameter which measures the strength of the shock. For example

$$V_\Sigma / \lambda_+ - 1 = O(\epsilon).$$

We assume that the flow behind the shock varies over a length scale of the order ϵ in the normal direction. This implies that

$$(1.3) \qquad \mathbf{Y}^0 = O(\epsilon), \; \mathbf{Y}^1 = O(1), \; \dots, \; \mathbf{Y}^k = O(\epsilon^{-k+1}).$$

We shall assume the formal expansion:

$$(1.4) \qquad \begin{aligned}
\mathbf{Y}^0 &= \epsilon \mathbf{Y}_1^0 + \epsilon^2 \mathbf{Y}_2^0 + \epsilon^3 \mathbf{Y}_3^0 + \cdots \\
\mathbf{Y}^1 &= \mathbf{Y}_0^1 + \epsilon \mathbf{Y}_1^1 + \epsilon^2 \mathbf{Y}_2^1 + \cdots \\
\mathbf{Y}^2 &= \epsilon^{-1} \mathbf{Y}_0^2 + \mathbf{Y}_1^2 + \epsilon \mathbf{Y}_2^2 + \cdots \\
&\quad \cdots \\
\mathbf{Y}^k &= \epsilon^{-k+1} \mathbf{Y}_0^k + \epsilon^{-k+2} \mathbf{Y}_1^k + \epsilon^{-k+3} \mathbf{Y}_2^k + \cdots \\
&\quad \cdots
\end{aligned}$$

$$(1.5) \qquad V_\Sigma = b_0 + \epsilon b_1 + \cdots$$

We further assume that the flow varies over a length scale of order ϵ^0 along directions tangent to the shock. This implies that the covariant derivatives along the surface do not change the order of magnitude of flow quantities. That is if

$$w = O(\epsilon^k) \quad \text{then} \quad \bar{\partial}_i w = O(\epsilon^k).$$

From (1.5) the Thomas derivative is expanded in ϵ as:

$$(1.6) \qquad \frac{\delta}{\delta t} = D + \epsilon b_1 n^i \partial_i + \cdots$$

where $D \equiv \partial_t + \lambda_+ n^i \partial_i$.

Let us define the following scalar quantities:

(1.7) $$\pi^1 = \mathbf{L}\mathbf{Y}^1 , \; \pi^2 = \mathbf{L}\mathbf{Y}^2 , \; \ldots , \; \pi^k = \mathbf{L}\mathbf{Y}^k , \; \ldots$$

We shall derive an infinite system of equations for these scalars. The procedure is the following. We take the jump in the system (1.1) and in its normal derivatives, make use of the compatibility conditions across the shock and insert the expansions (1.4) into the obtained equation as well as in the jump conditions (1.4) of Chapter 5. To the lowest order we obtain a system of ODE's for π^k (Eq. (1.28)).

First of all we insert the expansion (1.4) into the jump conditions (1.4) of Chapter 5, and make use of the following expansions:

(1.8) $$\mathbf{F}^\alpha(\mathbf{U}_-) = \mathbf{F}^\alpha_+ + \epsilon \mathcal{A}^\alpha_+ \mathbf{Y}^0_1 + \epsilon^2 \left\{ \frac{1}{2} \mathcal{B}^\alpha_+ \mathbf{Y}^0_1 \mathbf{Y}^0_1 + \mathcal{A}^\alpha_+ \mathbf{Y}^0_2 \right\} + \cdots$$
$$(\alpha = 0, \ldots, 3)$$

where $\mathcal{B}^\alpha \equiv \nabla_{\mathbf{U}} \mathcal{A}^\alpha$ and

$$\mathcal{B}^\alpha \mathbf{V} = \sum_{a=1}^N (\partial \mathcal{A}^\alpha / \partial U^a) V^a .$$

By equating the terms of lowest order in ϵ we get

(1.9) $$(\mathcal{A}^i_+ n_i - b_0 \mathcal{A}^0_+) \mathbf{Y}^0_1 = 0 .$$

This gives $b_0 = \lambda_+$ and $\mathbf{Y}^0_1 = \pi^0 \mathbf{R}$. Here π^0 is a quantity which determines the strength of the shock. To first order in ϵ it is related to the jump in the field by

(1.10) $$\mathbf{L}[\![\mathbf{U}]\!] = \epsilon \pi^0 + O(\epsilon^2) .$$

To first order in ϵ we get the correction to the shock speed:

(1.11) $$b_1 = \frac{\pi^0}{2} \mathbf{L} \mathcal{N}^i_+ n_i \mathbf{R} \mathbf{R}$$

where $\mathcal{N}^i \equiv \nabla_{\mathbf{U}} \mathcal{M}^i$.

Now we take the jump in the field equations (1.1) across the shock and use the first order compatibility relations (1.10–11) of Chapter 5. The resulting equation is:

(1.12) $$\frac{\delta \mathbf{Y}^0}{\delta t} + (\mathcal{M}^i_- n_i - V_\Sigma I) \mathbf{Y}^1 + \mathcal{M}^i_- \tilde\partial_i \mathbf{Y}^0 + [\![\mathcal{M}^i]\!](\partial_i \mathbf{U})_+ - [\![\mathbf{g}]\!] = 0 .$$

By inserting the expansion (1.4) we get, to zeroth order in ϵ,

(1.13) $$(\mathcal{M}^i_+ n_i - \lambda_+ I) \mathbf{Y}^1_0 = 0$$

hence $\mathbf{Y}_0^1 = C\mathbf{R}$ and, from

$$\mathbf{L}(\mathbf{Y}_0^1 + \epsilon\mathbf{Y}_1^1 + \cdots) = \pi^1$$

we obtain

$$C = \pi^1, \quad \mathbf{L}\mathbf{Y}_j^1 = 0, \quad j \geq 1.$$

To first order in ϵ, after left multiplication by $\mathbf{L}$, we obtain

(1.14) $\quad \{\partial_t + (\lambda_+ n^i + \mathbf{L}\mathcal{M}_+^j\mathbf{R}h_j^i)\partial_i\}\pi^0 + \beta\pi^0 + \dfrac{1}{2}\alpha\pi^0\pi^1 = 0$

where

(1.15) $\quad \begin{cases} \alpha \equiv \mathbf{L}\mathcal{N}_+^i n_i \mathbf{R}\mathbf{R}. \\[2mm] \beta \equiv \mathbf{L}D\mathbf{R} + \mathbf{L}\mathcal{M}_+^i\nabla_{\mathbf{U}}\mathbf{R}(\tilde{\partial}_i\mathbf{U})_+ + \mathbf{L}\mathcal{M}_+^i\dfrac{\partial\mathbf{R}}{\partial n^j}\chi_{ij} + \\[3mm] \qquad + \mathbf{L}\mathcal{N}_+^i\mathbf{R}(\partial_i\mathbf{U})_+ - \mathbf{L}(\nabla_{\mathbf{U}}g)_+\mathbf{R}. \end{cases}$

Note that the derivative of π^0 may be written as a total derivative along "rays":

(1.16) $\qquad\qquad\qquad \dfrac{d\pi^0}{dt} + \beta\pi^0 + \dfrac{\alpha}{2}\pi^0\pi^1 = 0$

on

(1.17) $\qquad\qquad\qquad\qquad \dfrac{dx^i}{dt} = \Lambda^i$

with $\dfrac{d}{dt} \equiv \partial_t + \Lambda^i\partial_i$, and $\Lambda^i \equiv \lambda_+ n^i + \mathbf{L}\mathcal{M}_+^j\mathbf{R}h_j^i$.

In order to derive the other equations of the system we shall need the following identities which follow from the compatibility relations (1.10–13) of Chapter 5.

(1.18) $\quad [\![n^{i_1}n^{i_2}\ldots n^{i_{k-1}}\partial_{i_1\ldots i_k}^k\mathbf{U}]\!] =$

$\qquad = n^{i_1}\ldots n^{i_{k-1}}\tilde{\partial}_{i_k}[\![\partial_{i_1\ldots i_{k-1}}^{k-1}\mathbf{U}]\!] + n_{i_k}\mathbf{Y}^k$

(1.19) $\quad n^{i_1}n^{i_2}\ldots n^{i_k}\dfrac{\delta}{\delta t}[\![\partial_{i_1\ldots i_k}^k\mathbf{U}]\!] =$

$\qquad = \dfrac{\delta\mathbf{Y}^k}{\delta t} - k\dfrac{\delta n^{i_k}}{\delta t}n^{i_1}\ldots n^{i_{k-1}}\tilde{\partial}_{i_k}[\![\partial_{i_1\ldots i_{k-1}}^{k-1}\mathbf{U}]\!]$

(1.20) $\quad n^{i_1}\ldots n^{i_k}[\![\partial_{ti_1\ldots i_k}^{k+1}\mathbf{U}]\!] =$

$\qquad = \dfrac{\delta\mathbf{Y}^k}{\delta t} - V_\Sigma\mathbf{Y}^{k+1} - k\dfrac{\delta n^{i_k}}{\delta t}n^{i_1}\ldots n^{i_{k-1}}\tilde{\partial}_{i_k}[\![\partial_{i_1\ldots i_{k-1}}^{k-1}\mathbf{U}]\!]$

(1.21) $\quad n^{i_1}\ldots n^{i_k}\tilde{\partial}_j[\![\partial_{i_1\ldots i_k}^k\mathbf{U}]\!] =$

$\qquad = \tilde{\partial}_j\mathbf{Y}^k - k\chi_j^{i_k}n^{i_1}\ldots n^{i_{k-1}}\tilde{\partial}_{i_k}[\![\partial_{i_1\ldots i_{k-1}}^{k-1}\mathbf{U}]\!]$

(1.22) $\quad n^{i_1}\ldots n^{i_k}\partial_{i_1\ldots i_{k-1}}(\partial_{i_k}\mathbf{U}\partial_{i_{k+1}}\mathbf{V}) =$

$\qquad = n^{i_1}\ldots n^{i_k}\displaystyle\sum_{j=1}^{k}\binom{k-1}{j-1}\partial_{i_1\ldots i_j}^j\mathbf{U}\partial_{i_{j+1}\ldots i_{k+1}}^{k-j+1}\mathbf{V}.$

Let us derive the evolution equation for π^k to the lowest order in ϵ. The jump in the k-th space derivative, saturated with $n^{i_1}\ldots n^{i_k}$, gives

$$(1.23) \qquad n^{i_1}\ldots n^{i_k}[\![\partial_{t i_1\ldots i_k}^{k+1}\mathbf{U} + \partial_{i_1\ldots i_k}^{k}(M^j\partial_j\mathbf{U}) - \partial_{i_1\ldots i_k}^{k}\mathbf{g}]\!] = 0.$$

The second term becomes, after some manipulations

$$(1.24) \qquad \begin{aligned} &n^{i_1}\ldots n^{i_k}[\![\partial_{i_1\ldots i_k}^{k}(M^j\partial_j\mathbf{U})]\!] = \\ &= P_1^k + P_2^k + \ldots + P_k^k + n^{i_1}\ldots n^{i_k}[\![M^j\partial_{i_1\ldots i_k j}^{k+1}\mathbf{U}]\!] \end{aligned}$$

where

$$(1.25) \qquad \begin{aligned} P_j^k &= n^{i_1}\ldots n^{i_k}[\![\partial_{i_1\ldots i_{k-1}}^{k-1}(M^s\partial_{i_{k-j+1}}\mathbf{U}\partial_{i_{k-j+2}\ldots i_{k s}}^{j-1}\mathbf{U})]\!] = \\ &= n^{i_2}\ldots n^{i_k}\,\mathcal{N}_+^s[\![\partial_{i_{k-j+1}}\mathbf{U}\partial_{i_{k-j+2}\ldots i_{k s}}^{j-1}\mathbf{U}]\!] + O(\epsilon^{-k+2}) \end{aligned}$$

as it is easy to check. Using identity (1.22) we have

$$(1.26) \quad \sum_{j=1}^{k} P_j^k = \mathcal{N}_+^s n_s \sum_{j=1}^{k}\sum_{s=1}^{k-j+1}\binom{k-1}{s-1}\mathbf{Y}^s\mathbf{Y}^{k-s+1} + k\mathcal{N}_+^s n_s n^r(\partial_r\mathbf{U})_+\mathbf{Y}^k$$
$$+ \mathcal{N}_+^s\mathbf{Y}^k(\partial_s\mathbf{U})_+ + O(\epsilon^{-k+2}).$$

Using identities (1.18–22) one obtains, to order ϵ^{-k}, after some algebra,

$$(1.27) \qquad (M_+^i n_i - \lambda_+ I)\mathbf{Y}_0^{k+1} = 0$$

which gives

$$\mathbf{Y}_0^{k+1} = \pi^{k+1}\mathbf{R},$$

and, to order ϵ^{-k+1}

$$(1.28) \qquad \frac{d\pi^k}{dt} + \alpha\sum_{p=1}^{k}\binom{k}{p}\pi^p\pi^{k-p+1} + (\beta + k\gamma)\pi^k + \frac{\alpha}{2}\pi^0\pi^{k+1} = 0$$

where

$$(1.29) \qquad \gamma \equiv \mathbf{L}\,\mathcal{N}_+^i n^j(\partial_j\mathbf{U})_+ n_i\mathbf{R}.$$

This constitutes an infinite set of equations (the last terms of the k-th equation contains π^{k+1}) for the propagation of the jump in the field and in the normal derivatives along the rays which satisfy equation (1.17). The system is supplemented with equations (1.17) and (2.9), (2.15) of Chapter 5 which describe the evolution of the geometry. These equations are not coupled with (1.28) and can be solved in advance. Within this approximation the geometry and the nonlinearity are not coupled and the shock propagates along the bicharacteristics, just as a weak discontinuity.

2   Geometrical interpretation and applications to acoustic shocks in a constant state

We recall the concept of "expansion" of the rays introduced in the first section of Chapter 4. The position of a point in a ray is function of initial position and time:

$$(2.1) \qquad x^i = x^i(x_0^i, t).$$

It is, of course,

$$\left(\frac{\partial x^i}{\partial t}\right)_{x_0^i = \text{const}} = \Lambda^i.$$

Let

$$(2.2) \qquad J \equiv \frac{D(x^i)}{D(x_0^i)}$$

be the jacobian of the transformation (2.1). J is the measure of a volume element carried along the rays at time t. The quantity $1/J \ dJ/dt$ represents the rate of expansion of a volume element which moves along the rays and it is related to the velocity field Λ^i by the relation (see Appendix 4)

$$(2.3) \qquad \frac{1}{J}\frac{dJ}{dt} = \partial_i \Lambda^i$$

where $\partial_i \Lambda^i$ is the spatial divergence of the field Λ^i and

$$(2.4) \qquad \partial_s \Lambda^r \equiv (\partial \Lambda_s / \partial n_k)\partial_k n^r + \partial \Lambda^r / \partial x^s.$$

($\partial/\partial x^s$ denotes the derivative with $\vec{n}$ constant).
Now we rewrite the expression of the coefficients of system (1.28) in order to give a geometrical interpretation. We make use of the following identities (see Appendix 5).

$$(2.5) \qquad L\mathcal{N}^i n_i R = \nabla_U \lambda$$

$$(2.6) \qquad LM^i \frac{\partial R}{\partial n_k}\chi_{ik} = \frac{1}{2}\frac{\partial^2 \lambda}{\partial n_i \partial n_k}\chi_{ik} + \frac{\partial \lambda}{\partial n_i}L\frac{\partial R}{\partial n_k}\chi_{ik}$$

$$(2.7) \qquad \frac{\partial^2 \lambda}{\partial n_r \partial n_s}\chi_{rs} = \frac{d}{dt}\log J - \frac{\partial \Lambda^i}{\partial x^i}.$$

The coefficients of the equations, (1.15), (1.29), can be written as

$$(2.8) \qquad \alpha = (\nabla_U \lambda)R$$

$$\beta = LDR + L\mathcal{N}_+^i R(\tilde{\partial}_i U)_+ - L(\nabla_U g)_+ R + LM_+^i \nabla_U R(\tilde{\partial}_i U)_+ +$$
$$+ \chi_{rs}\frac{\partial \lambda}{\partial n_r}L\frac{\partial R}{\partial n_s} + \frac{1}{2}\frac{d\log J}{dt} - \frac{1}{2}\frac{\partial \Lambda^k}{\partial x^k}$$

$$(2.9) \qquad \gamma = L\mathcal{N}_+^i n^k(\partial_k U)_+ n_i R.$$

As an application we study the propagation into a constant state by using the system (1.28) truncated at the second equation, i.e. we neglect the term containing π^2 in the transport equation for π^1. In this case we have:

$$\mathbf{U}_+ = \text{const}, \qquad \frac{\partial \lambda}{\partial x^i} = 0.$$

It follows $dn^i/dt = 0$, hence $\vec{n}$ is constant along the rays, which are therefore straight lines. Also

$$\mathbf{L}D\mathbf{R} + \chi_{rs}\frac{\partial \lambda}{\partial n_r}\mathbf{L}\frac{\partial \mathbf{R}}{\partial n_s} = \mathbf{L}\frac{d\mathbf{R}}{dt} = 0$$

because $d\vec{n}/dt = 0$ and $\mathbf{R}$ is a function of $\mathbf{U}_+$ and $\vec{n}$. It follows:

$$(2.10) \qquad\qquad \beta = \frac{1}{2}\frac{d}{dt}\log J, \qquad \gamma = 0.$$

The first two equations of (1.28) are

$$(2.11) \qquad \begin{aligned} \frac{d\pi^0}{dt} + \frac{1}{2}\alpha\pi^0\pi^1 + \frac{\pi^0}{2}\frac{d}{dt}\log J &= 0 \\ \frac{d\pi^1}{dt} + \alpha(\pi^1)^2 + \frac{\pi^1}{2}\frac{d}{dt}\log J &= 0 \end{aligned}$$

where α is a constant along the rays. Let

$$\psi \equiv 1 + \pi_0^1\int_0^t \frac{\alpha}{\sqrt{J}}d\tau$$

where π_0^1 is the initial value of π^1 along the ray. Then the solution of system (2.11) is

$$\pi^1 = \frac{\pi_0^1}{\sqrt{J}\psi}, \qquad \pi^0 = \frac{\pi_0^0}{\sqrt{J}\psi}$$

where π_0^0 is the initial value of π^0 along the ray.

For plane waves $J = 1$ and

$$\pi^0 = \frac{\pi_0^0}{(1 + \alpha\pi_0^1 t)^{1/2}}.$$

For spherical waves $J = (\tau/\tau_0)^2$, where $\tau_0 = \tau(0)$, and

$$(2.12) \qquad\qquad \pi^0 = \frac{\pi_0^0\tau_0}{\tau[1 + \alpha\pi_0^1\tau_0\log(\tau/\tau_0)]^{1/2}}.$$

For cylindrical waves $J = \tau/\tau_0$, hence

$$(2.13) \qquad\qquad \pi^0 = \frac{\pi_0^0}{(\tau/\tau_0)^{1/2}[1 + 2\pi_0^1\alpha\tau_0(\sqrt{\tau/\tau_0} - 1)]^{1/2}}.$$

For a linearly degenerate wave $\alpha = 0$ and the decay law is the same as for a linear wave.

Let us apply the results to ideal gasdynamics. For a right progressive acoustic shock propagating into a constant state, at rest, we have $\lambda^{(3)} = c_0$, where c_0 is the unperturbed sound speed. The rays are the straight lines:

$$\frac{dx^i}{dt} = c_0 n^i .$$

Then

$$\alpha = c_0 + \rho_0 \frac{dc_0}{d\rho_0} .$$

For a polytropic gas

$$\alpha = c_0 \frac{\gamma + 1}{2} ,$$

where γ is the polytropic constant. Thus one recovers exactly the usual formula for the decay law of a plane parallel shock obtained by using the shock-fitting method.

3 Relationship with weakly nonlinear geometrical optics

System (1.28) is an infinite system of ODE's and there is no generally valid way of truncating it. This system can be interpreted as a power series expansion of a single partial differential equation for a *generating function*. Let

$$(3.1) \qquad \pi(\theta, t) = \sum_{k=0}^{\infty} \pi^k(t) \frac{\theta^k}{k!} .$$

Then

$$(3.2) \qquad \begin{cases} \dfrac{\partial \pi}{\partial \theta} = \displaystyle\sum_{k=0}^{\infty} \pi^{k+1} \frac{\theta^k}{k!} , \\[2mm] \dfrac{\partial \pi}{\partial t} = \displaystyle\sum_{k=0}^{\infty} \frac{d\pi^k}{dt} \frac{\theta^k}{k!} , \end{cases}$$

and the Cauchy product of π and $\dfrac{\partial \pi}{\partial \theta}$ is:

$$(3.3) \qquad \pi \frac{\partial \pi}{\partial \theta} = \sum_{k=0}^{\infty} \left(\sum_{p=0}^{k} \binom{k}{p} \pi^p \pi^{k-p+1} \right) \frac{\theta^k}{k!} .$$

Let us multiply the k-th equation of (1.28) by $\dfrac{\theta^k}{k!}$, sum over k and make use of (3.1–3). We obtain the following PDE for $\pi(\theta, t)$:

$$(3.4) \qquad \frac{\partial \pi}{\partial t} + \left[\left(\pi - \frac{1}{2}\pi^0 \right) \alpha + \theta\gamma \right] \frac{\partial \pi}{\partial \theta} + \beta\pi = 0$$

where α, β and γ are defined by (2.5–7) and

$$\pi^0(t) = \pi(0, t).$$

Equation (3.4) describes the evolution of the shock strength. It must be supplemented with equations describing the geometry, (1.17), and (2.9), (2.15) of Chapter 5.

Let us suppose we have solved these equations and interpret α, β and γ as known functions of time.

We will consider the initial value problem for (3.4), with

(3.5) $$\pi(\xi, 0) = f(\xi).$$

At the initial time we known the field $\mathbf{U}$ in the region behind the shock and we can compute $\pi^k(0)$ defined in (1.7). The function $f(\xi)$ is defined by the power series:

(3.6) $$f(\xi) = \sum_{k=0}^{\infty} \pi^k(0) \frac{\xi^k}{k!}$$

which we will assume convergent for each ξ.

Equation (3.4) can be written in characteristic form:

(3.7) $$\begin{cases} \dfrac{d\pi}{dt} + \beta(t)\pi = 0 \\ \dfrac{d\theta}{dt} = \alpha(t)\left[\pi(\theta, t) - \dfrac{1}{2}\pi(0, t)\right] + \gamma(t)\theta. \end{cases}$$

Integrating the first equation gives

(3.8) $$\pi = f(\xi)\exp\left(-\int_0^t \beta(\tau)d\tau\right)$$

and substituting in the second yields

(3.9) $$\frac{d\theta}{dt} = \alpha(t)\left[f(\xi)\exp\left(-\int_0^t \beta(\tau)d\tau\right) - \frac{1}{2}\pi^0(t)\right] + \gamma(t)\theta = 0.$$

This linear equation can be integrated formally:

(3.10) $$\theta(\xi, t) = e^{\int_0^t \gamma(\tau)d\tau}\left\{\xi + \int_0^t \left[f(\xi)\alpha(\tau)e^{-\int_0^\tau [\beta(t')+\gamma(t')]dt'} \right.\right.$$
$$\left.\left. - \frac{1}{2}\pi^0(\tau)\alpha(\tau)e^{-\int_0^\tau \gamma(t')dt'}\right]d\tau\right\}.$$

The integration is formal because $\pi^0(t)$ is unknown. It is, indeed, the only unknown function we are interested in, representing the shock amplitude. Equation (3.10) with $\theta = 0$ is a relation between time t and the parameter ξ of the characteristics which terminates on the shock at time t. The relation has the form of an integral equation for $\xi(t)$

$$(3.11) \qquad \xi(t) + f(\xi(t)) \int_0^t F(\tau)d\tau - \frac{1}{2}\int_0^t f(\xi(\tau))F(\tau)d\tau = 0$$

where

$$F(t) \equiv \alpha(t)e^{-\int_0^t [\beta(\tau)+\gamma(\tau)]d\tau}.$$

Differentiating Eq. (3.11) with respect to time, we obtain an ordinary differential equation for ξ:

$$(3.12) \qquad \frac{d\xi}{dt} = -\frac{1}{2}\frac{f(\xi)F(t)}{1 + f'(\xi)\int_0^t F(\tau)d\tau}.$$

The solution of this equation, together with (3.8) gives the shock amplitude as a function of time.

It can be shown that this approach gives the some description of the shock evolution obtained by the Weakly Nonlinear Geometrical Optics (Russo and Hunter, 1991).

REFERENCES

– A. M. Anile, *Propagation of weak shock waves*, Wave Motion **6**, 571 (1984).

– A. M. Anile and G. Russo, *Generalized Wavefront Expansion I. Higher Order Corrections for the Propagation of Weak Shock Waves*, Wave Motion **8**, 243 (1986).

A. M. Anile and G. Russo, *Generalized Wavefront Expansion II: the Propagation of Step Shocks*, Wave Motion **10**, 3 (1988).

– S. Chandrasekhar, *On the decay of plane shock waves*, Ballistic Research Laboratories, Report N. 423, Aberdeen Proving Ground, Maryland (1943).

– J. K. Hunter and J. B. Keller, *Weakly Nonlinear High Frequency Waves*, Comm. Pure and Appl. Math., Vol. XXXVI, 547 (1983).

– L. D. Landau, *On shock waves at large distances from the place of their origin*, Soviet Journal of Physics **9**, 496 (1945).

– E. M. Maslov, *Propagation of shock waves in an isotropic, nonviscous gas*, J. Sov. Math. **13**, 119 (1980).

– P. Prasad, *Kinematics of a Multi-Dimensional Shock of Arbitrary Strenght in an Ideal Gas*, Acta Mechanica, 136, Springer-Verlag (1982).

– G. Russo and J. K. Hunter, *A transport equation for the evolution of shock amplitude along the rays*, proceedings of the VI International Conference on *Waves and Stability in Continuous Media*, Acireale, May 27 - June 1, 1991.

– G. B. Whitham, *Linear and Nonlinear Waves*, Wiley, New York (1974).

SMALL TIME ANALYSIS AND SHOCK STABILITY

chapter 7

Introduction

In the previous chapter we developed a simple ray theory for weak shock waves. Now we face the question of deriving a ray approach for shocks of arbitrary strength. The attempts by Whitham (1974) and Prasad (1982) are at best heuristic and it is not clear how to derive them on a formal basis. In this chapter we present an approach which is closely related to the one previously expounded for weak shocks. Essentially this approach is a small time analysis of the solution near the initial wavefront. As such it is of limited practical interest although in some cases it leads to results which coincide with those obtained by applying Whitham's theory (over which it has the advantage that the basic equations can be derived formally and not simply postulated). However there is one area in which the theory leads to quite interesting results and this is stability theory. One of the most useful concepts for studying shock wave stability is that of *corrugation stability*, which was defined intuitively by Whitham (1974). The concept of corrugation stability is that a plane front is stable if, after perturbing the wavefront, the shock speed decreases where the front is expanding and increases where the front is converging.

This concept of corrugation stability could lead to sufficient stability criteria if the relationship between the shock velocity and the wavefront rate of deformation were known (which is not the case in general). In order to obtain such a relationship Whitham resorted to geometrical shock dynamics (see also Liberman and Velikovich, 1986) which is only an approximate heuristic theory. However by employing the method which will be treated in this chapter Anile and Russo (1986, 1989) were able to put Whitham intuitive definition on a rigorous basis and then to obtain exact stability results for a class of suitable perturbations (sufficiently general to be of physical interest). In Sec. 1 and 2 we expound the fundamentals of the method, in one and several dimensions respectively. In Sec. 3 we use the method in order to give the definition of corrugation stability and in Sec. 4 we present applications to shocks stability in an ideal fluid and in a Van der Waals gas.

1 One dimensional analysis

Let us consider a system of conservation laws in one space dimension:

$$(1.1) \qquad \partial_t \mathbf{F}^0(\mathbf{U}) + \partial_x \mathbf{F}^1(\mathbf{U}) = \mathbf{f}(\mathbf{U})$$

and let us write it in the form

$$(1.2) \qquad \partial_t \mathbf{U} + M \partial_x \mathbf{U} = \mathbf{g}(\mathbf{U}).$$

Let $\Sigma(t)$ be a surface of discontinuity representing a k-shock and V_Σ its velocity. Then the jump relations (1.4) of Chapter 5 are:

$$(1.3) \qquad -V_\Sigma[\![\mathbf{F}^0]\!] + [\![\mathbf{F}^1]\!] = 0.$$

These constitute a set of N equations relating the field behind the shock to the field ahead and to the shock velocity.

For a k-shock we can solve (1.3) for the jump in the field (Jeffrey, 1976)

$$(1.4) \qquad [\![\mathbf{U}]\!] = \mathbf{G}(\mathbf{U}_+, V_\Sigma).$$

Let us define the quantities

$$(1.5) \qquad \pi^1 = \mathbf{L}_-\mathbf{Y}^1 \ldots, \pi^n = \mathbf{L}_-\mathbf{Y}^n, \ldots$$

where $\mathbf{L}$ is the left eigenvector corresponding to the k-th eigenvalue λ, which we denote without superscript, and $\mathbf{Y}^1, \ldots, \mathbf{Y}^n$, defined in Chapter 6, Eq. (1.2), denote the jump in the spatial derivative of the field.

We shall derive an equation for V_Σ involving π^1. We take the jump of (1.2) across $\Sigma(t)$ and use the first order compatibility condition (1.10) of Chapter 5, obtaining:

$$(1.6) \qquad (M_- - V_\Sigma I)\mathbf{Y}^1 + \mathbf{k}^0 \frac{\delta V_\Sigma}{\delta t} + \mathbf{H}^0 = 0$$

where

$$\mathbf{k}^0 \equiv \frac{\partial \mathbf{G}}{\partial V_\Sigma}$$

$$\mathbf{H}^0 \equiv (\nabla_{\mathbf{U}_+}\mathbf{G})\frac{\delta \mathbf{U}_+}{\delta t} + [\![M]\!](\partial_x \mathbf{U})_+ - [\![\mathbf{g}]\!].$$

In the sequel we shall denote $\mathbf{L}_-$ simply as $\mathbf{L}$.
Multiplying (1.6) by $\mathbf{L}$ and assuming

$$\mathbf{L}\mathbf{k}^0 \neq 0$$

yields an equation for $\delta V_\Sigma/\delta t$

$$(1.7) \qquad \frac{\delta V_\Sigma}{\delta t} = \mathcal{G}^\circ(\mathbf{U}_+, \partial \mathbf{U}_+, V_\Sigma) + \Gamma(\mathbf{U}_+, V_\Sigma)\pi^1$$

where

(1.8)
$$\mathcal{G}^0 \equiv -\frac{\mathbf{L}\mathbf{H}^0}{\mathbf{L}\mathbf{k}^0},$$
$$\Gamma \equiv -\frac{\lambda - V_\Sigma}{\mathbf{L}\mathbf{k}^0}$$

and we denote by

$$\{\partial \mathbf{U}_+\} \equiv \{(\partial_t \mathbf{U})_+, (\partial_x \mathbf{U})_+\}.$$

Substituting (1.7) into (1.6) and solving for $\mathbf{Y}^1$ we obtain

(1.9)
$$\mathbf{Y}^1 = \hat{\mathbf{Y}}^1(\mathbf{U}_+, \partial \mathbf{U}_+, V_\Sigma) + \pi^1 \mathbf{k}^1(\mathbf{U}_+, V_\Sigma)$$

where

(1.10)
$$\hat{\mathbf{Y}}^1 \equiv (\mathcal{M}_- - V_\Sigma I)^{-1}\left(\frac{\mathbf{k}^0 \mathbf{L}}{\mathbf{L}\mathbf{k}^0} - I\right)\mathbf{H}^0$$
$$\mathbf{k}^1 \equiv (\lambda_- - V_\Sigma)(\mathcal{M}_- - V_\Sigma I)^{-1}\frac{\mathbf{k}^0}{\mathbf{L}\mathbf{k}^0}\pi^1.$$

This procedure can be iterated.
Let us take the x-derivative of (1.2) on both sides of Σ, make the difference and use the compaibility relation to obtain:

(1.11) $$\frac{\delta \mathbf{Y}^1}{\delta t} + [\nabla_\mathbf{U}\mathcal{M}\,\partial_x\mathbf{U}\,\partial_x\mathbf{U} + \mathcal{M}\,\partial_x^2\mathbf{U}] - V_\Sigma[\partial_x^2\mathbf{U}] - [\nabla_\mathbf{U}g\,\partial_x\mathbf{U}] = 0$$

This can be written as:

(1.12) $$\frac{\delta \mathbf{Y}^1}{\delta t} + (\mathcal{M}_- - V_\Sigma I)\mathbf{Y}^2 + [\mathcal{M}](\partial_x^2\mathbf{U})_+ + \mathcal{N}_-(\partial_x\mathbf{U})_+\mathbf{Y}^1$$

$$+ \mathcal{N}_-\mathbf{Y}^1(\partial_x\mathbf{U})_+ + \mathcal{N}_-\mathbf{Y}^1\mathbf{Y}^1 - (\nabla_\mathbf{U}g)_-\mathbf{Y}^1$$

$$- [\nabla_\mathbf{U}g](\partial_x\mathbf{U})_+ + [\mathcal{N}](\partial_x\mathbf{U})_+(\partial_x\mathbf{U})_+ = 0$$

where $\mathcal{N} \equiv \nabla_\mathbf{U}\mathcal{M}$.
Multiplying (1.12) by $\mathbf{L}$ yields the transport equation for π^1

(1.13) $$\frac{\delta \pi^1}{\delta t} - \frac{\delta \mathbf{L}}{\delta t}\mathbf{Y}^1 + \mathbf{L}[\mathcal{M}](\partial_x^2\mathbf{U})_+ + \mathbf{L}\mathcal{N}_-(\partial_x\mathbf{U})_+\mathbf{Y}^1$$

$$+ \mathbf{L}\mathcal{N}_-\mathbf{Y}^1\mathbf{Y}^1 - \mathbf{L}(\nabla_\mathbf{U}g)_-\mathbf{Y}^1 + \mathbf{L}\mathcal{N}_-\mathbf{Y}^1(\partial_x\mathbf{U})_+$$

$$- \mathbf{L}[\nabla_\mathbf{U}g](\partial_x\mathbf{U})_+ + \mathbf{L}[\mathcal{N}](\partial_x\mathbf{U})_+(\partial_x\mathbf{U})_+ + (\lambda_- - V_\Sigma)\pi^2 = 0$$

Proceeding in a similar way it is possible to derive an infinite system of the form

$$
\text{(1.14)} \quad
\begin{cases}
\dfrac{\delta V_\Sigma}{\delta t} = \mathcal{G}^\circ(\mathbf{U}_+, \partial\mathbf{U}_+, V_\Sigma) - \dfrac{\lambda_- - V_\Sigma}{Lk^0}\pi^1 \\[3mm]
\dfrac{\delta \pi^1}{\delta t} = \mathcal{G}^1(\mathbf{U}_+, \partial\mathbf{U}_+, \partial^2\mathbf{U}_+, V_\Sigma, \pi^1) - (\lambda_- - V_\Sigma)\pi^2 \\[3mm]
\qquad\vdots \\[3mm]
\dfrac{\delta \pi^n}{\delta t} = \mathcal{G}^n(\mathbf{U}_+, \partial\mathbf{U}_+, \dots, \partial^{n+1}\mathbf{U}_+, V_\Sigma, \pi^1, \dots, \pi^n) \\[2mm]
\qquad\qquad - (\lambda_- - V_\Sigma)\pi^{n+1} \\[3mm]
\qquad\vdots \\[3mm]
\dfrac{\delta x_s}{\delta t} = V_\Sigma
\end{cases}
$$

and the jump in the space derivatives, $\mathbf{Y}^n$, have the form

$$
\mathbf{Y}^n = \hat{\mathbf{Y}}^n(\mathbf{U}_+, \partial\mathbf{U}_+, \dots, \partial^n\mathbf{U}_+, V_\Sigma, \pi^1, \dots, \pi^{n-1}) + \pi^n \mathbf{k}^n(\mathbf{U}_+, V_\Sigma),
$$
$$
n = 0, 1, \dots
$$

If the solution is known at time $t = 0$, then it is possible to evaluate all the terms $\pi^1, \pi^2, \dots$ at $t = 0$ and, from (1.14) one can build the Taylor expansion of the shock position:

$$
\text{(1.15)} \qquad x_s(t) = x_s(0) + V_\Sigma(0)t + \frac{1}{2}\left(\frac{\delta V_\Sigma}{\delta t}\right)_0 t^2 + \frac{1}{6}\left(\frac{\delta^2 V_\Sigma}{\delta t^2}\right)_0 t^3 + \cdots
$$

where $\delta^2 V_\Sigma/\delta t^2$ is obtained by differentiating the first equation and using the second, and so on.

This expansion describes the propagation of the shock for short times. Let us consider the first equation of the system, (1.7). If there are cases in which $\Gamma\pi^1$ vanishes then the equation gives an exact description of the evolution of the shock. This is the case, for example, of an intermediate discontinuity, as described in Chapter 5.

If $(\lambda_- - V_\Sigma)\pi^1$ is small compared to the other terms then we get an approximate equation for the propagation of the shock:

$$
\text{(1.16)} \qquad \frac{\delta V_\Sigma}{\delta t} + \frac{1}{L(\partial\mathbf{G}/\partial V_\Sigma)}\left(\mathbf{L}\nabla_{\mathbf{U}_+}\mathbf{G}\frac{\delta\mathbf{U}_+}{\delta t} + \mathbf{L}[\![\mathbf{M}]\!](\partial_x\mathbf{U})_+ - \mathbf{L}[\![\mathbf{g}]\!]\right) = 0.
$$

We apply this expression to the propagation of a shock in gas dynamics into a non uniform athmosphere. We consider a polytropic gas and a static unperturbed field

$$
u_0 = 0, \; \rho_0(x), \; p_0(x).
$$

The field equations are given of the form (1.2) with

(1.17) $\qquad \mathbf{U} = \begin{pmatrix} \rho \\ u \\ p \end{pmatrix}, \ M = \begin{pmatrix} u & \rho & 0 \\ 0 & u & 1/\rho \\ 0 & \gamma p & u \end{pmatrix}, \ \mathbf{g} = \begin{pmatrix} 0 \\ -g \\ 0 \end{pmatrix}$

where γ is the polytropic constant and g is the gravitational constant. The solution of the jump relations are given by (Whitham, 1974)

$$\rho = \rho_0 \frac{(\gamma+1)M^2}{(\gamma-1)M^2 + 2},$$

(1.18) $$u = c_0 \frac{2(M^2 - 1)}{(\gamma+1)M},$$

$$p = p_0 \frac{2\gamma M^2 - \gamma + 1}{\gamma + 1}$$

where $M \equiv V_\Sigma / c_0$ is the Mach number, the quantitias with the subscript zero are evaluated in the impertubed field and the quantitias without subscript are evaluated behind the shock.

Equation (1.16) becomes

(1.19) $\quad (1 + 2\mu + 1/M^2) \dfrac{1}{M^2 - 1} \dfrac{1}{c_0} \dfrac{\delta M}{\delta t} + \left[1 + 2\mu + \dfrac{2}{\gamma M^2}(\mu - 1) \right] \dfrac{1}{c_0} \dfrac{dc_0}{dx}$

$\qquad + \left[\mu + \dfrac{1}{\gamma M^2}(\mu - 1) \right] \dfrac{1}{\rho_0} \dfrac{d\rho_0}{dx} = 0$

where

(1.20) $$\mu \equiv \left(\frac{(\gamma - 1)M^2 + 2}{2\gamma M^2 - \gamma + 1} \right)^{1/2}.$$

This equation gives a relation between the variation of the Mach number and the state ahead of Σ. For an isothermal athmosphere $c_0 = $ const and the equation is integrated to:

(1.21) $$\log \bar{\rho}_0 / \rho_0 = \int \mathcal{F}(\mu) d\mu + \text{const}$$

where

(1.22) $\quad \mathcal{F}(\mu) = \dfrac{\gamma(\gamma+1)\mu[2(1 - \gamma)\mu^2 - (\gamma+1)\mu + \gamma - 3]}{(1 - \mu)(2\gamma\mu^2 - \gamma + 1)[\gamma(\gamma+1)\mu^3 - 2\gamma\mu^2 + (\gamma+1)\mu + \gamma - 1]}$

In the case of strong shocks, i.e. $M \gg 1$, equation (1.19) simplifies to:

(1.23) $$\frac{1}{V_\Sigma} \frac{\delta V_\Sigma}{\delta t} + \beta \frac{1}{\rho_0} \frac{\delta \rho_0}{\delta t} = 0$$

where

(1.24) $$\frac{1}{\beta} = 2 + \left(\frac{2\gamma}{\gamma - 1} \right)^{1/2}$$

which gives $V_\Sigma \propto \rho_0^{-\beta}$. A similar rule which relates the Mach number to the state ahead has been obtained by Whitham (1974), using a slightly different approach. The relation that he obtained coincides with (1.23–24) in the limit case of strong shocks.

2 Multidimensional case

It is possible to extend the above approach to the propagation of a shock Σ in more than one space dimension.

Let us consider the full system in $\mathbb{R}^4$:

$$(2.1) \qquad \partial_t \mathbf{U} + \mathcal{M}^i \partial_i \mathbf{U} = \mathbf{g}.$$

Taking the jump of this system across Σ and making use of the compatibility relations (1.10–11) of Chapter 5 we get:

$$(2.2) \qquad \frac{\partial \mathbf{G}}{\partial V_\Sigma} \frac{\delta V_\Sigma}{\delta t} + \mathbf{H}^0 + (\mathcal{M}^i_- n_i - V_\Sigma I) \mathbf{Y}^1 = 0$$

where

$$(2.3) \qquad \mathbf{H}^0 \equiv -\frac{\partial \mathbf{G}}{\partial n_i} \tilde{\partial}_i V_\Sigma + \nabla_{\mathbf{U}_+} \mathbf{G} \frac{\delta \mathbf{U}_+}{\delta t} + \mathcal{M}^i_- \mathbf{G}' \tilde{\partial}_i V_\Sigma$$

$$+ \mathcal{M}^j_- \frac{\partial \mathbf{G}}{\partial n_i} \chi_{ij} + \mathcal{M}^i_- \nabla_{\mathbf{U}_+} \mathbf{G} \tilde{\partial}_i \mathbf{U}_+ + [\![\mathcal{M}^i]\!] (\partial_i \mathbf{U})_+ - [\![\mathbf{g}]\!]$$

depends on quantities assigned on Σ and $\mathbf{G}' \equiv \partial \mathbf{G}/\partial V_\Sigma$.

The symbols used have been introduced in Chapter 5.

Multiplying (2.2) by $\mathbf{L} \equiv \mathbf{L}_-$ we obtain

$$(2.4) \qquad \mathbf{L}\mathbf{G}' \frac{\delta V_\Sigma}{\delta t} + \mathbf{L}\mathbf{H}^0 + (\lambda_- - V_\Sigma)\pi^1 = 0.$$

This equation gives a relation between the acceleration of the shock surface, the quantities defined on Σ, and π^1. If $\lambda_- = V_\Sigma$ (i.e. for an intermediate discontinuity) this equation can be interpreted as a PDE for V_Σ as a function of t and two parameters along the surface.

If V_Σ does not coincide with an eigenvalue of λ_- then we can solve (2.2) for $\mathbf{Y}^1$:

$$(2.5) \qquad \mathbf{Y}^1 = \mathcal{P}_-^{-1} \left(\frac{\mathbf{L}\mathbf{H}^0}{\mathbf{L}\mathbf{G}'} - \mathbf{H}^0 + (\lambda_- - V_\Sigma)\pi^1 \frac{\mathbf{G}'}{\mathbf{L}\mathbf{G}'} \right)$$

where

$$\mathcal{P} \equiv \mathcal{M}^i n_i - V_\Sigma I.$$

$\mathbf{Y}^1$ can be expressed as a function of V_Σ, $\vec{n}$, the unperturbed state ahead of the shock, and π^1. For propagation into a constant state, π^1 is invariant under a transformation of field variables. More precisely, if we perform a regular transformation of field variables,

$$\mathbf{U} = \Psi(\tilde{\mathbf{U}}')$$

then the field equations (2.1) can be written as

$$\partial_t \tilde{\mathbf{U}}' + \tilde{\mathcal{M}}^i \partial_i \tilde{\mathbf{U}} = \tilde{\mathbf{g}}$$

where

$$\tilde{M}^i \equiv C^{-1} M^i C, \quad \tilde{g} \equiv C^{-1} g, \quad C \equiv \nabla_{\tilde{U}} \Psi.$$

The quantity π^1 transforms according to

$$\tilde{\pi}^1 = \pi^1 + L C_- [\![C^{-1}]\!] n^i (\partial_i U)_+ .$$

If **R** has the same dimensions as **U** (and they are both of the same order of magnitude) then π^1 has the dimension of the inverse of length and represents the spatial rate of variation of the field behind the shock.

We shall now rewrite (2.4) in a form which is more suitable for the applications. By differentiating the jump conditions (1.3) we obtain the following identities

(2.6)
$$P_-(I + \nabla_{U_+} G) = P_+$$
$$P_- G' = [\![F^0]\!]$$
$$P_- \frac{\partial G}{\partial n_j} = -[\![F^j]\!].$$

Making use of these identities and of the formulas (see Appendix A6)

(2.7)
$$L P^{-1} = \frac{L}{\lambda - V_\Sigma}, \quad L M^i P^{-1} = L P^{-1} M^i$$

we write (2.4) in the following form:

(2.8)
$$L [\![F^0]\!] \frac{\delta V_\Sigma}{\delta t} + (L [\![F^i]\!] + L M_-^i [\![F^0]\!]) \tilde{\partial}_i V_\Sigma - L M_-^i [\![F^j]\!] \chi_{ij}$$
$$+ (\lambda_- - V_\Sigma) L [\![M^i]\!] (\partial_i U)_+ - L(M_-^i - \lambda_- n^i) [\![M^j n_j]\!] (\partial_i U)_+$$
$$- L [\![M^j n_j]\!] \frac{\delta U_+}{\delta t} - (\lambda_- - V_\Sigma) L [\![g]\!] + (\lambda_- - V_\Sigma)^2 \pi^1 = 0 .$$

This expression does not require the explicit solutions of the jump conditions and is in a form more suitable for algebraic manipulators. In order to derive the equation for the evolution of π^1, we consider the derivative of (2.1) with respect to x^j, take the jump across Σ, make use of the compatibility relations (1.10–13) of Chapter 5 and left multiply by **L**. We obtain

(2.9)
$$\frac{\delta \pi^1}{\delta t} - \frac{\partial L}{\partial V_\Sigma} \frac{\delta V_\Sigma}{\delta t} Y^1 + L \tilde{\partial}_i G \tilde{\partial}^i V_\Sigma + L N_-^i n_i Y^1 Y^1 + L N_-^i Y^1 \tilde{\partial}_i G$$
$$+ L N_-^i Y^1 (\partial_i U)_+ + L N_-^i n^j (\partial_j U)_+ \tilde{\partial}_i G + L N_-^i n_i n^j (\partial_i U)_+ Y^1$$
$$+ L [\![N^i]\!] n^j (\partial_j U)_+ (\partial_i U)_+ + L M^i \tilde{\partial}_i Y^1 - L M_-^i \chi_i^k \tilde{\partial}_k G$$
$$+ (\lambda_- - V_\Sigma) \pi^2 = 0$$

where

$$\tilde{\partial}_i G = \nabla_U G (\tilde{\partial}_i U)_+ + G' \tilde{\partial}_i V_\Sigma + \frac{\partial G}{\partial n_j} \chi_{ij}$$

and $\delta V_\Sigma / \delta t$ is obtained from the previous equation.

The term $\tilde{\partial}_i \mathbf{Y}^1$ is evaluated from the expression of $\mathbf{Y}^1$ obtained from (2.5). Equations (2.4) and (2.9) are of the form

(2.10)
$$\frac{\delta V_\Sigma}{\delta t} = \mathcal{F}^0(\mathbf{U}_+, V_\Sigma, n^i, (\partial_i \mathbf{U})_+, \tilde{\partial}_i V_\Sigma, \tilde{\partial}_i n^j, \pi^1)$$

(2.11)
$$\frac{\delta \pi^1}{\delta t} = \mathcal{F}^1(\mathbf{U}_+, V_\Sigma, n^i, (\partial_i \mathbf{U})_+, \tilde{\partial}_i V_\Sigma, \tilde{\partial}_j n^i, \pi^1,$$
$$(\partial_{ij}^2 \mathbf{U})_+, \tilde{\partial}_{ij}^2 V_\Sigma, \tilde{\partial}_{ij} n^s, \tilde{\partial}_i \pi^1, \pi^2).$$

By iterating the method we obtain an infinite system of the form

$$\frac{\delta V_\Sigma}{\delta t} = \mathcal{F}^0(Q_0, Q_1)$$

$$\frac{\delta \pi^1}{\delta t} = \mathcal{F}^1(Q_0, Q_1, Q_2)$$

(2.12)
$$\vdots$$

$$\frac{\delta \pi^n}{\delta t} = \mathcal{F}^n(Q_0, Q_1, \ldots, Q_{n+1})$$

$$\vdots$$

where

$$Q_0 \equiv \{\mathbf{U}_+, V_\Sigma, n^i\}$$
$$Q_1 \equiv \{(\partial_i \mathbf{U})_+, \tilde{\partial}_i V_\Sigma, \tilde{\partial}_j n^i, \pi^1\}$$
$$Q_2 \equiv \{(\partial_{ij}^2 \mathbf{U})_+, \tilde{\partial}_{ij}^2 V_\Sigma, \tilde{\partial}_{ij}^2 n^s, \tilde{\partial}_i \pi^1, \pi^2\}$$

(2.13)
$$\vdots$$

$$Q_n \equiv \{(\partial_{i_1 \ldots i_n}^n \mathbf{U})_+, \tilde{\partial}_{i_1 \ldots i_n}^n V_\Sigma, \tilde{\partial}_{i_1 \ldots i_n}^n n^s,$$
$$\tilde{\partial}_{i_1 \ldots i_{n-1}} \pi^1, \tilde{\partial}_{i_1, \ldots, i_{n-2}} \pi^2, \ldots, \tilde{\partial}_i \pi^n, \pi^{n+1}\}.$$

From (2.11) it is possible to compute higher order time derivatives of V_Σ. Formally:

(2.14)
$$\frac{\delta^2 V_\Sigma}{\delta t^2} = \frac{\partial \mathcal{F}^0}{\partial Q_0} \frac{\delta Q_0}{\delta t} + \frac{\partial \mathcal{F}^0}{\partial Q_1} \frac{\delta Q_1}{\delta t}.$$

The derivatives $(\delta Q_0 / \delta t)$ are computed by using the first equation of (2.11) and the transport equation for n^i. The term $(\delta Q_1 / \delta t)$ contains the derivatives

$$\frac{\delta}{\delta t} \tilde{\partial}_i V_\Sigma, \; \frac{\delta}{\delta t} \chi_{ij}, \; \frac{\delta \pi^1}{\delta t}.$$

The last one is computed by using the second equation of (2.11). In order to compute the first two terms we make use of the commutation relations (Appendix A3)

(2.15)
$$\frac{\delta}{\delta t} \tilde{\partial}_i - \tilde{\partial}_i \frac{\delta}{\delta t} = (n_i \delta^{ks} \tilde{\partial}_s V_\Sigma - V_\Sigma \chi_i^k) \tilde{\partial}_k.$$

Then we compute the derivatives

$$\tilde{\partial}_i(\delta V_\Sigma/\delta t) \,,\, \tilde{\partial}_i(\delta n^j/\delta t)$$

form (2.10) and Eq. (1.15) of Chapter 5.

We obtain expressions which depend on $\mathbf{U}_+\,, V_\Sigma\,, n^i$ and their first and second derivatives and on π^1, $\tilde{\partial}_i\pi^1$, π^2.

This procedure can be iterated and expressions for

$$\frac{\delta Q_2}{\delta t}\,,\, \frac{\delta Q_3}{\delta t}\,,\cdots$$

can be obtained in a similar way. Each expression is of the form:

$$\frac{\delta Q_k}{\delta t} = \mathcal{G}^k(Q_0\,,Q_1\,,\cdots\,,Q_{k+1})\,.$$

Making use of these expressions and of the chain rule for the derivative it is possible to compute higher order derivatives of the shock velocity at time $t = 0$. Once the terms

$$\frac{\delta^k V_\Sigma}{\delta t^k} \quad\text{and}\quad \frac{\delta^k n^i}{\delta t^k}$$

have been computed, the Taylor series for a point on the shock is obtained:

$$(2.16)\qquad x_s^i(t) = x_s^i(0) + \left(\frac{\delta x^i}{\delta t}\right)_0 t + \frac{1}{2}\left(\frac{\delta^2 x^i}{\delta t^2}\right)_0 t^2 + \cdots$$

with

$$\frac{\delta x^i}{\delta t} = V_\Sigma n^i$$

$$(2.17)\qquad \frac{\delta^2 x^i}{\delta t^2} = \frac{\delta V_\Sigma}{\delta t}n^i + V_\Sigma\frac{\delta n_i}{\delta t}$$

$$\vdots$$

3   Propagation into a constant state and definition of corrugation stability

For propagation into a constant state equation (2.8) reduces to:

$$(3.1)\qquad \mathbf{L}[\mathbf{F}^0]\frac{\delta V_\Sigma}{\delta t} + (\mathbf{L}[\mathbf{F}^i] + \mathbf{L}\mathcal{M}_-^i[\mathbf{F}^0])\tilde{\partial}_i V_\Sigma$$
$$- \mathbf{L}\mathcal{M}_-^i[\mathbf{F}^j]\chi_{ij} + (\lambda_- - V_\Sigma)^2\pi^1 = 0\,.$$

Let us consider now the constant shock solution:

$$(3.2)\qquad \mathbf{U}(x^i,t) = \begin{cases} \mathbf{U}_1 & \text{for } x^1 - \overset{\circ}{V}_\Sigma t < 0 \\ \mathbf{U}_0 & \text{for } x^1 - \overset{\circ}{V}_\Sigma t \geq 0 \end{cases}$$

This is a plane shock moving with constant velocity and separating two constant states $\mathbf{U}_1$ and $\mathbf{U}_0$.

We shall consider a class of perturbation of this solution for which, at the initial time $t = 0$,

(i) $V_\Sigma|_{t=0} = \overset{\circ}{V}_\Sigma$ on $\Sigma(0)$,

(ii) the field ahead remains constant,

(iii) the field behind is such that $\pi^1 = 0$.

Note that this is an admissible perturbation and that $\mathbf{U}_-$ and $(\partial_i \mathbf{U})_-$ are uniquely determined by the shape of the perturbed shock (normal to the wavefront). Let A be the cross sectional area of an infinitesimal tube of orthogonal rays (normal to the wavefront). The rate of variation of A is related to the curvature of the surface:

$$\frac{1}{V_\Sigma A} \frac{\delta A}{\delta t} = \chi_i^i \,.$$

We say that the plane fronted shock is unstable if

(3.3)
$$\frac{1}{A} \frac{\delta A}{\delta t} \frac{\delta V_\Sigma}{\delta t} > 0$$

that is if a small perturbation in the shock surface tends to grow.

For a given shock configuration condition (3.3) can be checked by using (3.1). In two dimensions it is

$$\chi_{ij} = (1/r) h_{ij}\,,$$

where $h_{ij} = \delta_{ij} - n_i n_j$ and r is the radius of curvature of Σ and is positive for an expanding surface, therefore condition (3.3) becomes:

(3.4)
$$\frac{\mathbf{L}\mathcal{M}_-^i[\![\mathbf{F}^j]\!] h_{ij}}{\mathbf{L}[\![\mathbf{F}^0]\!]} = \frac{\mathbf{L}\mathcal{M}_-^i[\![\mathbf{F}^j]\!] \delta_{ij}}{\mathbf{L}[\![\mathbf{F}^0]\!]} - \lambda V_\Sigma > 0\,.$$

This expression for the instability condition is valid also in three dimensions in cases in which the term

$$\mathbf{L}\mathcal{M}_-^i[\![\mathbf{F}^j]\!] \chi_{ij}$$

depends only on the mean curvature χ_i^i.

In many cases of physical interest the quantity

$$\mathbf{L}[\![\mathbf{F}^i]\!] + \mathbf{L}\mathcal{M}_-^i[\![\mathbf{F}^0]\!]$$

is proportional to n^i.

In this case a broader class of perturbations can be considered, because condition (i) is no longer required.

Both these properties are true in the case of classical and relativistic fluid dynamics.

4   Application to gas dynamics

The Euler equations of gas dynamics can be written in the conservation form (1.1)
of Chapter 5 with

$$\mathbf{U} = \begin{pmatrix} \rho \\ \vec{u} \\ E \end{pmatrix}, \quad \mathbf{F}^0 = \begin{pmatrix} \rho \\ \rho\vec{u} \\ \varepsilon \end{pmatrix},$$

(4.1)

$$\mathbf{F}^i = \begin{pmatrix} \rho u^i \\ \rho\vec{u}u^i + P\hat{e}^i \\ (\varepsilon + P)u^i \end{pmatrix}$$

where $\hat{e}^i$ is the row or column unit vector of the i axis, E is the internal energy per
unit mass, and $\varepsilon = \rho(\frac{1}{2}|\vec{u}|^2 + E)$ the total energy per unit volume. We suppose
that pressure is related to density and energy by an equation of state

$$P = P(\rho, E).$$

The system can be written in the form (1.3) of Chapter 5, with

(4.2)
$$\mathcal{M}^i = \begin{pmatrix} u^i & \rho\hat{e}^i & 0 \\ \dfrac{P'_\rho}{\rho}\hat{e}^i & u^i I & \dfrac{P'_E}{\rho}\hat{e}^i \\ 0 & \dfrac{P}{\rho}\hat{e}^i & u^i \end{pmatrix}$$

where I is the 3×3 identity matrix, $P'_\rho = \partial P/\partial\rho$, $P'_E = \partial P/\partial E$.
We consider the stability of a plane acoustic shock corresponding to the eigenvalue

(4.3)
$$\lambda = \vec{u} \cdot \vec{n} + \varepsilon$$

where $c \equiv (\partial P/\partial\rho)_s^{1/2}$ is the sound speed. A left eigenvector of the matrix $\mathcal{M}^i n_i$
corresponding to the eigenvalue (4.3) is

(4.4)
$$\mathbf{L} = (P'_\rho, \rho c\vec{n}, P'_E).$$

The terms appearing in (2.4) are:

(4.5)
$$\mathbf{L}\frac{\partial \mathbf{G}}{\partial V_\Sigma} = \rho c\frac{d\hat{u}}{dV_\Sigma} + \frac{d\hat{P}}{dV_\Sigma}$$

(4.6)
$$\mathbf{L}\mathcal{M}^k_-\frac{\partial \mathbf{G}}{\partial V_\Sigma} = \left(\rho c\frac{d\hat{u}}{dV_\Sigma} + \frac{d\hat{P}}{dV_\Sigma}\right)(u^k + cn^k)$$

(4.7)
$$\mathbf{L}\frac{\partial \mathbf{G}}{\partial n_i} = \rho c u n^i$$

(4.8)
$$\mathbf{L}\mathcal{M}^k_-\frac{\partial \mathbf{G}}{\partial n^i} = u c\rho(c\delta^k_i + u^k n_i)$$

where

$$(4.9) \qquad \frac{d\cdot}{dV_\Sigma} \equiv \left(\frac{\partial\cdot}{\partial V_\Sigma}\right)_{U_+,\vec{n}=\text{const}}$$

is the derivative along the Hugoniot curve, $u \equiv [\![u^i]\!]n_i$, and all the quantities are evaluated behind Σ.

It is convenient to express all the quantities in the rest frame of the unperturbed fluid ahead of the shock.

In this case $u^k = un^k$ and therefore the term proportional to $\tilde{\partial}_i V_\Sigma$ in equation (2.4) vanishes.

The instability condition (3.4) becomes:

$$(4.10) \qquad AV_\Sigma \frac{\delta V_\Sigma}{\delta A} = \frac{\rho u c^2}{\rho c \dfrac{d\hat{u}}{dV_\Sigma} + \dfrac{d\hat{P}}{dV_\Sigma}} < 0 \,.$$

The jump conditions across Σ, solved with respect to ρ, P and E, are:

$$(4.11) \qquad \rho = \frac{\rho_0 V_\Sigma}{V_\Sigma - u}$$

$$(4.12) \qquad P = P_0 + \rho_0 u V_\Sigma$$

$$(4.13) \qquad E = E_0 + \frac{P_0}{\rho_0}\frac{u}{V_\Sigma} + \frac{1}{2}u^2 \,.$$

Here the subscript 0 denotes the unperturbed state. Differentiating (4.12) and substituting in (4.10), the instability condition becomes:

$$(4.14) \qquad \phi \equiv \left(1 + \frac{C_s}{V_\Sigma - u}\right)\frac{V_\Sigma}{u}\frac{d\hat{u}}{dV_\Sigma} + 1 > 0 \,.$$

This condition is expressed in terms of the shape of the Hugoniot curve in the (u, V_Σ) plane. By differentiation of relations (4.11–13) it is possible to express $(d\hat{u}/dV_\Sigma)$ in terms of thermodynamical quantities (Appendix A7):

$$(4.15) \qquad \frac{V_\Sigma}{u}\frac{d\hat{u}}{dV_\Sigma} = \frac{1 + M^2 - M^2(1/\rho_0 - 1/\rho)P'_\epsilon}{1 - M^2} \,.$$

Substituting this expression into (4.14) the condition for instability can be written in the form:

$$(4.16) \qquad F_s \equiv 1 + M - M^2\left(\frac{1}{\rho_0} - \frac{1}{\rho}\right)P'_E < 0$$

where $M = (V_\Sigma - u)/C_s$ is the Mach number behind the shock. We remark here that this same condition appears when considering the stability of the shock against linear perturbations (Gardner and Kruskal, 1963). In this case (4.16) is a necessary and sufficient condition for the existence of unstable modes.

Now we apply condition (4.16) to some specific cases to check whether the stability requirements are satisfied.

First note that the stability requirement is of the form

$$(4.17) \qquad\qquad \mathcal{F}(U_+, U_-) < 0$$

that is the inequality involves the field values on both sides of the shock. These values are related by the jump conditions (4.11–13) and therefore, in order to check (4.17) it is necessary to solve the jump conditions.

Now we derive a simpler condition which does not involve the shock amplitude and which is simpler to apply.

PROPOSITION (4.1). *If the pressures are positive and the shock is compressive and*

$$(4.18) \qquad\qquad P'_\rho > 0$$

then condition (4.16) cannot be satisfied.

PROOF. From (4.11–12) it follows:

$$(4.19) \qquad\qquad \rho^2 (V_\Sigma - u)^2 = \frac{P - P_0}{1/\rho_0 - 1/\rho}.$$

From the first law of thermodynamics:

$$T ds = dE - \frac{P}{\rho^2} d\rho$$

therefore

$$(4.20) \qquad\qquad c^2 = \left(\frac{\partial P}{\partial \rho}\right)_s = P'_\rho + \frac{P}{\rho^2} P'_E.$$

Making use of (4.19-20) we obtain the following expression for F_s:

$$F_s = M + \frac{P_0}{P} + \frac{P - P_0}{P} P'_\rho,$$

which is positive. $\qquad\qquad\qquad\qquad\qquad\qquad\qquad\qquad\qquad\qquad$ □

In the proof, as well as in stating the equivalence of (4.14) with (4.16) we assumed that the flow is subsonic behind the shock, i.e. $M < 1$.

In the case of a polytropic gas the equation of state is

$$P = (\gamma - 1)\rho E$$

with $\gamma > 1$. It is

$$P'_\rho = (\gamma - 1)E > 0$$

and therefore a plane shock in a polytropic gas is stable against linear perturbations. Let us consider now a Van der Waals gas.

The equation of state in nondimensional variables are (Chap. 1, Eq. (1.24–25)).

(4.21)
$$(q + 3r^2)(3/r - 1) = 8\Theta$$

(4.22)
$$\eta = -3r + f(\Theta)$$

where $r = 1/V$, q, η and Θ are nondimensional density, pressure, energy density and temperature. Condition (4.18) becomes:

(4.23)
$$\left(\frac{\partial q}{\partial V}\right)_\eta = \frac{3[\tau + f'(\Theta)(q - q_m(V))/V]}{f'(\Theta)(3 - V)} > 0$$

where $q_m(V) = 3V^2 - 2V^3$ is the locus of the maxima and minima of isothermal lines. In the region $q < q_m(V)$ equations (4.21–22) are no more valid because the condition of mechanical stability,

$$\left(\frac{\partial P}{\partial \rho}\right)_E > 0$$

is not satisfied and a single phase state cannot exist. Outside this region condition (4.23) is satisfied, because

$$f'(\Theta) = (8/3)C_V/\overline{R} > 0.$$

Therefore in a Van der Waals gas thermodynamical stability implies the linear stability of shock waves.

Other considerations about the stability of shock waves in a Van der Waals gas can be found in (Russo, 1990).

Stability condition (4.16) can be extended to relativistic gas dynamics. This subject was treated by Russo and Anile (1987) and by Russo (1988).

REFERENCES

- A. M. Anile and G. Russo, *Corrugation stability for plane relativistic shock waves*, Phys. Fluids **29**, 2847 (1986).
- A. M. Anile and G. Russo *Corrugation stability of magnetohydrodynamic shock waves*, in *Nonlinear Wave Motion*, Alan Jeffrey Ed., Longman, New York (1989).
- C. S. Gardner and M. D. Kruskal, *Stability of Plane Magneto hydrodynamic Shocks*, Phys. Fluids **7**, 700 (1964).
- A. Jeffrey, *Quasilinear Hyperbolic Systems and Waves*, Pitman, London (1976).
- M. A. Liberman and A. L. Velikovich, *Physics of Shock Waves in Gases and Plasmas*, Springer, Berlin (1986).
- P. Prasad, *Kinematics of a Multi-Dimensional Shock of Arbitrary Strenght in an Ideal Gas*, Acta Mechanica, 136, Springer-Verlag (1982).
- G. Russo and A. M. Anile, *Stability properties of relativistic shock waves: Basic results*, Phys. Fluids **30**, 2406 (1987).
- G. Russo, *Stability properties of relativistic shock waves: Applications*, The Astrophysical Journal **334**, 702 (1988).
- G. Russo, *Some remarks on the stability of shock waves*, Meccanica **25**, 83 (1990).
- G. B. Whitham, *Linear and Nonlinear Waves*, Wiley, New York (1974).

RAY METHODS FOR NONLINEAR DISPERSIVE WAVES

chapter 8

Introduction

In Chapter 2 Sec. 3 we introduced the perturbation-reduction method in order to treat the one-dimensional propagation of far fields in nonlinear dissipative and/or dispersive media. The KdV or Burgers' equations have also been derived. In this chapter we will generalize these methods to several dimensions by using the ray method, introduced in Chapter 5.

The first paper using ray methods for nonlinear equations is that of Y. Choquet-Bruhat (1969) which extended the method previously introduced for the linear case by J. Leray (1961), L. Gårding, T. Kotake, and J. Leray (1964), Kline and Kay (1965) among others. These methods use an asympotic expansion which has its origin in the WKB method and which has been applied to linear systems by P. Lax (1957), D. Ludwig (1960), and many others.

P. Germain (1971) extended these methods to treat the propagation of waves in both nonlinear and dissipative media. The multidimensional dispersive case, limited to shallow water waves, was treated by Prasad and Revindran (1977). The general formulation of the multidimensional methods for nonlinear dispersive and/or dissipative media is due to S. Giambó, A. Greco, P. Pantano (1979) and D. Fusco (1979). Hunter (1985) extended the method to treat non-resonant interaction of several dissipative or dispersive waves. These methods have been also successfully applied to ion-acoustic wave propagation by S. Giambó and P. Pantano (1982). They deduced a general KdV equation which, in particular cases, reduces to the cylindrical or spherical KdV. These theoretical results have received several experimental corroborations.

Recently the mutiple scale case has been studied by S. Giambó and P. Pantano (1989) who obtained a series of model equations describing the anelastic collisions between solitons.

Another description of cylindrical and spherical solitons is given by Infeld and Rowlands (1990).

1 Perturbation-reduction methods in several dimensions: derivation of the generalized KdV or Burgers equations

When we consider three dimensional wave propagation various new effects arise due to the wavefront geometry. In this case we have several interacting phenomena, including nonlinearity, the geometry and, of course, dissipation and/or dispersion. However the far field concept, introduced in Section 2.3, can be suitably extended in order to obtain model equations for multi-dimensional waves as well.

We consider the following system of equations

(1.1)
$$\mathcal{A}^\alpha \frac{\partial \mathbf{U}}{\partial x^\alpha} + \left[\sum_{I=1}^{S} \prod_{J=1}^{P} \left(\mathcal{H}_J^{I\alpha} \frac{\partial}{\partial x^\alpha} \right) \right] \mathbf{U} = 0,$$

satisfied by a field $\mathbf{U}$ with N components. Here $\alpha = 0, 1, 2, 3$ and the Einstein convention on the repeated indices is assumed. Also, $\mathcal{A}^\alpha$ and $\mathcal{H}_J^{I\alpha}$ are $N \times N$ matrices whose elements are analytic function of the field $\mathbf{U}$ in a domain D of $\mathbb{R}^N$, containing the unperturbed state $\mathbf{U}_0$. Then we can write the following expansions

(1.2)
$$\begin{cases} \mathcal{A}^\alpha = \mathcal{A}_0^\alpha + (\nabla_\mathbf{U} \mathcal{A}^\alpha)_0 (\mathbf{U} - \mathbf{U}_0) + \cdots \\ \mathcal{H}_J^{I\alpha} = \mathcal{H}_{J0}^{I\alpha} + (\nabla_\mathbf{U} \mathcal{H}_J^{I\alpha})_0 (\mathbf{U} - \mathbf{U}_0) + \cdots \end{cases}$$

where

$$\mathcal{A}_0^\alpha = \mathcal{A}^\alpha(\mathbf{U}_0), \qquad \mathcal{H}_{J0}^{I\alpha} = \mathcal{H}_J^{I\alpha}(\mathbf{U}_0)$$

For the sake of simplicity we assume that $\mathcal{A}_0^0 = I$ and that the matrix $\mathcal{A}_0^n = \mathcal{A}_0^i n_i$ $(i = 1, 2, 3)$, where $\vec{n}$ is an arbitrary vector, has N linearly independent eigenvectors corresponding to real eigenvalues. We suppose that a simple eigenvalue λ_0 exists. We want to study the evolution of a wave train taking into account the effects of nonlinearity, dispersion and the wavefront geometry. Let ϵ be the amplitude of the perturbation. Then, typically, nonlinear terms in Eq. (1.1) will be of order ϵ^2/L where L is the length scale for nonlinearity to be important (unless, due to some degeneracy, quadratic nonlinearities are absent). If we ignore the dispersive and dissipative terms in Eq. (1.1) we have the high-frequency approximation of Chapter 4. Since we are interested in dispersive waves we must consider those terms, which are of order ϵ/L^P. Therefore, in order for dispersion/dissipation to balance nonlinearity one must have

$$L \sim \epsilon^{-a}, \qquad a = \frac{1}{P-1}.$$

Therefore the length scale over which nonlinearity is important and is balanced by dispersion/dissipation is $L \sim \epsilon^{-a}$.

We want the wavefront geometry to vary on the length scale L and in a locally isotropic way (there are no preferred directions on the front). Let $\varphi(x^\alpha) = \text{const}$ describe surfaces. Then φ must enter in the problem through the variable $\xi = \epsilon^a \varphi$. We introduce the variables ξ and η^α.

(1.3)
$$\begin{cases} \eta^\alpha = \epsilon^{a+1} x^\alpha \\ \xi = \frac{1}{\epsilon} \varphi(\epsilon^{a+1} x^\alpha) \qquad a = (P-1)^{-1} \end{cases}$$

We assume that the field $\mathbf{U}$ can be expanded in a formal asymptotic series in terms of a small parameter ϵ

(1.4)
$$\mathbf{U} = \mathbf{U}_0 + \epsilon \mathbf{U}_1 + \epsilon^2 \mathbf{U}_2 + \cdots$$

where the unperturbed field $\mathbf{U}_0$ is assumed to be constant.

As usual (see Chapter 2, Sec. 3) we consider $\mathbf{U}_1, \mathbf{U}_2 \ldots$ as functions of ξ and η^α.

Then

$$\frac{\partial}{\partial x^\alpha} = \epsilon^{a+1}\frac{\partial}{\partial \eta^\alpha} + \epsilon^a \varphi_\alpha \frac{\partial}{\partial \xi}$$

with

$$\varphi_\alpha \equiv \frac{\partial \varphi}{\partial \eta^\alpha} \, .$$

Multiplying by ϵ^{-a}, system (1.1) becomes

(1.5) $$\epsilon \mathcal{A}^\alpha \frac{\partial \mathbf{U}}{\partial \eta^\alpha} + \mathcal{A}^\alpha \varphi_\alpha \frac{\partial \mathbf{U}}{\partial \xi} + \epsilon \left[\sum_{I=1}^S \prod_{J=1}^P \mathcal{H}_J^{I\alpha} \left(\epsilon \frac{\partial}{\partial \eta^\alpha} + \varphi_\alpha \frac{\partial}{\partial \xi} \right) \right] \mathbf{U} = 0 \, .$$

Using expansions (1.2) and (1.4), Eq. (1.5) becomes a power series in ϵ. Equating the coefficients of the resulting series termwise to zero, we obtain, to first order in ϵ

(1.6) $$\mathcal{A}_0^\alpha \varphi_\alpha \frac{\partial \mathbf{U}_1}{\partial \xi} = 0$$

and, to second order

(1.7) $$\mathcal{A}_0^\alpha \left(\frac{\partial \mathbf{U}_1}{\partial \eta^\alpha} + \varphi_\alpha \frac{\partial \mathbf{U}_2}{\partial \xi} \right) + \varphi_i (\nabla_{\mathbf{U}} \mathcal{A}^i)_0 \mathbf{U}_1 \frac{\partial \mathbf{U}_1}{\partial \xi}$$

$$+ \left[\sum_{I=1}^S \prod_{J=1}^P \mathcal{H}_{J0}^{I\alpha} \varphi_\alpha \right] \frac{\partial^P \mathbf{U}_1}{\partial \xi^P} = 0 \qquad (i = 1, 2, 3) \, .$$

Let, as usual,

(1.8) $$n_i = \frac{\varphi_i}{|\nabla \varphi|}; \qquad \lambda = -\frac{\varphi_0}{|\nabla \varphi|} \, .$$

Equation (1.6) can be rewritten as

$$(\mathcal{A}_0^n - \lambda I)\frac{\partial \mathbf{U}_1}{\partial \xi} = 0 \, ,$$

where $\mathcal{A}_0^n = \mathcal{A}_0^i n_i$.

This equation admits a nontrivial solution for $\partial \mathbf{U}_1/\partial \xi$ if λ is an eigenvalue of $\mathcal{A}_0^n$. If we choose a simple eigenvalue λ_0 equation (1.8) describes the phase φ. Furthermore, from (1.6)

(1.9) $$\mathbf{U}_1 = \pi(\eta^\alpha, \xi)\mathbf{R}_0(\mathbf{U}_0, n_i) \, ,$$

where $\mathbf{R}_0$ is the right eigenvector of A_0^n corresponding to the eigenvalue λ_0. By substituting (1.9) in (1.7) and multiplying to the left by $\mathbf{L}_0$ (which is the left eigenvector of A_0^n corresponding to λ_0), we obtain

$$(1.10) \qquad \mathbf{L}_0 A_0^\alpha \left(\frac{\partial \pi}{\partial \eta^\alpha} \mathbf{R}_0 + \pi \frac{\partial \mathbf{R}_0}{\partial \eta^\alpha} \right) + \mathbf{L}_0 [(\nabla_\mathbf{U} A^n)_0 \mathbf{R}_0] \mathbf{R}_0 |\nabla\varphi| \pi \frac{\partial \pi}{\partial \xi}$$

$$+ \mathbf{L}_0 \left[\sum_{I=1}^S \prod_{J=1}^P (-\lambda_0 \mathcal{H}_{J0}^{I0} + n_i \mathcal{H}_{J0}^{Ii}) \right] \mathbf{R}_0 |\nabla\varphi|^P \frac{\partial^P \pi}{\partial \xi^P} = 0 \,.$$

The phase φ satisfies the equation

$$(1.11) \qquad\qquad \psi_0 \equiv \psi(\mathbf{U}_0, \varphi_\alpha) = 0 \,,$$

where

$$(1.12) \qquad\qquad \psi(\mathbf{U}, \varphi_\alpha) \equiv \varphi_0 + \lambda |\nabla\varphi| \,.$$

As in Chapter 4, the rays are given by

$$(1.13) \qquad \begin{cases} \dfrac{d\eta^\alpha}{d\sigma} = \dfrac{\partial\psi_0}{\partial\varphi_\alpha} \\[2mm] \dfrac{d\varphi_\alpha}{d\sigma} = -\dfrac{\partial\psi_0}{\partial\eta^\alpha} \end{cases} \,.$$

We can write (Boillat, 1965)

$$(1.14) \qquad\qquad \mathbf{L}_0 A_0^\alpha \mathbf{R}_0 = \frac{\partial\psi_0}{\partial\varphi_\alpha} \mathbf{L}_0 \mathbf{R}_0$$

and

$$(1.15) \qquad\qquad \mathbf{L}_0 A_0^\alpha \frac{\partial\mathbf{R}_0}{\partial\eta^\alpha} = \mathbf{L}_0 \mathbf{R}_0 \frac{d}{d\sigma} \log \tilde{\theta}$$

where $\tilde{\theta} = \sqrt{J}$, J being the Jacobian of the transformation between the two values of η^α at two different σ's, with σ parameter along the rays.

Hence, equation (1.10) can be written as

$$(1.16) \qquad \frac{\partial\pi}{\partial\sigma} + \pi \frac{\partial}{\partial\sigma} \log\tilde{\theta} + \alpha\pi \frac{\partial\pi}{\partial\xi} + \beta \frac{\partial^P \pi}{\partial\xi^P} = 0$$

where

$$(1.17) \qquad\qquad \alpha = \frac{\mathbf{L}_0 [(\nabla_\mathbf{U} A^n)_0 \mathbf{R}_0] \mathbf{R}_0}{\mathbf{L}_0 \mathbf{R}_0} |\nabla\varphi|$$

and

$$(1.18) \qquad \beta = \frac{\mathbf{L}_0 \left[\displaystyle\sum_{I=1}^S \prod_{J=1}^P (n_i \mathcal{H}_{J0}^{Ii} - \lambda_0 \mathcal{H}_{J0}^{I0}) \right] \mathbf{R}_0}{\mathbf{L}_0 \mathbf{R}_0} |\nabla\varphi|^P \,.$$

Equation (1.16) describes the balance between nonlinearity, dispersion or dissipation, and geometrical effects.

When $P = 2$, equation (1.16) is the generalized Burgers' equation; when $P = 3$ it is the generalized KdV equation.

If we let $w = \tilde{\theta}\pi$ and $d\tau = \alpha/\tilde{\theta}\,d\sigma$, equation (1.16) becomes

(1.19)
$$\frac{\partial w}{\partial \tau} + w\frac{\partial w}{\partial \xi} = f(\tau)\frac{\partial^P w}{\partial \xi^P}$$

where $f(\tau) = -\beta\tilde{\theta}/\alpha$. Equation (1.19) is a KdV or Burgers equation with variable coefficients.

By means of equation $(1.13)_1$ we can define, as in Chapter 4, the ray velocity $\vec{\Lambda}$ by

(1.20)
$$\vec{\Lambda} = \frac{\partial \psi}{\partial(\nabla\varphi)} = \frac{\partial \lambda}{\partial \vec{n}}.$$

If the propagation occurs in an isotropic constant state, we have

(1.21)
$$\vec{\Lambda} = \lambda_0\vec{n}$$

i.e. the rays are, at every instant, orthogonal to the spatial surfaces and the waves are "parallel".

In this case the explicit form for $\tilde{\theta}$ is (Boillat, 1965)

(1.22)
$$\tilde{\theta} = \sqrt{\lambda_0^2 K_0\sigma^2 + 2\lambda_0 k_0\sigma + 1},$$

where k_0 and K_0 are the mean and total Gauss curvatures, respectively.

If the initial surfaces have spherical symmetry, $K_0 = k_0^2$ and $\tilde{\theta} = (1 + \lambda_0 k_0\sigma) = r$, where r is the ratio of the radius of the spatial surfaces at the instant σ and initial radius, then

$$\frac{\partial}{\partial \sigma}\log\tilde{\theta} = \lambda_0 k_0/r$$

and equation (1.16) becomes

(1.23)
$$\frac{\partial \pi}{\partial r} + \frac{\pi}{r} + \alpha'\pi\frac{\partial \pi}{\partial \xi} + \beta'\frac{\partial^P \pi}{\partial \xi^P} = 0,$$

where

$$\alpha' = \frac{\alpha}{k_0\lambda_0} \quad \text{and} \quad \beta' = \frac{\beta}{k_0\lambda_0}.$$

Equation (1.23) is the "spherical" KdV equation when $P = 3$ (Maxon and Viecelli, 1974 a-b) and the "spherical" Burgers equation when $P = 2$ (Crighton, 1979).

When the initial surface has cylindrical symmetry $K_0 = 0$, $\tilde{\theta} = \sqrt{1 + \lambda_0 k_0\sigma}$ and the relative radius is

$$r = 1 + \lambda_0 k_0\sigma,$$

then equation (1.16) becomes

(1.24)
$$\frac{\partial \pi}{\partial r} + \frac{1}{2r}\pi + \alpha'\pi\frac{\partial \pi}{\partial \xi} + \beta'\frac{\partial^P \pi}{\partial \xi^P} = 0.$$

Equation (1.24) is known as "cylindrical" KdV or Burgers' equation for $P = 3$ or $P = 2$, respectively.

In the case $P = 3$ equations (1.23) and (1.24) were first derived by Maxon and Viecelli in studying the ion-acoustic waves in a collisionless plasma.

2 Solutions of the generalized KdV equation

Equations (1.16) or (1.19) and their particular forms (1.23) and (1.24) arise in several physical media such as: collisionless plasmas (Maxon and Viecelli, 1974a, 1974b, Giambò and Pantano, 1982), a case which we will treat in detail in Section 3; shallow water (Prasad and Revindran, 1977, Johnson, 1980, Bartuccelli, Brugarino and Pantano, 1983); and inhomogeneous media (Asano and Ono, 1971, Asano, 1974, Martina and Santini, 1978).

These equations describe a balance between a nonlinear term (tending to steepen the wave profile), a dispersive or dissipative term and a geometrical linear term. This last term tends to decrease or increase the wave amplitude according to whether

$$\dot{\tilde{\theta}} \equiv \frac{d\tilde{\theta}}{d\sigma} > 0 \quad \text{or} \quad \frac{d\tilde{\theta}}{d\sigma} < 0,$$

i.e. whether the waves are exploding or imploding, respectively. Leibovich and Randall (1979) have demonstrated that, for imploding waves and for $\dot{\tilde{\theta}}/\tilde{\theta} \to \infty$, the dominant term is increasing and has a solitonic shape. For exploding waves the dominant solitonic shape was first observed by Ott and Sudan (1970, 1971). It is interesting to note that, in this case, on a large scale the dominant term is the geometrical one which modifies the wavefront shape.

Equations (1.16) and (1.19) are very difficult to study in general. Equations (1.23) and (1.24) have been numerically integrated by Maxon and Viecelli. They observe the soliton-like pulses with variable amplitude; for these solitons the usual relations between amplitude, width and velocity are retained. Another interesting observation is the presence of a residue behind the soliton.

Variable Solitons

In general, analytical results do not exist for the generalized KdV equation, but some information can be obtained for particular cases. For example the cylindrical KdV equation can be reduced to KdV by a suitable transformation and therefore it is integrable (Lugovtzov and Lugovtzov, 1969, Hirota, 1979).

To investigate in detail the effects of geometry on soliton propagation we consider the "quasi-plane" waves, i.e. we suppose that

$$\dot{\tilde{\theta}} \simeq \delta, \tag{2.1}$$

where δ is a "small" parameter. Assuming that the solution of (1.16) has the shape of a soliton (Kaup and Newell, 1978, Karpman and Maslov 1977):

$$\pi_s(\xi, \sigma) = 2z^2 \operatorname{sech}^2 z(\xi - \overline{\xi}), \tag{2.2}$$

where $\overline{\xi}$ is the position of the soliton, allowing z to vary with σ and using the energy conservation law, it is found that

$$\frac{dz}{d\sigma} = -\frac{2}{3}\delta z \tag{2.3}$$

and

(2.4)
$$\frac{d\bar{\xi}}{d\sigma} = -4z^2 + 0(\delta).$$

Under the assumption (2.2), the conservation of mass and momentum are not satisfied, and a new term $\tilde{\pi}$ must be added to π_s. Then the solution of the generalized KdV equation is given by

(2.5)
$$\pi = \pi_s + \tilde{\pi}$$

where π_s is obtained from (2.2) setting

(2.6)
$$\begin{cases} z = z_0 \tilde{\theta}^{-2/3} \\ \bar{\xi} = -4z_0^2 \tilde{\theta}^{-4/3} + 0(\delta) \end{cases}$$

and $\tilde{\pi}$ assumes a very complicated form (Kaup and Newell, 1978).

It can be noticed that a shelf of height $-\delta/3z$ is created between $\xi = 0$ and $\bar{\xi} = \xi$. This is a "residue".

From (2.5–6) we notice that, similar to the cases of cylindrical or spherical waves, the quasi-plane waves evolve retaining the dominant solitonic shape and the relationships among amplitude, velocity and depth hold. The shape is modified during the propagation and depends on the local curvature for each ray. The amplitude and velocity are increasing or decreasing if the waves are locally imploding or expanding.

Behind the soliton a negative or positive shelf appears if the waves are locally expanding or imploding, respectively. The modification of the solitonic shape and the shelf appear to be a characteristic of the three-dimensional propagation, and are due to the geometrical effects on the initial soliton.

If we have other geometrical effects, such as shallow water of variable depth (Brugarino and Pantano, 1981, Carbonaro, Floris and Pantano, 1983), or plasma inhomogeneities, the shelf may disappear for a particular geometry, although the soliton shape is still modified.

3   Applications to plasmas

Let us consider a collisionless plasma, without a magnetic field. We assume that the ions form a cold fluid, described by equation (4.15) of Chapter 1 without the collision term,

(3.1)
$$\frac{\partial \vec{w}_1}{\partial t} + (\vec{w}_1 \cdot \nabla)\vec{w}_1 - \frac{e\vec{E}}{m_1} = 0$$

and by the mass conservation equation

(3.2)
$$\frac{\partial \rho_1}{\partial t} + \nabla \cdot (\rho_1 \vec{w}_1) = 0.$$

The electrons form an isothermal gas and are therefore described by equation (4.16) of Chapter 1 without the collision term

(3.3)
$$\frac{m_2}{m_1}\left\{\frac{\partial \vec{w}_2}{\partial t} + (\vec{w}_2 \cdot \nabla)\vec{w}_2\right\} + \frac{m_2}{m_1}\frac{(a_2)^2}{\rho_2}\nabla\rho_2 + \frac{e\vec{E}}{m_1} = 0$$

where

$$(a_2)^2 = \frac{k_B T}{m_2} = \text{const.}$$

Let $\bar{a}^2 = k_B T/m_1$ be the isothermal ion-acoustic speed, then

(3.4)
$$\frac{m_2}{m_1}\frac{d\vec{w}_2}{dt} + \bar{a}^2\frac{\nabla\rho_2}{\rho_2} + \frac{e\vec{E}}{m_1} = 0.$$

Neglecting the electron inertia term, one obtains

(3.4')
$$\bar{a}^2\frac{\nabla\rho_2}{\rho_2} + \frac{e\vec{E}}{m_1} = 0$$

and, substituting into equation (3.1), this gives

(3.5)
$$\frac{\partial \vec{w}_1}{\partial t} + (\vec{w}_1 \cdot \nabla)\vec{w}_1 + \bar{a}^2\frac{\nabla\rho_2}{\rho_2} = 0.$$

The equation for ρ_2 is obtained as follows.
Let us consider Eq. $(4.14)_1$ of Chapter 1,

(4.14')
$$\frac{\rho_2 e_2}{m_2} = -\frac{\rho_1 e_1}{m_1} + \epsilon_0 \nabla \cdot \vec{E}.$$

Multiplying it by $\vec{w}_1$ and taking the divergence we have

$$\frac{e_2}{m_2}\nabla \cdot (\rho_2\vec{w}_1) = -\frac{e_1}{m_1}\nabla \cdot (\rho_1\vec{w}_1) + \epsilon_0 \nabla \cdot (\vec{w}_1\nabla \cdot \vec{E}).$$

By adding to this equation the time derivative of (4.14') we obtain

$$\frac{\partial \rho_2}{\partial t} + \nabla \cdot (\rho_2\vec{w}_1) = \epsilon_0\frac{m_2}{e_2}\nabla \cdot \left[\frac{\partial \vec{E}}{\partial t} + (\vec{w}_1\nabla \cdot \vec{E})\right].$$

Making use of (3.4'), the equation can be written as

(3.6)
$$\frac{\partial \rho_2}{\partial t} + \nabla \cdot (\rho_2\vec{w}_1) =$$

$$= \epsilon_0\frac{m_2 m_1}{e^2}\nabla \cdot \left[\frac{\partial}{\partial t}\left(\bar{a}^2\frac{\nabla\rho_2}{\rho_2}\right) + \left(\vec{w}_1\nabla \cdot \left(\bar{a}^2\frac{\nabla\rho_2}{\rho_2}\right)\right)\right].$$

If we let

(3.7)
$$\bar{a}^2 = \frac{1}{M^2}$$

and

(3.8)
$$l_D^2 = \epsilon_0 \frac{m_2 m_1}{e^2} \bar{a}^2 \, ,$$

then (3.6) and (3.5) become

(3.9)
$$\frac{\partial \rho_2}{\partial t} + \nabla \cdot (\rho_2 \vec{w}_1) = l_D^2 \nabla \cdot \left[\frac{\partial}{\partial t} \frac{\nabla \rho_2}{\rho_2} + \left(\vec{w}_1 \nabla \cdot \left(\frac{\nabla \rho_2}{\rho_2} \right) \right) \right] ,$$

(3.10)
$$\frac{\partial \vec{w}_1}{\partial t} + (\vec{w}_1 \cdot \nabla) \vec{w}_1 + \frac{1}{M^2} \frac{\nabla \rho_2}{\rho_2} = 0 \, ,$$

respectively. Eqs. (3.9) and (3.10) constitute a system in the unknowns ρ_2 and $\vec{w}_1$.

The method described in the first section can be applied to the Eqs. (3.9–10) describing a collisionless plasma, that we rewrite in the following form

(3.11)
$$\begin{cases} \dfrac{\partial \rho}{\partial t} + w_k \dfrac{\partial \rho}{\partial x^k} + \rho \dfrac{\partial w_k}{\partial x^k} - l_D^2 \dfrac{\partial}{\partial x^k} \left(w_k \dfrac{\partial}{\partial x^i} + \delta_{ki} \dfrac{\partial}{\partial t} \right) \dfrac{1}{\rho} \dfrac{\partial \rho}{\partial x^i} = 0 \\[2mm] \dfrac{\partial w_k}{\partial t} + w_j \dfrac{\partial w_k}{\partial x^j} + \dfrac{1}{M^2} \dfrac{1}{\rho} \dfrac{\partial \rho}{\partial x^k} = 0 \end{cases}$$

where we omit, for simplicity, the subscripts of ρ_2 and $\vec{w}_1$.

Let us expand the field variables:

(3.12)
$$\begin{cases} \rho = \rho_0 + \epsilon \rho_1 + \epsilon^2 \rho_2 + \cdots \\ \vec{w} = \epsilon \vec{w}_1 + \epsilon^2 \vec{w}_2 + \cdots . \end{cases}$$

For the ion-acoustic wave $\lambda_0 = 1/M$, and Eq. (1.9) becomes

(3.13)
$$\begin{cases} \rho_1 = M \pi \\ \vec{w}_1 = (\pi / \rho_0) \vec{n} \end{cases} .$$

The transport equation is

(3.14)
$$\frac{\partial \pi}{\partial \sigma} + \pi \frac{\partial}{\partial \sigma} \log \tilde{\theta} + \alpha \pi \frac{\partial \pi}{\partial \xi} + \beta \frac{\partial^3 \pi}{\partial \xi^3} = 0 \, ,$$

where, as in the one-dimensional case,

$$\alpha = \frac{1}{M \rho_0} \quad \text{and} \quad \beta = \frac{l_D}{2 M \rho_0}.$$

Experiments

The first experiments on multidimensional solitons in a collisionless plasma were performed by Hershkowitz and Romesser in (1974). In these experiments they used a DP plasma with a cylindrical symmetry.

They observed imploding pulses that, for small differences, have the same characteristics as one-dimensional solitons (the square root of the maximum amplitude, multiplied by the width is a constant; the velocity of pulse is a function of the maximum amplitude of the soliton; they retain their shapes under collisions).

Expanding cylindrical solitons, which are more difficult to observe because of complicated double plasma technical properties, were observed between 1976 and 1977 by Chen and Schott (1976, 1977). Hershkowitz and Romesser observed these cylindrical solitons after the solitons crossed the center and diverge. Chen and Scott observed solitons that directly diverge from their source of excitation.

These experiments are performed in a spherical plasma machine. Waves are excited with a cylindrical sonda. Applying negative pulses to the internal conductor of the sonda, ion-acoustic solitons are excited. Their velocity is smaller than that of plane solitons and their thickness increases little by little as they turn away from the sonda.

Cylindrical divergent solitons, as well as their basic properties like the dependence of the width and the thickness on the velocity, were observed by Ze (1979). This article presents one the first experiments in which soliton theory is used in order to check the validity of an experimental technique.

In a very interesting paper, Nakamura, Ooyama and Ogino (1980) confirmed previous results on spherical and cylindrical solitons and they were able to demonstrate the solitons' stability in collisions and the corresponding phase shifts (which are more difficult to observe than in the one-dimensional case). Even the residue is clearly in evidence.

4   Derivation of the generalized KP equation: application to plasmas

In this section we shall consider the propagation of waves subject to small transversal disturbances. These effects are described with the introduction of an additional phase, ϕ, and with a different scaling of the field (Hunter, 1986).

We look for solutions of system (1.1) of the following form:

$$(4.1) \qquad \mathbf{U} = \mathbf{U}_0 + \epsilon \mathbf{U}_1 + \epsilon^{3/2} \mathbf{U}_2 + \epsilon^2 \mathbf{U}_3 + \epsilon^{5/2} \mathbf{U}_4 + \dots$$

where $\mathbf{U}_0$ is an unperturbed constant solution and $\mathbf{U}_1, \mathbf{U}_2, \dots$ depend on the variables

$$(4.2) \qquad \begin{cases} \xi = \epsilon^{-1} \varphi(\epsilon^{a+1} x^\alpha) \\ \varsigma = \epsilon^{-1/2} \phi(\epsilon^{a+1} x^\alpha) \\ \eta = \epsilon^{a+1} x^\alpha \end{cases}$$

where $a = (P-1)^{-1}$. The derivative with respect to x^α becomes:

$$\frac{\partial}{\partial x^\alpha} = \epsilon^a \varphi_\alpha \frac{\partial}{\partial \xi} + \epsilon^{a+\frac{1}{2}} \phi_\alpha \frac{\partial}{\partial \varsigma} + \epsilon^{a+1} \frac{\partial}{\partial \eta^\alpha},$$

where $\varphi_\alpha = \partial\varphi/\partial\eta^\alpha, \phi_\alpha = \partial\phi/\partial\eta^\alpha$.

The matrices in (1.1) are expanded as

$$\mathcal{A}^\alpha = \mathcal{A}_0^\alpha + \epsilon\nabla\mathcal{A}_0^\alpha \cdot \mathbf{U}_1 + \epsilon^{3/2}\nabla\mathcal{A}_0^\alpha \cdot \mathbf{U}_2 + \cdots$$

$$\mathcal{H}_J^\alpha = \mathcal{H}_{J0}^\alpha + \epsilon\nabla\mathcal{H}_{J0}^\alpha \cdot \mathbf{U}_1 + \cdots$$

By inserting this expansion into (1.1) and equating the coefficients of ϵ^{a+1}, $\epsilon^{a+3/2}$ and ϵ^{a+2} to zero we obtain, respectively:

$$(4.3) \qquad \mathcal{A}_0^\alpha \varphi_\alpha \frac{\partial \mathbf{U}_1}{\partial \xi} = 0,$$

$$(4.4) \qquad \mathcal{A}_0^\alpha \phi_\alpha \frac{\partial \mathbf{U}_1}{\partial \varsigma} + \mathcal{A}_0^\alpha \varphi_\alpha \frac{\partial \mathbf{U}_2}{\partial \xi} = 0,$$

$$(4.5) \qquad \mathcal{A}_0^\alpha \varphi_\alpha \frac{\partial \mathbf{U}_3}{\partial \xi} + (\nabla\mathcal{A}_0^\alpha \cdot \mathbf{U}_1)\varphi_\alpha \frac{\partial \mathbf{U}_1}{\partial \xi} + \mathcal{A}_0^\alpha \phi_\alpha \frac{\partial \mathbf{U}_2}{\partial \varsigma} + \mathcal{A}_0^\alpha \frac{\partial \mathbf{U}_1}{\partial \eta^\alpha}$$

$$+ \sum_{I=1}^{S} \prod_{J=1}^{P} \mathcal{H}_{J0}^{I\alpha}\varphi_\alpha \frac{\partial^P \mathbf{U}_1}{\partial \xi^P} = 0.$$

Equation (4.3) admits non trivial solution if

$$\det(\mathcal{A}_0^\alpha \varphi_\alpha) = 0.$$

This is the eikonal equation for the phase φ. A non trivial solution for the field is

$$\mathbf{U}_1 = \pi \mathbf{R}_0,$$

where $\pi(\xi, \varsigma, \eta^\alpha)$ is a scalar function to be determined and $\mathbf{R}_0$ is the right nullvector of the matrix $\mathcal{A}_0^\alpha \varphi_\alpha$.

The solvability condition for Eq. (4.4) is

$$(4.6) \qquad \mathbf{L}_0 \phi_\alpha \mathcal{A}_0^\alpha \mathbf{R}_0 = 0,$$

where $\mathbf{L}_0$ is the left nullvector of $\mathcal{A}_0^\alpha \varphi_\alpha$.

This is the transversality condition i.e. the transverse variable ϕ is constant along the rays associated with the phase φ.

A solution of (4.4) is given by

$$(4.7) \qquad \mathbf{U}_2 = \chi \mathbf{S}_0,$$

where $\chi(\xi, \varsigma, \eta^\alpha)$ is a scalar amplitude such that

$$(4.8) \qquad \chi_\xi - \pi_\varsigma = 0,$$

and $\mathbf{S}_0$ is a solution of

(4.9) $$\varphi_\alpha \mathcal{A}_0^\alpha \mathbf{S}_0 + \phi_\alpha \mathcal{A}_0^\alpha \mathbf{R}_0 = 0.$$

This equation has a solution because of the solvability condition (4.6).

An equation for π and σ is obtained as solvability condition for Eq. (4.5).

Taking the scalar product of (4.5) with $\mathbf{L}_0$ we have

(4.10) $$\mathbf{L}_0 \mathcal{A}_0^\alpha \frac{\partial}{\partial \eta^\alpha}(\pi \mathbf{R}_0) + \mathbf{L}_0 \mathcal{A}_0^\alpha \phi_\alpha \mathbf{S}_0 \frac{\partial \chi}{\partial \varsigma}$$
$$+ \mathbf{L}_0(\varphi_\alpha \nabla \mathcal{A}_0^\alpha \mathbf{R}_0)\mathbf{R}_0 \pi \frac{\partial \pi}{\partial \xi} + \mathbf{L}_0 \left(\sum_{I=1}^{S} \prod_{J=1}^{P} \mathcal{H}_{J0}^{I\alpha} \varphi_\alpha \mathbf{R}_0 \right) \frac{\partial^P \pi}{\partial \xi^P} = 0.$$

This equation has the form:

(4.11) $$\frac{\partial \pi}{\partial \sigma} + \pi \frac{\partial}{\partial \sigma} \log \tilde{\theta} + \alpha \pi \frac{\partial \pi}{\partial \xi} + \beta \frac{\partial^P \pi}{\partial \xi^P} + \gamma \frac{\partial \chi}{\partial \varsigma} = 0,$$

where

$$\mathbf{L}_0 \mathbf{R}_0 \frac{\partial}{\partial \sigma} = \mathbf{L}_0 \mathcal{A}_0^\alpha \mathbf{R}_0 \frac{\partial}{\partial \eta^\alpha}$$

is the derivative along the rays, and the coefficients α, β and γ are given by:

$$\begin{cases} \alpha = \dfrac{\mathbf{L}_0(\varphi_\alpha \nabla \mathcal{A}_0^\alpha \mathbf{R}_0)\mathbf{R}_0}{\mathbf{L}_0 \mathbf{R}_0}, \\[2em] \beta = \dfrac{\mathbf{L}_0 \displaystyle\sum_{I=1}^{S} \prod_{J=1}^{P} H_{J0}^{I\alpha} \varphi_\alpha \mathbf{R}_0}{\mathbf{L}_0 \mathbf{R}_0}, \\[2em] \gamma = \dfrac{\mathbf{L}_0 \phi_\alpha \mathcal{A}_0^\alpha \mathbf{S}_0}{\mathbf{L}_0 \mathbf{R}_0}. \end{cases}$$

and $\tilde{\theta} = \sqrt{J}$ is the divergence of the rays.

By taking the derivative of (4.11) with respect to ξ and making use of (4.8) one obtains

(4.12) $$\frac{\partial}{\partial \xi} \left[\frac{\partial}{\partial \sigma} \pi + \pi \frac{\partial}{\partial \sigma} \log \tilde{\theta} + \alpha \pi \frac{\partial \pi}{\partial \xi} + \beta \frac{\partial^P \pi}{\partial \xi^P} \right] + \gamma \frac{\partial^2 \pi}{\partial \varsigma^2} = 0.$$

The rays and the expansion coefficient $\tilde{\theta}$ are determined as in Section 1 by the system (1.13).

Finally, if we set $W = \tilde{\theta} \pi$ and introduce the variable τ defined along the rays by the relation $d\tau = \alpha/\tilde{\theta} \, d\sigma$ we obtain for W the following equation

(4.13) $$\frac{\partial}{\partial \xi} \left(\frac{\partial W}{\partial \tau} + W \frac{\partial W}{\partial \xi} + \tilde{\beta} \frac{\partial^P W}{\partial \xi^P} \right) + \tilde{\gamma} \frac{\partial^2 W}{\partial \varsigma^2} = 0,$$

with

$$\tilde{\beta} = \frac{\beta}{\alpha} \tilde{\theta}, \qquad \tilde{\gamma} = \frac{\gamma \tilde{\theta}}{\alpha}.$$

For $P = 2$ and $P = 3$ equation (4.13) is, respectively, the generalized Burgers or Kadometsev-Petviashvilii (KP) equation in two dimensions. The equation (4.13) describes the most general propagation subject to weak transversal perturbations in one direction with respect to the surface of propagation.

The generalized KP equation

If we consider the case $P = 3$, (4.13) is just a generalized KP equation. For this equation general analytic solutions do not exist, but some information can be obtained in particular cases.

We consider now propagation into a constant isotropic state.

For plane symmetry θ is a constant and equation (4.12) is similar to KP equation.

When cylindrical symmetry is imposed, a cylindrical coordinate r' can be defined as $r' = r_0 + \sigma\lambda_0$, and equations (4.12) becomes, for expanding waves

(4.14)
$$\frac{\partial}{\partial\xi}\left(\frac{\partial f}{\partial r} + \frac{1}{2r}f + f\frac{\partial f}{\partial\xi} + \frac{1}{2}\frac{\partial^3 f}{\partial\xi^3}\right) + \frac{1}{2}\frac{\partial^2 f}{\partial\mu^2} = 0,$$

and for implosive waves

(4.15)
$$\frac{\partial}{\partial\xi}\left(\frac{\partial f}{\partial r} - \frac{1}{2r}f + f\frac{\partial f}{\partial\xi} + \frac{1}{2}\frac{\partial^3 f}{\partial\xi^3}\right) + \frac{1}{2}\frac{\partial^2 f}{\partial\mu^2} = 0,$$

where

$$r = 2\frac{\beta}{\lambda_0}r', \quad \mu = \sqrt{\frac{\beta}{\gamma}}\varsigma, \quad f = \frac{\alpha}{2\beta}\pi.$$

Equation (4.14) is the cylindrical form of the KP equation. The quantity in parenthesis, when equated to zero, is the well-known cylindrical KdV equation. This equation describes the propagation of cylindrical waves subject to small transverse disturbances. The corresponding KP equation for spherical waves is

(4.16)
$$\frac{\partial}{\partial\xi}\left(\frac{\partial f}{\partial r} + \frac{f}{r} + f\frac{\partial f}{\partial\xi} + \frac{1}{2}\frac{\partial^3 f}{\partial\xi^3}\right) + \frac{1}{2}\frac{\partial^2 f}{\partial\mu^2} = 0.$$

These equations can be also deduced directly from the generalized KdV equation (1.16).

Equation (4.14) can be transformed with the substitution

(4.17)
$$\begin{cases} f = W(\hat{\xi}, r), \\ \hat{\xi} = \xi + \frac{1}{2}\frac{\mu^2}{r}. \end{cases}$$

After an integration in $\hat{\xi}$, provided that $W \to 0$ as $|\hat{\xi}| \to \infty$, the following equation is obtained for expanding waves:

(4.18)
$$\frac{\partial W}{\partial r} + \frac{1}{r}W + W\frac{\partial W}{\partial\hat{\xi}} + \frac{1}{2}\frac{\partial^3 W}{\partial\hat{\xi}^3} = 0$$

Eq. (4.15) for implosive waves becomes

(4.19)
$$\frac{\partial W}{\partial r} + W\frac{\partial W}{\partial\hat{\xi}} + \frac{1}{2}\frac{\partial^3 W}{\partial\hat{\xi}^3} = 0.$$

Equation (4.18) is the spherical KdV equation and (4.19) the usual KdV equation.

Applications to plasmas

We consider now the application of these methods to a collisionless plasma.

We look for a solution of (3.11) describing a perturbation of a given constant solution

$$\mathbf{U}_0 = \begin{pmatrix} \rho_0 \\ \vec{0} \end{pmatrix}$$

expressed as an asymptotic series of a small parameter ϵ, of the form (4.1), where $\mathbf{U}_1, \mathbf{U}_2, \ldots$ depend on the variables $\xi, \varsigma, \eta^\alpha$ defined in (4.2).

For the ion-acoustic waves

$$\lambda_0 = \frac{1}{M},$$

where M is defined in (3.7).

We now look for solutions in which the phase φ does not depend on x^3. In this case, a particular solution of (4.6) and (4.9) is

$$\phi = x^3, \qquad \mathbf{S}_0 = \left(0, 0, 0, \frac{|\nabla\varphi|}{\rho_0}\right)^T.$$

The transport equation for π is given by (4.11) with

$$\alpha = \frac{|\nabla\varphi|}{M\rho_0}, \qquad \beta = \frac{|\nabla\varphi|^3}{2}\frac{l_D^2}{M\rho_0}, \qquad \gamma = \frac{|\nabla\varphi|}{2M}.$$

Some experimental observations on anelastic collisions between solitons are available. The first experiments were made by Ze, Hershkowitz and Lonngren (1980) who used a large multidipole plasma device to study the collision of two expanding spherical ion-acoustic solitons.

They observe the formation of a new wave originating from the anelastic interaction of the two waves and they describe the main features of this phenomenon. The qualitative behaviour is in good agreement with the solutions of the KP equation.

These first results are confirmed for plane interacting solitons by Folkes, Ikezi and Davis (1980), and by Nagasawa, Shimizu and Nishida (1981). The experimental results are in agreement with the theoretical results predicted by the KP model. A large number of other papers for multi-dimensional interaction of ion acoustic solitons confirm substantially the phenomenon of anelastic collisions. See Tsukubayashi and Nakamura (1981), Khazei, Bulson and Lonngren (1982), Tsukabayashi, Nakamura, Kako and Lonngren (1983), Gabl, Bulson and Lonngren (1984), Gabl, Tsikis and Lonngren (1984), Gabl and Lonngren (1984).

5   Derivation of the generalized nonlinear Schrödinger equation

In this section we study the nonlinear modulation of a dispersive wave in several space dimensions, by suitably generalizing the approach of Asano (1974) presented in Chapter 2.

The phenomenon in governed by cubic nonlinearities because it can be described as a nonlinear self-interaction with four wave resonant interactions. A justification of the choice of the independent variables and of their relative order of magnitude is given in the framework of resonant wave interactions in Sec. (10.1). A similar case was treated in Section 2.3 for one dimensional propagation.

Let us start with the quasilinear system.

$$(5.1) \qquad A^\alpha \frac{\partial \mathbf{U}}{\partial x^\alpha} + \mathbf{B}(\mathbf{U}) = 0; \qquad A^\alpha = A^\alpha(\mathbf{U}).$$

We introduce the following set of variables and derivatives:

$$\text{very slow variables}: \hat{x}^\alpha = \epsilon^2 x^\alpha; \quad \partial_{,\hat{\alpha}} \equiv \frac{\partial}{\partial \hat{x}^\alpha}; \quad f_{,\hat{\alpha}} \equiv \frac{\partial f}{\partial \hat{x}^\alpha};$$

$$\text{slow variables}: X^\alpha = \epsilon x^\alpha; \quad \partial_{,\alpha} \equiv \frac{\partial}{\partial X^\alpha}; \quad f_{,\alpha} \equiv \frac{\partial f}{\partial X^\alpha};$$

also we introduce

$$\text{four phase}: \xi^\alpha = \xi^\alpha(X^\beta).$$

Let $\mathbf{U}^{(0)} = \mathbf{U}^{(0)}(\hat{x}^\alpha)$ be the unperturbed state, satisfying

$$(5.2) \qquad \qquad \mathbf{B}(\mathbf{U}^{(0)}) = 0.$$

$\mathbf{U}^{(0)}$ is an approximate solution of equation (5.1) to the order ϵ^2.

We make the following *ansatz*:

$$(5.3) \qquad \qquad \mathbf{U} = \mathbf{U}^{(0)} + \sum_{a=1}^{\infty} \epsilon^a \mathbf{U}^{(a)}$$

$$(5.4) \qquad \qquad \mathbf{U}^{(a)} = \sum_{l=-\infty}^{+\infty} \mathbf{U}_l^{(a)}(\xi^\alpha, \hat{x}^\alpha) e^{il\theta}$$

with

$$\theta = \frac{1}{\epsilon^2} \hat{\theta}(\hat{x}^\alpha) \quad \text{(main phase) and} \quad k_\alpha = \frac{\partial \theta}{\partial x^\alpha} = \partial_{\hat{\alpha}} \hat{\theta}.$$

The reality of $\mathbf{U}$ implies

$$(5.5) \qquad \qquad \mathbf{U}_{-l}^{(a)*} = \mathbf{U}_l^{(a)}.$$

Moreover, we have

$$\frac{\partial}{\partial x^\alpha} \mathbf{U} = \epsilon \sum_{l=-\infty}^{+\infty} ilk_\alpha \mathbf{U}_l^{(1)} e^{il\theta}$$

$$(5.6) \qquad + \epsilon^2 \left[\partial_{\hat{\alpha}} \mathbf{U}^{(0)} + \sum_{l=-\infty}^{+\infty} e^{il\theta} \left(ilk_\alpha \mathbf{U}_l^{(2)} + \frac{\partial \mathbf{U}_l^{(1)}}{\partial \xi^\mu} \xi^\mu_{,\alpha} \right) \right]$$

$$+ \epsilon^3 \left[\sum_{l=-\infty}^{+\infty} e^{il\theta} \left(ilk_\alpha \mathbf{U}_l^{(3)} + \partial_{\hat{\alpha}} \mathbf{U}_l^{(1)} + \frac{\partial \mathbf{U}_l^{(2)}}{\partial \xi^\mu} \xi^\mu_{,\alpha} \right) \right] + 0(\epsilon^4)$$

(5.7) $\mathbf{B}(\mathbf{U}) = \epsilon(\nabla B)_0 \mathbf{U}^{(1)} + \epsilon^2 \left[(\nabla B)_0 \mathbf{U}^{(2)} + \frac{1}{2}(\nabla\nabla B)_0 \mathbf{U}^{(1)}\mathbf{U}^{(1)} \right]$

$$+ \epsilon^3 \left[(\nabla B)_0 \mathbf{U}^{(3)} + (\nabla\nabla B)_0 \mathbf{U}^{(1)}\mathbf{U}^{(2)} \right.$$

$$\left. + \frac{1}{6}(\nabla\nabla\nabla B)_0 \mathbf{U}^{(1)}\mathbf{U}^{(1)}\mathbf{U}^{(1)} \right] + 0(\epsilon^4),$$

(5.8) $A^\alpha(\mathbf{U}) = A_0^\alpha + \epsilon(\nabla A^\alpha)_0 \mathbf{U}^{(1)}$

$$+ \epsilon^2 \left[(\nabla A^\alpha)_0 \mathbf{U}^{(2)} + \frac{1}{2}(\nabla\nabla A^\alpha)_0 \mathbf{U}^{(1)}\mathbf{U}^{(1)} \right] + 0(\epsilon^3),$$

where $\nabla \equiv \nabla_\mathbf{U}$ when applied to N dimensional vectors or matrices that are function of $\mathbf{U}$. To the first order in ϵ, Equation (5.1) yields

$$A_0^\alpha \sum_{l=-\infty}^{+\infty} ilk_\alpha e^{il\theta} \mathbf{U}_l^{(1)} + (\nabla B)_0 \sum_{l=-\infty}^{+\infty} \mathbf{U}_l^{(1)} e^{il\theta} = 0$$

and, by equating the Fourier components termwise,

(5.9) $\mathcal{W}_l \mathbf{U}_l^{(1)} = 0, \quad \text{with} \quad \mathcal{W}_l \equiv ilk_\alpha A_0^\alpha + (\nabla B)_0.$

We shall consider modulation of the first harmonic and therefore assume that

(5.10) $\det \mathcal{W}_l \neq 0 \qquad |l| \neq 1,$

(5.11) $\det \mathcal{W}_1 = 0 \qquad \text{(dispersion relation)}.$

From (5.9) one obtains

(5.12) $\mathbf{U}_1^{(1)} = \pi(\xi^\alpha, \hat{x}^\alpha) \mathbf{R}(\hat{x}^\alpha),$

we assume that k_α is simple root of (5.11) and $\mathbf{R}$ is the corresponding nullvector of $\mathcal{W}_1$. Also, from (5.11) we have

(5.13) $\mathbf{U}_l^{(1)} = 0 \quad \text{for} \quad |l| \neq 1.$

To the second order in ϵ, equation (5.1) yields

(5.14) $A_0^\alpha \partial_\alpha \mathbf{U}^{(0)} + \sum_{l=-\infty}^{+\infty} e^{il\theta} \left\{ \mathcal{W}_l \mathbf{U}_l^{(2)} + A_0^\alpha \xi_{,\alpha}^\mu \frac{\partial \mathbf{U}_l^{(1)}}{\partial \xi^\mu} + \right.$

$$\left. + \sum_{r=-\infty}^{+\infty} \left[irk_\alpha(\nabla A^\alpha)_0 + \frac{1}{2}(\nabla\nabla B)_0 \right] \mathbf{U}_{l-r}^{(1)} \mathbf{U}_r^{(1)} \right\} = 0.$$

The term with $l = 1$ yields

(5.15) $$\mathcal{W}_1 \mathbf{U}_1^{(2)} + \mathcal{A}_0^\alpha \, \xi_{,\alpha}^\mu \mathbf{R} \frac{\partial \pi}{\partial \xi^\mu} = 0.$$

Let $\mathbf{L}$ be the left nullvector of $\mathcal{W}_1$. Multiplying (5.15) by $\mathbf{L}$ gives

(5.16) $$\mathbf{L} \mathcal{A}_0^\alpha \mathbf{R} \xi_{,\alpha}^\mu \frac{\partial \pi}{\partial \xi^\mu} = 0.$$

We choose the variables ξ^μ satisfying

(5.17) $$\mathbf{L} \mathcal{A}_0^\alpha \mathbf{R} \xi_{,\alpha}^\mu = 0.$$

The rays (corresponding to the group speed) for a wavelet of the form (5.14) are given by the tangent vector

$$\Lambda^\alpha \equiv \mathbf{L} \mathcal{A}_0^\alpha \mathbf{R}$$

Hence (5.17) implies that

(5.18) $$\Lambda^\alpha \xi_{,\alpha}^\mu = 0$$

and ξ^μ is given by four linearly independent solutions of (5.16). In particular, introducing the rays in the X^α space,

$$\frac{dX^\alpha}{d\tau} = \Lambda^\alpha,$$

if $\Lambda^0 \neq 0$, we have

$$\frac{dX^i}{dX^0} = \frac{\Lambda^i}{\Lambda^0}, \qquad i = 1, 2, 3.$$

Since $\Lambda^\alpha = \Lambda^\alpha(\hat{x}^\mu)$ it follows

$$X^i = \frac{\Lambda^i}{\Lambda^0} X^0 + X_0^i \quad \text{with} \quad X_0^i = \text{constant}$$

and a solution of (5.18) is

$$\xi^i = X^i - \frac{\Lambda^i}{\Lambda^0} X^0; \qquad \xi^0 = 0.$$

The general solution of equation (5.15) is of the following form

(5.19) $$\mathbf{U}_1^{(2)} = \psi \mathbf{R} + \mathbf{Z}, \qquad \psi = \psi(\xi^\alpha, \hat{x}^\alpha)$$

where ψ is arbitrary and $\mathbf{Z}$ is a particular solution.

Let us try a solution of the form

$$\mathbf{Z} = \beta_\alpha \frac{\partial \mathbf{R}}{\partial k_\alpha}.$$

Now, from $W_1\mathbf{R} = 0$ it follows that

$$\frac{\partial W_1}{\partial k_\alpha}\mathbf{R} + W_1\frac{\partial \mathbf{R}}{\partial k_\alpha} = 0.$$

Hence from $W_1 = ik_\alpha \mathcal{A}_0^\alpha + (\nabla \mathbf{B})_0$, one has

$$\frac{\partial W_1}{\partial k_\alpha} = i\mathcal{A}_0^\alpha$$

and therefore

$$W_1\frac{\partial \mathbf{R}}{\partial k_\alpha} = -i\mathcal{A}_0^\alpha\mathbf{R}.$$

Finally $W_1\mathbf{Z} = \beta_\alpha W_1\dfrac{\partial \mathbf{R}}{\partial k_\alpha} = -i\beta_\alpha \mathcal{A}_0^\alpha\mathbf{R}$, whence

$$-i\beta_\alpha \mathcal{A}_0^\alpha\mathbf{R} + \frac{\partial \pi}{\partial \xi^\mu}\xi^\mu_{,\alpha}\mathcal{A}_0^\alpha\mathbf{R} = 0$$

and we obtain

(5.20)
$$\mathbf{Z} = -i\frac{\partial \pi}{\partial \xi^\mu}\xi^\mu_{,\alpha}\frac{\partial \mathbf{R}}{\partial k_\alpha}.$$

The term with $l = 0$ yields

(5.21)
$$\mathcal{A}_0^\alpha\partial_{,\hat{\alpha}}\mathbf{U}^{(0)} + W_0\mathbf{U}_0^{(2)}+$$
$$|\pi|^2\left[ik_\alpha(\nabla \mathcal{A}^\alpha)_0 + \frac{1}{2}(\nabla\nabla\mathbf{B})_0\right](\mathbf{R}\mathbf{R}^* + \mathbf{R}^*\mathbf{R}) = 0,$$

whence

(5.22)
$$\begin{cases} \mathbf{U}_0^{(2)} = |\pi|^2\mathbf{R}_0^{(2)} + \mathbf{V}, \text{ with} \\ \mathbf{R}_0^{(2)} = -W_0^{-1}\left[ik_\alpha(\nabla \mathcal{A}^\alpha)_0 + \frac{1}{2}(\nabla\nabla\mathbf{B})_0\right](\mathbf{R}\mathbf{R}^* + \mathbf{R}^*\mathbf{R}) \\ \mathbf{V} = -W_0^{-1}\mathcal{A}_0^\alpha\partial_\alpha\mathbf{U}^{(0)} \end{cases}$$

For $l = 2$ we obtain

$$W_2\mathbf{U}_2^{(2)} + \left[ik_\alpha(\nabla \mathcal{A}^\alpha)_0 + \frac{1}{2}(\nabla\nabla\mathbf{B})_0\right]\mathbf{U}_1^{(1)}\mathbf{U}_1^{(1)} = 0$$

(5.23)
$$\begin{cases} \mathbf{U}_2^{(2)} = \pi^2\mathbf{R}_2^{(2)}, \text{ with} \\ \mathbf{R}_2^{(2)} = -W_2^{-1}\left[ik_\alpha(\nabla \mathcal{A}^\alpha)_0 + \frac{1}{2}(\nabla\nabla\mathbf{B})_0\right](\mathbf{R}\mathbf{R}). \end{cases}$$

For $|l| \geq 3$ one obtains $\mathbf{U}_l^{(2)} = 0$. To the 3rd order in ϵ, equation (5.1) yields

$$\sum_{l=-\infty}^{+\infty} e^{il\theta} \left\{ W_l \mathbf{U}_l^{(3)} + A_0^\alpha \partial_{,\dot\alpha} \mathbf{U}_l^{(1)} + A_0^\alpha \frac{\partial \mathbf{U}_l^{(2)}}{\partial \xi^\mu} \xi_{,\alpha}^\mu + (\nabla A^\alpha)_0 \mathbf{U}_l^{(1)} \partial_{,\dot\alpha} \mathbf{U}^{(0)} \right\} +$$

$$+ (\nabla A^\alpha) \sum_{l=-\infty}^{+\infty} e^{il\theta} \left\{ \sum_{n=-\infty}^{+\infty} \mathbf{U}_{l-n}^{(1)} \left(ink_\alpha \mathbf{U}_n^{(2)} + \frac{\partial \mathbf{U}_n^{(1)}}{\partial \xi^\mu} \xi_{,\alpha}^\mu \right) \right.$$

$$+ \sum_{n=-\infty}^{+\infty} (\nabla A^\alpha)_0 \mathbf{U}_{l-n}^{(2)} (ink_\alpha) \mathbf{U}_n^{(1)} + \frac{1}{2} (\nabla \nabla A^\alpha)_0 \sum_{q,n=-\infty}^{+\infty} iqk_\alpha \mathbf{U}_{l-n-q}^{(1)} \mathbf{U}_n^{(1)} \mathbf{U}_q^{(1)} +$$

$$\left. + (\nabla \nabla B)_0 \sum_{n=-\infty}^{+\infty} \mathbf{U}_{l-n}^{(1)} \mathbf{U}_n^{(2)} + \frac{1}{6} (\nabla \nabla \nabla B)_0 \sum_{q,n=-\infty}^{+\infty} \mathbf{U}_{l-n-q}^{(1)} \mathbf{U}_n^{(1)} \mathbf{U}_q^{(1)} \right\} = 0$$

which gives

$$(5.24) \qquad W_l \mathbf{U}_l^{(3)} + A_0^\alpha \partial_{,\dot\alpha} \mathbf{U}_l^{(1)} + A_0^\alpha \frac{\partial \mathbf{U}_l^{(2)}}{\partial \xi^\mu} \xi_{,\alpha}^\mu + (\nabla A^\alpha)_0 \mathbf{U}_l^{(1)} \partial_{\dot\alpha} \mathbf{U}^{(0)}$$

$$+ \sum_{n=-\infty}^{+\infty} \left[ink_\alpha (\nabla A^\alpha)_0 \mathbf{U}_{l-n}^{(1)} \mathbf{U}_n^{(2)} \right.$$

$$+ (\nabla A^\alpha)_0 \xi_{,\alpha}^\mu \mathbf{U}_{l-n}^{(1)} \frac{\partial \mathbf{U}_n^{(1)}}{\partial \xi^\mu} ink_\alpha (\nabla A^\alpha)_0 \mathbf{U}_{l-n}^{(2)} \mathbf{U}_n^{(1)}$$

$$\left. + (\nabla \nabla B)_0 \mathbf{U}_{l-n}^{(1)} \mathbf{U}_n^{(2)} \right] + \sum_{q=-\infty}^{+\infty} \sum_{n=-\infty}^{+\infty} \left[iqk_\alpha \frac{1}{2} (\nabla \nabla A^\alpha)_0 \right.$$

$$\left. + \frac{1}{6} (\nabla \nabla \nabla B)_0 \right] \mathbf{U}_{l-n-q}^{(1)} \mathbf{U}_n^{(1)} \mathbf{U}_q^{(1)} = 0.$$

For $l = 1$ we obtain

$$(5.25) \qquad W_1 \mathbf{U}_1^{(3)} + A_0^\alpha \partial_{,\dot\alpha} \mathbf{U}_1^{(1)} + A_0^\alpha \frac{\partial \mathbf{U}_1^{(2)}}{\partial \xi^\mu} \xi_{,\alpha}^\mu + (\nabla A^\alpha)_0 \mathbf{U}_1^{(1)} \partial_{,\dot\alpha} \mathbf{U}^{(0)} +$$

$$+ 2ik_\alpha (\nabla A^\alpha)_0 \mathbf{U}_{-1}^{(1)} \mathbf{U}_2^{(2)} + ik_\alpha (\nabla A^\alpha)_0 \mathbf{U}_0^{(2)} \mathbf{U}_1^{(1)}$$

$$ik_\alpha (\nabla A^\alpha)_0 \mathbf{U}_2^{(2)} \mathbf{U}_{-1}^{(1)} + (\nabla \nabla B)_0 (\mathbf{U}_1^{(1)} \mathbf{U}_0^{(2)} + \mathbf{U}_{-1}^{(1)} \mathbf{U}_2^{(2)})$$

$$+ \left[\frac{1}{2} iqk_\alpha (\nabla \nabla A^\alpha)_0 + \frac{1}{6} (\nabla \nabla B)_0 \right] (\mathbf{U}_{-1}^{(1)} \mathbf{U}_1^{(1)} \mathbf{U}_1^{(1)}$$

$$+ \mathbf{U}_1^{(1)} \mathbf{U}_{-1}^{(1)} \mathbf{U}_1^{(1)} + \mathbf{U}_1^{(1)} \mathbf{U}_1^{(1)} \mathbf{U}_{-1}^{(1)}) = 0.$$

Left multiplication by $\mathbf{L}$ yields

$$(5.26) \qquad \Lambda^\alpha \partial_{,\dot\alpha} \pi - i\xi_{,\beta}^\nu \xi_{,\alpha}^\mu \mathbf{L} A_0^\alpha \frac{\partial \mathbf{R}}{\partial k_\beta} \frac{\partial^2 \pi}{\partial \xi^\mu \partial \xi^\nu} + M|\pi|^2 \pi + N\pi = 0$$

with

$$(5.27) \quad \begin{cases} M = \mathbf{L}ik_\alpha(\nabla \mathcal{A}^\alpha)_0[2\mathbf{R}^*\mathbf{R}_2^{(2)} + \mathbf{R}_0^{(2)}\mathbf{R} - \mathbf{R}_2^{(2)}\mathbf{R}^*] \\ \quad + \mathbf{L}(\nabla\nabla\mathbf{B})_0(\mathbf{R}\mathbf{R}_0^{(2)} + \mathbf{R}^*\mathbf{R}_2^{(2)}) + \mathbf{L}[1/2(\nabla\nabla\mathcal{A}^\alpha)_0 \\ \quad + 1/6(\nabla\nabla\nabla\mathbf{B})_0](\mathbf{R}^*\mathbf{R}\mathbf{R} + \mathbf{R}\mathbf{R}^*\mathbf{R} + \mathbf{R}\mathbf{R}\mathbf{R}^*) \\ N = \mathbf{L}[\mathcal{A}_0^\alpha \partial_{\dot{a}}\mathbf{R} + (\nabla \mathcal{A}^\alpha)_0\mathbf{R}(\partial_{\dot{a}}\mathbf{U}^{(0)}) \\ \quad + ik_\alpha(\nabla \mathcal{A}^\alpha)_0\mathbf{V}\mathbf{R} + (\nabla\nabla\mathbf{B})_0\mathbf{R}\mathbf{V}, \end{cases}$$

Eq. (5.26) is a generalized nonlinear Schrödinger equation.

6 Derivation of the KP equation for magnetosonic waves

The field equations

If we consider a collisionless cold plasma, the field equations (4.6) and (4.9) of Chapter 1 become

$$(6.1) \qquad \frac{\partial \rho_\alpha}{\partial t} + \nabla \cdot (\rho_\alpha \vec{w}_\alpha) = 0,$$

$$(6.2) \qquad \rho_\alpha \left(\frac{\partial \vec{w}_\alpha}{\partial t} + (\vec{w}_\alpha \cdot \nabla)\vec{w}_\alpha \right) - \rho_\alpha \vec{F}_\alpha = 0,$$

where

$$\vec{F}_\alpha = \frac{e_\alpha}{m_\alpha}(\vec{E} + \vec{w}_\alpha \times \vec{B}).$$

To equations (6.1-2), we must add Maxwell equations (4.13) of Chapter 1

$$(6.3) \qquad \begin{cases} \nabla \cdot \vec{B} = 0 \\ \epsilon\nabla \cdot \vec{E} = \dfrac{e_1\rho_1}{m_1} + \dfrac{e_2\rho_2}{m_2} \\ \nabla \times \vec{E} = -\dfrac{\partial \vec{B}}{\partial t} \\ \nabla \times \vec{B} = \mu \left(\dfrac{\rho_1 e_1}{m_1}\vec{w}_1 + \dfrac{\rho_2 e_2}{m_2}\vec{w}_2 \right) + \mu\epsilon\dfrac{\partial \vec{E}}{\partial t} \end{cases}$$

We can normalize the system (6.1-3) by introducing the following characteristic quantities:
   n_0^*: characteristic numerical density
   u_0^*: characteristic speed
   L_0^*: characteristic length
   B_0^*: characteristic magnetic field
   where the characteristic frequency and electric field are
   $\omega_0^* = u_0^*/L_0^*, \quad E_0^* = u_0^*B_0^*$

We also introduce the nondimensional parameters R_i, R_e, R_{pe} ionic-cyclotron, electron-cyclotron, normalized electronic plasma frequencies defined as

$$R_i = \frac{\omega_i}{\omega_0^*}, \quad R_e = \frac{\omega_e}{\omega_0^*}, \quad R_{pe} = \frac{\omega_{pe}}{\omega_0^*}$$

and the Mach number defined as

$$M_A = \frac{u_0^*}{V_A}$$

where V_A is the Alfvén speed.

Then equation (6.1) remains the same, but (6.2) and (6.3) are rewritten as

(6.4)
$$\frac{\partial \vec{w}_1}{\partial t} + (\vec{w}_1 \cdot \nabla)\vec{w}_1 = R_i[\vec{E} + \vec{w}_1 \times \vec{B}],$$

(6.5)
$$\frac{\partial \vec{w}_2}{\partial t} + (\vec{w}_2 \cdot \nabla)\vec{w}_2 = -R_e[\vec{E} + \vec{w}_2 \times \vec{B}],$$

(6.6)
$$\nabla \times \vec{B} = \left(\frac{u_0^*}{c}\right)^2 \frac{\partial \vec{E}}{\partial t} + M_A \frac{R_i R_e}{R_i + R_e}[\rho_1 \vec{w}_1 - \rho_2 \vec{w}_2],$$

(6.7)
$$\nabla \times \vec{E} = -\frac{\partial \vec{B}}{\partial t},$$

(6.8)
$$\nabla \cdot \vec{B} = 0,$$

(6.9)
$$\nabla \cdot \vec{E} = \frac{R_{pe}^2}{R_e}(n_1 - n_2).$$

If we consider waves for which $u_0^*/c \ll 1$ the first term of (6.6) on the RHS can be neglected. Equation (6.8) holds at all times if it is satisfied initially as one can see by considering the divergence of (6.7). We also suppose that the plasma is quasi-neutral, i.e. $n_1 \approx n_2 = n$.

Derivation of the KP equation

If we eliminate $\vec{E}$ and $\vec{w}_2$ by (6.1) and (6.4–9) we obtain the following equations for n, the electron density, $\vec{v}$, the velocity of ionic component and $\vec{B}$, the magnetic field

$$(6.10) \quad \begin{cases} \dfrac{\partial n}{\partial t} + \operatorname{div}(n\vec{v}) = 0, \\[2mm] M_A^2 \dfrac{d\vec{v}}{dt} = \dfrac{1}{n} \nabla \times \vec{B} \times \vec{B} + \dfrac{1}{R_e} \left\{ \left(\dfrac{1}{n}(\nabla \times \vec{B}) \cdot \nabla \right) \vec{v} \right. \\[2mm] \qquad\qquad + \dfrac{d}{dt}\left(\dfrac{1}{n} \nabla \times \vec{B} \right) \Big\} \\[2mm] \qquad\qquad - \dfrac{1}{M_A^2 R_i R_e}\left(1 + \dfrac{m_e}{m_1} \right) \left(\dfrac{1}{n}(\nabla \times \vec{B}) \cdot \nabla \right)\left(\dfrac{1}{n}\nabla \times \vec{B} \right), \\[2mm] \dfrac{\partial \vec{B}}{\partial t} = \nabla \times (\vec{v} \times \vec{B}) - \dfrac{1}{R_i}\nabla \times \left(\dfrac{d\vec{v}}{dt} \right), \end{cases}$$

By supposing that the field variables $n, \vec{v} = (u, v, w)$ and $\vec{B} = (B_x, B_y, B_z)$ are functions of x, z and t, system (6.10) can be written as

$$\frac{\partial n}{\partial t} + n\frac{\partial u}{\partial x} + u\frac{\partial n}{\partial x} + n\frac{\partial w}{\partial z} + w\frac{\partial n}{\partial z} = 0,$$

$$\frac{du}{dt} - \frac{1}{M_A^2 n}\left(-B_y\frac{\partial B_y}{\partial x} - B_z\frac{\partial B_z}{\partial x} + B_z\frac{\partial B_x}{\partial z} \right)$$

$$- \frac{1}{M_A^2 R_e n}\left(-\frac{\partial B_y}{\partial z}\frac{\partial u}{\partial x} + \frac{\partial B_y}{\partial x}\frac{\partial u}{\partial z} \right) + \frac{1}{M_A^2 R_e}\frac{d}{dt}\left(\frac{1}{n}\frac{\partial B_y}{\partial z} \right)$$

$$+ \frac{1}{M_A^2 R_i R_e n}\left(1 + \frac{m_e}{m_i} \right)\left(\frac{\partial B_y}{\partial z}\frac{\partial}{\partial x}\left(\frac{1}{n}\frac{\partial B_y}{\partial z} \right) \right.$$

$$\left. - \frac{\partial B_y}{\partial x}\frac{\partial}{\partial z}\left(\frac{1}{n}\frac{\partial B_y}{\partial z} \right) \right) = 0,$$

$$(6.11) \quad \frac{dv}{dt} - \frac{1}{M_A^2 n}\left(B_z\frac{\partial B_y}{\partial z} + B_x\frac{\partial B_y}{\partial x} \right) - \frac{1}{M_A^2 R_e n}\left(-\frac{\partial B_y}{\partial z}\frac{\partial v}{\partial z} \right.$$

$$\left. + \frac{\partial B_y}{\partial x}\frac{\partial v}{\partial z} \right) - \frac{1}{M_A^2 R_e}\frac{d}{dt}\left(\frac{1}{n}\frac{\partial B_x}{\partial z} \right) + \frac{1}{M_A^2 R_e}\frac{d}{dt}\left(\frac{1}{n}\frac{\partial B_z}{\partial x} \right) +$$

$$+ \frac{1}{M_A^4 R_i R_e n}\left(1 + \frac{m_e}{m_i} \right)\left(-\frac{\partial B_y}{\partial z}\frac{\partial}{\partial x}\left(\frac{1}{n}\frac{\partial B_x}{\partial z} \right) \right.$$

$$+ \frac{\partial B_y}{\partial z}\frac{\partial}{\partial x}\left(\frac{1}{n}\frac{\partial B_z}{\partial x} \right) + \frac{\partial B_y}{\partial x}\frac{\partial}{\partial z}\left(\frac{1}{n}\left(\frac{\partial B_x}{\partial z} - \frac{\partial B_z}{\partial x} \right) \right) \Big) = 0,$$

$$\frac{dw}{dt} - \frac{1}{M_A^2 n}\left(-B_y\frac{\partial B_y}{\partial z} - B_x\frac{\partial B_x}{\partial z} + B_x\frac{\partial B_z}{\partial x}\right)$$

$$-\frac{1}{M_A^2 R_e n}\left(-\frac{\partial B_y}{\partial z}\frac{\partial w}{\partial x} + \frac{\partial B_y}{\partial x}\frac{\partial w}{\partial z}\right) - \frac{1}{M_A^2 R_e}\frac{d}{dt}\left(\frac{1}{n}\frac{\partial B_y}{\partial x}\right)$$

$$+\frac{1}{M_A^2 R_i R_e n}\left(1 + \frac{m_e}{m_i}\right)\left(-\frac{\partial B_y}{\partial z}\frac{\partial}{\partial x}\left(\frac{1}{n}\frac{\partial B_y}{\partial x}\right)\right.$$

$$\left.+\frac{\partial B_y}{\partial x}\frac{\partial}{\partial z}\left(\frac{1}{n}\frac{\partial B_y}{\partial x}\right)\right) = 0,$$

$$\frac{\partial B_x}{\partial t} + \frac{\partial}{\partial z}(wB_x - uB_z) - \frac{1}{R_i}\frac{\partial}{\partial z}\left(\frac{dv}{dt}\right) = 0,$$

$$\frac{\partial B_y}{\partial t} - \frac{\partial}{\partial t}(vB_z - wB_y) + \frac{\partial}{\partial x}(uB_y - vB_x)$$

$$+\frac{1}{R_i}\frac{\partial}{\partial z}\left(\frac{du}{dt}\right) - \frac{1}{R_i}\frac{\partial}{\partial x}\left(\frac{dw}{dt}\right) = 0,$$

$$\frac{\partial B_z}{\partial t} - \frac{\partial}{\partial x}(wB_x - uB_z) + \frac{1}{R_i}\frac{\partial}{\partial x}\left(\frac{dv}{dt}\right) = 0.$$

As usual, by using a generalized form of the perturbation-reduction methods, we look now for a solution of this system describing a perturbation of a given constant solution which can be expressed with the following asymptotic series in terms of a small parameter ϵ:

(6.12)
$$\begin{cases} n = 1 + \epsilon n_1 + \epsilon^2 n_2 + \cdots, \\ u = \epsilon u_1 + \epsilon^2 u_2 + \cdots, \\ v = \epsilon v_1 + \epsilon^2 v_2 + \cdots, \\ w = \epsilon^{3/2} w_1 + \epsilon^{5/2} w_2 + \cdots, \\ B_x = \cos\theta, \\ B_y = \sin\theta + \epsilon B_{y_1} + \epsilon^2 B_{y_2} + \cdots, \\ B_z = \epsilon^{3/2} B_{z_1} + \epsilon^{5/2} B_{z_2} + \cdots, \end{cases}$$

where θ is the angle between the x-axis and the magnetic field $\vec{B}_0 = (\cos\theta, \sin\theta, 0)$.

We suppose also that the quantities appearing in (6.12) are functions of the following stretched variables:

$$\begin{cases} \xi = \epsilon^{1/2}(x - \lambda t), \\ \eta = \epsilon z, \\ \tau = \epsilon^{3/2}t, \end{cases}$$

where λ is a constant to be determined at the next stage. Then, by equating to zero the coefficients of the obtained series in ϵ, system (6.11) becomes, to order

$\epsilon^{3/2}$:

$$\begin{cases} -\lambda \dfrac{\partial n_1}{\partial \xi} + \dfrac{\partial u_1}{\partial \xi} = 0, \\[2mm] -\lambda \dfrac{\partial u_1}{\partial \xi} + \dfrac{\sin \theta}{M_A^2} \dfrac{\partial B_{y1}}{\partial \xi} = 0, \\[2mm] -\lambda \dfrac{\partial v_1}{\partial \xi} - \dfrac{\cos \theta}{M_A^2} \dfrac{\partial B_{y1}}{\partial \xi} = 0, \\[2mm] -\lambda \dfrac{\partial B_{y1}}{\partial \xi} + \sin \theta \dfrac{\partial u_1}{\partial \xi} - \cos \theta \dfrac{\partial v_1}{\partial \xi} = 0, \end{cases}$$

to order ϵ^2:

(6.13)
$$\begin{cases} -\lambda \dfrac{\partial w_1}{\partial \xi} + \dfrac{\sin \theta}{M_A^2} \dfrac{\partial B_{y1}}{\partial \eta} - \dfrac{\cos \theta}{M_A^2} \dfrac{\partial B_{z1}}{\partial \xi} + \dfrac{\lambda}{M_A^2 R_e} \dfrac{\partial^2 B_{y1}}{\partial \xi^2} = 0, \\[2mm] -\lambda \dfrac{\partial B_{z1}}{\partial \xi} - \cos \theta \dfrac{\partial w_1}{\partial \xi} - \dfrac{\lambda}{R_i} \dfrac{\partial^2 v_1}{\partial \xi^2} = 0, \end{cases}$$

to order $\epsilon^{5/2}$:

(6.14)
$$\begin{cases} -\lambda \dfrac{\partial n_2}{\partial \xi} + \dfrac{\partial u_2}{\partial \xi} + \dfrac{\partial u_1}{\partial \tau} + n_1 \dfrac{\partial u_1}{\partial \xi} + u_1 \dfrac{\partial n_1}{\partial \xi} + \dfrac{\partial w_1}{\partial \eta} = 0 \\[3mm] -\lambda \dfrac{\partial u_2}{\partial \xi} + \dfrac{\sin \theta}{M_A^2} \dfrac{\partial B_{y2}}{\partial \xi} + \dfrac{\partial u_1}{\partial \tau} + u_1 \dfrac{\partial u_1}{\partial \xi} + \dfrac{1}{M_A^2} B_{y1} \dfrac{\partial B_{y1}}{\partial \xi} \\[3mm] \qquad - \lambda n_1 \dfrac{\partial u_1}{\partial \xi} - \dfrac{\lambda}{M_A^2 R_e} \dfrac{\partial^2 B_{y1}}{\partial \xi \partial \eta} = 0, \\[3mm] -\lambda \dfrac{\partial v_2}{\partial \xi} - \dfrac{\cos \theta}{M_A^2} \dfrac{\partial B_{y2}}{\partial \xi} + \dfrac{\partial v_1}{\partial \tau} + u_1 \dfrac{\partial v_1}{\partial \xi} - \lambda n_1 \dfrac{\partial v_1}{\partial \xi} \\[3mm] \qquad - \dfrac{\lambda}{M_A^2 R_e} \dfrac{\partial^2 B_{z1}}{\partial \xi^2} = 0, \\[3mm] -\lambda \dfrac{\partial B_{y2}}{\partial \xi} + \sin \theta \dfrac{\partial u_2}{\partial \xi} - \cos \theta \dfrac{\partial v_2}{\partial \xi} + \dfrac{\partial B_{y1}}{\partial \tau} + \sin \theta \dfrac{\partial w_1}{\partial \eta} \\[3mm] \qquad + u_1 \dfrac{\partial B_{y1}}{\partial \xi} + B_{y1} \dfrac{\partial u_1}{\partial \xi} - \dfrac{\lambda}{R_i} \dfrac{\partial^2 u_1}{\partial \xi \partial \eta} + \dfrac{\lambda}{R_i} \dfrac{\partial^2 w_2}{\partial \xi^2} = 0. \end{cases}$$

We have a nontrivial solution of system (6.13) when $\lambda = 1/M_A$, corresponding to acoustic waves, and

(6.15)
$$\begin{pmatrix} n_1 \\ u_1 \\ v_1 \\ B_{y1} \end{pmatrix} = \pi(\xi, \eta, \tau) \begin{pmatrix} 1 \\ 1/M_A \\ -(\cos \theta / \sin \theta)(1/M_A) \\ 1/\sin \theta \end{pmatrix},$$

where π is a function of ξ, η, τ to be determined at the next stage.

By using (6.15), and multiplying system (6.14) by the left eigenvector $\mathbf{L} = (0, M_A \sin \theta, -M_A \cos \theta, 1)$, we obtain the following evolution equation for π:

(6.16)
$$\frac{\partial}{\partial \xi} \left(\frac{\partial \pi}{\partial \tau} + \alpha \pi \frac{\partial \pi}{\partial \xi} + \beta \frac{\partial^3 \pi}{\partial \xi^3} \right) + \frac{1}{3} \alpha \frac{\partial^2 \pi}{\partial \eta^2} = 0,$$

where

$$\alpha = \frac{3}{2M_A}, \quad \beta = \frac{1}{2M_A^3} \left(\frac{1}{R_i} - \frac{1}{R_e} \right) \left(\frac{\cot^2 \theta}{R_e} - \frac{1}{\sin^2 \theta R_i} \right).$$

Equation (6.16) is the KP equation.

REFERENCES

- N. Asano, H. Ono, *Nonlinear dispersive or dissipative waves in inhomogeneous media*, J. Phys. Soc. Jpn. **31**, 1830 (1971).
- N. Asano, *Wave propagation in non uniform media*, Suppl. Prog. Theor. Phys. **55**, 52 (1974).
- M. Bartucelli, T. Brugarino, P. Pantano, *Two-dimensional Burgers equation*, Lett. Nuovo Cimento **36**, 200 (1983).
- G. Boillat, *La propagation des ondes*, Ganthier-Villars, Paris (1965).
- T. Brugarino, P. Pantano, *Two-Dimensional Solitons in Shallow Water of Variable Depth*, Phys. Lett. **86 A**, 478 (1981).
- P. Carbonaro, R. Floris, P. Pantano, *Cylindrical Solitons in Shallow Water of Variable Depth*, Il Nuovo Cimento **6 C**, 133 (1983).
- T. Chen, L. Schott, *Observation of diverging cylindrical solitons excited with a probe*, Phys. Lett. **58 A**, 459 (1976).
- T. Chen, L. Schott, *Excitation of ion-acoustic waves with probes*, Phys. Fluids **20**, 844 (1977).
- Y. Choquet, Bruhat, *Ondes Asymptotiques et Approchées pour des Systéms d'Equations aux Dérivées partielles non lineaires*, J. Math. et Appl. **48**, 117 (1969).
- D. G. Crighton, *Model equations of nonlinear acoustics*, Ann. Rev. Fluid Mech. **17**, 11 (1979).
- O. A. Folkes, H. Ikezi, R. Davis, *Two-dimensional Interaction of Ion-acoustic Solitons*. Phy. Rev. Lett. **45**, 902 (1980).
- D. Fusco, *An asymptotic method for nonlinear magnetosonic waves in an isothermal plasma with a finite conductivity*, Proc. Roy. Soc. Ediburgh, **82 A**, 103 (1979).
- E. F. Gabl, J. M. Bulson, K. E. Lonngren, *Exitation of the Two-dimensional Ion-acoustic Soliton*, Phys. Fluids **27**, 269 (1984).
- E. F. Gabl, E. K. Tsikis, K. E. Lonngren, *Resonant Interaction of Ion-acoustic Solitons in Three-dimensional revisited*, Phys. Fluids **27**, 704 (1984).
- E. F. Gabl, K. E. Lonngren, *On the Oblique Collision of Unequal Amplitude Ion-acoustic Solitons in a Field-free Plasma*, Phys. Fluids **100 A**, 153 (1984).
- Garding, T. Kotake, J. Leray, *Uniformisation et Solution du Probléme de Cauchy Linéaire*, Bull. Soc. Math. Fr. **92**, 263 (1964).
- P. Germain, *Progressive waves*, centenaire de Prandl (1971).
- S. Giambò, A. Greco, P. Pantano, *Sur la méthode pertubative et réductive à n-dimensions: le cas général*, C. R. Acad. Sc., Paris, **289 A**, 553 (1979).
- S. Giambò, P. Pantano, *Three-dimensional ion-acoustic waves in a collisionless plasma*, Lett. Nuovo Cimento **34**, 380 (1982).
- S. Giambò, P. Pantano, *Modified Kodometsev-Petviashvili equation for ion-acoustic waves*, Rend. Circolo Mat. Palermo, II, XXXVIII, 79 (1989).
- N. Hershkowitz, T. Romesser, *Observations of ion-acoustic cylindrical solitons*, Phys. Rev. Lett. **32**, 1581 (1974).

– R. Hirota, *Exact solutions to the equation describing cylindrical soltons*, Phys. Lett. **71A**, 393 (1979).

– J. K. Hunter, *A ray method for slowly modulated nonlinear waves*, SIAM J. Appl. Math., **45**, 735 (1985).

– J. K. Hunter, *Transverse diffraction and singular rays*, SIAM J. Appl. Math. **75**, 187 (1986).

– E. Infeld and G. Rowlands, *Nonlinear Waves, Solitons and Chaos*, Cambridge Unversity Press, Cambridge (1990).

– R. S. Johnson, *Water waves and Korteweg-de Vries equations*, J. Fluid Mech. **97**, 701 (1980).

– V. I. Karpman, E. M. Maslov, *A perturbation theory for the Korteweg de Vries equation*, Phys. Lett. **60 A**, 307 (1977).

– D. J. Kaup, A. C. Newell, *Solitons as particles, oscillations, and in slowly changing media: a singular perturbation theory*, Proc. Roy. Soc. London **361 A**, 413 (1978).

– M. Khazei, J. M. Bulson, K. E. Lonngren, *Resonant Interaction of Ion-acoustic Solitons in Three-dimensions*, Phys. Fluids **25**, 759 (1982).

– M. Kline, I. W. Kay, *Eletromagnetic Theory and Geometrical Optics*, Pure and Applied Mathematics, XII, Interscience (1965).

– P. Lax, *Asymptotic Solutions of Oscillatory Initial Value Problem.* Duke Math. J. **24**, 627 (1957).

– S. Leibovich, J. D. Randall, *On soliton amplification*, Phys. Fluids **22**, 2289 (1979).

– J. Leray, *Particules et Singularités de Ondes.* Cahiers de Physique **15**, 373 (1961).

– D. Ludwig, *Exact and Asymptotic Solutions of the Cauchy Problem*, Comm. on Pure and Appl. Math. **13**, 473 (1960).

– A. A. Lugovtzov and B. A. Lugovtzov, *Dinamika sploshnoy sryedi, I.*, Nauka, Novosbirsk (1969).

– L. Martina, P. Santini, *Propagation of ion-acoustic waves in cold inhomogeneous plasmas*, Lett. Nuovo Cimento **29**, 513, (1978).

– S. Maxon, J. Viecelli, *Spherical solitons*, Phys. Rev. Lett. **32**, 4 (1974)$_a$.

– S. Maxon, J. Viecelli, *Cylindrical solitons*, Phys. Fluids **17**, 1614 (1974)$_b$.

– T. Nagasawa, M. Shimizu, Y. Nishida, *Strong Interaction of Plane Ion-Acoustic Solitons*, Phys. Lett. **87 A**, 37 (1981).

– A. Nakamura, M. Ooyama, T. Ogino, *Observation of spherical ion-acoustic solitons*, Phys. Rev. Lett. **45**, 1565 (1980).

– E. Ott, R. N. Sudan, *Damping of Solitary Waves*, Phys. Fluids **13**, 1432 (1970).

– E. Ott, *Damping of a Large Amplitude Ion Acoustic Solitary Wave*, Phys. Fluids **14**, 748 (1971).

– P. Prasad, R. Ravindran, *A Theory of Non-Linear Waves in Multi-Dimensions: with Special Reference to Surface Water Waves*, J. Inst. Math. Applics. **20**, 9 (1977).

- I. Tsukabayashi, Y. Nakamura, *Resonant Interaction of Cylindrical Ion-acoustic Solitons*, Phys. Lett. **85 A**, 151 (1981).
- I. Tsukabayashi, Y. Nakamura, F. Kako, K. E. Lonngren, *Oblique Collision of Cylindrical Outgoing Ion-acoustic Solitons*, Phys. Fluids **26**, 790 (1983).
- F. Ze, *Excitation of spherical ion-acoustic solitons with a conducting proble*, Phys. Fluids **22**, 1554 (1979).
- F. Ze, N. Hershkowitz, K. E. Lonngren, *Oblique Collision of Ion-acoustic Solitons*, Phys. Fluids **23**, 1155 (1980).

RAY METHODS FOR NONLINEAR DISSIPATIVE WAVES

Introduction

In Chapter 8 we have demonstrated that, in various physical systems, equations of the following type arise

$$(0.1) \qquad \frac{\partial u}{\partial \sigma} + u \frac{\partial u}{\partial \xi} = f(\sigma, u, u_\xi, u_{\xi\xi}, \ldots) .$$

Here u is the field variable and f represents terms due to various phenomena as inhomogeneities, dispersion, dissipation, geometry and others. In this chapter we consider nonlinear dissipative systems in which this equation arises. In Sec. 1 we recall some properties of generalized Burgers equation. In Sec. 2–3 we derive the two-dimensional and generalized Burgers equation for some dissipative systems. Finally, in Sec. 4 we derive a modified KdV for a collisional plasma.

1   The generalized Burgers equation

In Chapter 8 we deduced a particular form of (0.1), equation (1.16):

$$(1.1) \qquad \frac{\partial \pi}{\partial \sigma} + \pi \frac{\partial}{\partial \sigma} \log \tilde{\theta} + \alpha \pi \frac{\partial \pi}{\partial \xi} + \beta \frac{\partial^2 \pi}{\partial \xi^2} = 0 .$$

This equation is known as *generalized Burgers equation*. As in Section 1 of Chapter 8 we can write (1.1) in the cylindrical or spherical form as

$$(1.2) \qquad \frac{\partial \pi}{\partial r} + \frac{1}{2r} \pi + \pi \frac{\partial \pi}{\partial \xi} + \frac{\partial^2 \pi}{\partial \xi^2} = 0$$

$$(1.3) \qquad \frac{\partial \pi}{\partial r} + \frac{1}{r} \pi + \pi \frac{\partial \pi}{\partial \xi} + \frac{\partial^2 \pi}{\partial \xi^2} = 0 .$$

These equations arise in various physical media, such as the three-dimensional propagation of acoustic waves in thermoviscous fluids (Brugarino and Pantano, 1980, Crighton 1979), radiative gas dynamics (Palumbo, 1983) and magnetogasdynamics (Sedov and Nariboli, 1978, Bartuccelli, 1983), waves in thermostatic media (Rizun and Engel'breckt, 1975), one-dimensional propagation in nonhomogeneous media (Asano, 1970), etc.

Burgers equation ($\tilde{\theta} = 1$) can be reduced to the linear heat equation by the Cole-Hopf transform (Whitham, 1974). Nimmo and Crighton (1982) showed that no such linearizing Backlünd transforms exist for the generalized Burgers equation (1.1). Solutions can be obtained only in special cases (Sachdev, 1987).

For the *cylindrical* Burgers equation some similarity solutions have been obtained in (Brugarino, De Pascale, and Pantano, 1982).

2   Acoustic waves in a thermoviscous fluid

We consider the balance equations (1.1–3) of Chapter 1:

$$(2.1) \qquad \frac{\partial \rho}{\partial t} + \frac{\partial}{\partial x^i}(\rho u^i) = 0 \,,$$

$$(2.2) \qquad \frac{\partial (\rho u_i)}{\partial t} + \frac{\partial}{\partial x^j}(\rho u^i u^j - t^{ij}) = f^i \,,$$

$$(2.3) \qquad \frac{\partial}{\partial t}\left(\frac{1}{2}\rho|\vec{u}|^2 + \rho E\right) + \frac{\partial}{\partial x^j}\left\{\left(\frac{1}{2}\rho|\vec{u}|^2 + \rho E\right)u^j - t^{ij}u_i + q^j\right\} = f^i u_i \,.$$

We suppose that $f^i = 0$. We consider Equations (2.11), (2.5), (2.6), (2.12), (2.13) of Chapter 1,

$$(2.4) \qquad q_i = -k\frac{\partial T}{\partial x^i}$$

$$(2.5) \qquad t_{ij} = -P\delta_{ij} - \tilde{\pi}_{ij}$$

$$(2.6) \qquad \tilde{\pi}_{ij} = \pi_{ij} + \frac{\pi}{3}\delta_{ij}$$

$$(2.7) \qquad \pi_{ij} = -2\eta e_{ij}$$

$$(2.8) \qquad \pi = -3\varsigma\theta$$

where

$$e_{ij} = \frac{1}{2}\left(\frac{\partial u_i}{\partial x^j} + \frac{\partial u_j}{\partial x^i}\right) - \frac{1}{3}\delta_{ij}\frac{\partial u^k}{\partial x^k}$$

and

$$\theta = \frac{1}{3}\frac{\partial u^i}{\partial x^i} \,.$$

Equation (2.6) can be written as

$$\tilde{\pi}_{ij} = -\frac{\varsigma}{3}\delta_{ij}\frac{\partial u_k}{\partial x^k} - \eta\left(\frac{\partial u_i}{\partial x^j} + \frac{\partial u_j}{\partial x^i} - \frac{2}{3}\delta_{ij}\frac{\partial u_k}{\partial x^k}\right) \,.$$

By using (2.5) and (2.4), Equations (2.1–3) become

$$(2.9) \quad \begin{cases} \dfrac{\partial \rho}{\partial t} + \dfrac{\partial}{\partial x^i}(\rho u_i) = 0 \\[2mm] \dfrac{\partial(\rho u_i)}{\partial t} + \dfrac{\partial}{\partial x^j}[\rho u^i u^j + P\delta_{ij} + \tilde{\pi}_{ij}] = 0 \\[2mm] \dfrac{\partial}{\partial t}\left(\dfrac{1}{2}\rho|\vec{u}|^2 + \rho E\right) + \dfrac{\partial}{\partial x^j}\left\{ \left(\dfrac{1}{2}\rho|\vec{u}|^2 + \rho E + P\right)u^j + \tilde{\pi}_{ij}u_i - k\dfrac{\partial T}{\partial x^j} \right\} = 0 \end{cases}$$

We will consider now a perfect polytropic gas, i.e.

$$P = \overline{R}\rho T, \quad E = \frac{P}{\rho(\gamma - 1)}, \quad \overline{R} = C_P - C_v, \quad \gamma = \frac{C_P}{C_V} = \text{const.}$$

where C_P and C_V are specific heats at constant pressure and volume respectively.

Now, following the method developed in Sec. 1 of Chap. 8 we look for a solution $\rho, \vec{u}, P$, of system (2.1–3), which describes a perturbation of a given constant solution $\rho_0, 0, P_0$, expressed as an asymptotic series in terms of a parameter ϵ:

$$(2.10) \quad \begin{cases} \rho = \rho_0 + \epsilon\rho_1 + \epsilon^2\rho_2 + \dots \\ \vec{u} = \epsilon\vec{u}_1 + \epsilon^2\vec{u}_2 + \dots \\ P = P_0 + \epsilon P_1 + \epsilon^2 P_2 + \dots \end{cases}$$

where

$$(2.11) \qquad \rho_k = \rho_k(\eta^\alpha, \xi) \quad \vec{u}_k = \vec{u}_k(\eta^\alpha, \xi), \quad P_k = P_k(\eta^\alpha, \xi),$$

$k = 1, 2, \dots$ with

$$(2.12) \qquad \begin{cases} \eta^\alpha = \epsilon^2 x^\alpha \\ \xi = \dfrac{1}{\epsilon}\phi(\epsilon^2 x^\alpha) \end{cases} \qquad \alpha = 0, 1, 2, 3, \quad x^0 = t.$$

By inserting (2.10) in (2.9), and using (2.11–12), to the first order we obtain:

$$(2.13) \qquad \begin{cases} -\lambda\dfrac{\partial\rho_1}{\partial\xi} + \rho_0\vec{n}\cdot\dfrac{\partial\vec{u}_1}{\partial\xi} = 0 \\[2mm] -\lambda\dfrac{\partial\vec{u}_1}{\partial\xi} + \dfrac{\vec{n}}{\rho_0}\dfrac{\partial P_1}{\partial\xi} = 0 \\[2mm] \gamma P_0\vec{n}\cdot\dfrac{\partial\vec{u}_1}{\partial\xi} - \lambda\dfrac{\partial P_1}{\partial\xi} = 0 \end{cases}$$

where, as usually, we have put:

$$(2.14) \qquad \lambda = -\frac{\phi_0}{|\nabla\phi|} \quad \text{and} \quad \vec{n} = \frac{\nabla\phi}{|\nabla\phi|},$$

and ∇ denotes the gradient with respect to η_i ($i = 1, 2, 3$).

We obtain nontrivial solutions $\rho_1, \vec{u}_1, P_1$, of (2.13) by choosing

$$\lambda = \left[\gamma \frac{P_0}{\rho_0}\right]^{1/2} .$$

In this case we can take

(2.15) $$\rho_1 = \rho_1(\eta^\alpha, \xi), \quad \vec{u}_1 = \vec{n}\lambda_0\rho_0^{-1}\rho_1(\eta^\alpha, \xi),$$

$$P_1 = \gamma\frac{P_0}{\rho_0}\rho_1(\eta^\alpha, \xi)$$

ρ_1 being an arbitrary function of its arguments. The second order terms give:

(2.16)
$$
\left\{
\begin{aligned}
& \frac{\partial \rho_1}{\partial \eta^0} + \rho_0(\nabla \cdot \vec{u}_1) + |\nabla\phi|\left\{-\lambda\frac{\partial P_2}{\partial \xi} + \rho_0\vec{n}\cdot\frac{\partial \vec{u}_2}{\partial \xi}\right\} \\
& +|\nabla\phi|\left\{u_{1n}\frac{\partial \rho_1}{\partial \xi} + \rho_1\vec{n}\cdot\frac{\partial \vec{u}_1}{\partial \xi}\right\} = 0 \\[2mm]
& \frac{\partial \vec{u}_1}{\partial \eta^0} + \frac{1}{\rho}\nabla P_1 + |\nabla\phi|\left\{-\lambda\frac{\partial \vec{u}_2}{\partial \xi} + \frac{1}{\rho}\vec{n}\frac{\partial P_2}{\partial \xi}\right\} \\
& +|\nabla\phi|\left\{u_{1n}\frac{\partial \vec{u}_1}{\partial \xi} - \frac{\rho_1}{\rho_0}\vec{n}\frac{\partial P_1}{\partial \xi}\right\} \\
& -|\nabla\phi|^2\left\{\frac{\eta}{\rho_0}\frac{\partial^2 \vec{u}_1}{\partial \xi^2} + \frac{\nu}{\rho_0}\vec{n}\left(\vec{n}\cdot\frac{\partial^2 \vec{u}_1}{\partial \xi^2}\right)\right\} = 0 \\[2mm]
& \gamma P_0(\nabla \cdot \vec{u}_1) + \frac{\partial P_1}{\partial \eta^0} + |\nabla\phi|\left\{\gamma P_0\vec{n}\cdot\frac{\partial \vec{u}_2}{\partial \xi} - \lambda\frac{\partial P_2}{\partial \xi}\right\} \\
& +|\nabla\phi|\left\{\gamma P_1\vec{n}\cdot\frac{\partial \vec{u}_1}{\partial \xi} - u_{1n}\frac{\partial P_1}{\partial \xi}\right\} \\
& +|\nabla\phi|^2\left\{k(\gamma-1)\overline{R}^{-1}\frac{P_0}{\rho_0^2}\frac{\partial^2 \rho_1}{\partial \xi^2} - k(\gamma-1)\frac{\overline{R}^{-1}}{\rho_0}\frac{\partial^2 P_1}{\partial \xi^2}\right\} = 0
\end{aligned}
\right.
$$

with

$$u_{1n} = \vec{u}_1 \cdot \vec{n}, \quad \nu = \eta + \frac{\varsigma}{9}.$$

The algebraic compatibility condition for the existence of the derivatives with respect to ξ of $\rho_2, \vec{u}_2, P_2$, taking into account Eqs. (2.15), gives us the following equation for ρ_1:

(2.17) $$\frac{\partial \rho_1}{\partial \sigma} + \rho_1\frac{\partial}{\partial \sigma}\log \tilde{\theta} + a|\nabla\phi|\rho_1\frac{\partial \rho_1}{\partial \xi} = b|\nabla\phi|^2\frac{\partial^2 \rho_1}{\partial \xi^2},$$

$$a = \frac{1}{2}(1+\gamma)\frac{\lambda_0}{\rho_0}, \quad b = \frac{1}{2\rho_0}[\overline{R}^{-1}k(\gamma-1)^2\gamma^{-1} + (\eta+\nu)].$$

3   Two-dimensional Burgers equation

We consider now the propagation of acoustic waves in a viscous fluid. The field equations are (2.1)

$$\frac{\partial \rho}{\partial t} + \nabla \cdot (\rho \vec{u}) = 0$$

and (2.2) that we can write as

$$\frac{\partial \vec{u}}{\partial t} + (\vec{u} \cdot \nabla)\vec{u} = -\frac{1}{\rho}\nabla P + \frac{\eta}{\rho}\Delta \vec{u} + \frac{1}{\rho}\left(\frac{\varsigma}{3} + \frac{\eta}{3}\right)\nabla(\nabla \cdot \vec{u})$$

where Δ is the Laplacian operator. We suppose that the flow is isentropic. For a polytropic gas

$$P = K\rho^\gamma$$

where K and γ are constants.

If we apply the generalized form of the perturbation-reduction method introduced in Section 3 of Chapter 8, we obtain a 2-dimensional form of the Burgers equation.

We obtain

$$\lambda_0 = \sqrt{\gamma\frac{P_0}{\rho_0}} \,,$$

corresponding to acoustic waves. As in the case of ion-acoustic waves, we consider the particular geometry:

$$n_2 = n_3 = 0, \qquad \phi = y.$$

Then $u_{1x} = \lambda\rho_1/\rho_0$, $u_{1y} = 0$, $\rho_2 = u_{2x} = 0$ and

$$\frac{\partial u_{2y}}{\partial \xi} = \left(\frac{\lambda_0}{\rho_0}\right)\frac{\partial \rho_1}{\partial \eta}.$$

The compatibility equation can be written as

$$\frac{\partial}{\partial \xi}\left(\frac{\partial \rho_1}{\partial \sigma} + a\rho_1\frac{\partial \rho_1}{\partial \xi} + b\frac{\partial^2 \rho_1}{\partial \xi^2}\right) + c\frac{\partial^2 \rho_1}{\partial \varsigma^2} = 0$$

where

$$a = \frac{1}{2}(1+\gamma)\frac{\lambda_0}{\rho_0}, \quad b = \frac{1}{2\rho_0}\left(\frac{4}{3}\eta + \varsigma\right), \quad c = \frac{\lambda_0}{2}.$$

A suitable transformation of ρ_1 and τ reduces this equation to the standard form

(3.1)
$$\frac{\partial}{\partial \xi'}\left(\frac{\partial \psi}{\partial \sigma} + \psi\frac{\partial \psi}{\partial \xi'} + \frac{\partial^2 \psi}{\partial \xi'^2}\right) + \delta\frac{\partial^2 \psi}{\partial \varsigma^2} = 0,$$

where

$$\psi = \frac{a}{\sqrt{b}}\rho_1, \quad \xi = \sqrt{b}\xi', \quad \delta = \sqrt{b}c.$$

This equation is the *genuinely two-dimensional Burgers equation* (2-D Burgers) and it corresponds to the KP equation, obtained when dispersive media are considered.

We call equation (3.1) *genuinely* 2-D Burgers to distinguish it from the Burgers equation in several dimensions (1.1).

4 Collisional plasma

In this section we derive an unusual evolution equation (4.17) for sound waves in a collisional plasma. For large collision times, this equation reduces to KdV equation with a small fourth order dissipative term. We consider a ionized plasma composed of cold ions ($T_i = 0$) and warm isothermal electrons ($T_e = $ cost $\neq 0$); we neglect any external magnetic fields. We take into account electron-ion collisions. In this case, we can use the equations (4.14), (4.6), (4.15), (4.16) of Chapter 1, by considering a proton-electron plasma,

(4.1)
$$\begin{cases} \epsilon_0 \nabla \cdot \vec{E} = e(n_i - n_e) \\ \nabla \times \vec{E} = 0 \\ \epsilon_0 \dfrac{\partial \vec{E}}{\partial t} + e(n_i \vec{w}_i - n_e \vec{w}_e) = 0 \end{cases}$$

where n_i and n_e are the numerical densities of protons and electrons, respectively.

(4.2)
$$\begin{cases} \dfrac{\partial n_i}{\partial t} + \nabla \cdot (n_i \vec{w}_i) = 0 \\ \dfrac{\partial n_e}{\partial t} + \nabla \cdot (n_e \vec{w}_e) = 0 \end{cases},$$

(4.3)
$$\begin{cases} \dfrac{\partial \vec{w}_i}{\partial t} + (\vec{w}_i \cdot \nabla)\vec{w}_i - \dfrac{e}{m_i}\vec{E} = -\dfrac{m_e}{m_i \tau}\dfrac{n_e}{n_i}(\vec{w}_i - \vec{w}_e) \\ \dfrac{m_e}{m_i}\left\{ \dfrac{\partial \vec{w}_e}{\partial t} + (\vec{w}_e \cdot \nabla)\vec{w}_e \right\} + \dfrac{m_e}{m_i}\dfrac{(a_2)^2}{n_e}\nabla n_e + \dfrac{e}{m_i}\vec{E} = \dfrac{m_e}{\tau m_i}(\vec{w}_i - \vec{w}_e) \end{cases}$$

If we let

$$R_p^2 = \frac{e}{m_i}, \qquad N = \frac{m_e}{m_i \tau}, \qquad \frac{1}{M^2} = \frac{m_e}{m_i}(a_2)^2$$

and $\epsilon_0 = 1$, $e = 1$ in appropriate units, then (4.1–3) can be written as

(4.4)
$$\nabla \times \vec{E} = 0,$$

(4.5)
$$\frac{\partial \vec{E}}{\partial t} = n_e \vec{w}_e - n_i \vec{w}_i,$$

(4.6)
$$\nabla \cdot \vec{E} = n_i - n_e,$$

(4.7)
$$\frac{\partial n_i}{\partial t} + \nabla \cdot (n_i \vec{w}_i) = 0,$$

(4.8)
$$\frac{\partial n_e}{\partial t} + \nabla \cdot (n_e \vec{w}_e) = 0,$$

(4.9)
$$\frac{\partial \vec{w}_i}{\partial t} + (\vec{w}_i \cdot \nabla)\vec{w}_i = R_p^2 \vec{E} + N\frac{n_e}{n_i}(\vec{w}_e - \vec{w}_i),$$

(4.10)
$$\frac{m_e}{m_i}\left[\frac{\partial \vec{w}_e}{\partial t} + (\vec{w}_e \cdot \nabla)\vec{w}_e \right] = -R_p^2 \vec{E} - N(\vec{w}_e - \vec{w}_i) - \frac{1}{M^2 n_e}\nabla n_e,$$

where R_p is the oscillation frequency of the ionic plasma, M the Mach number and N the collision parameter. All the quantities are suitably normalized. For simplicity the ions are assumed to be monovalent.

In this case the electron inertia (the term on the LHS of (4.10)) can be neglected. The Mach number can be expressed in terms of R_p and the Debye length l_D as $M = (R_p l_D)^{-1}$.

By taking the divergence of (4.5) we see that among the equations (4.5–8), only three are independent.

By eliminating n_i and $\vec{w}_e$ from the equations (4.5–10), we obtain

(4.11)
$$\frac{\partial n}{\partial t} + \nabla \cdot (n\vec{w}) = 0,$$

(4.12)
$$\frac{\partial \vec{w}}{\partial t} + (\vec{w} \cdot \nabla)\vec{w} = \frac{1}{n}\left\{ R_p^2 \vec{E}(\nabla \cdot \vec{E}) + \frac{1}{M^2}\nabla(\nabla \cdot \vec{E} - n) \right\},$$

(4.13)
$$\frac{\partial \vec{E}}{\partial t} + \vec{w}\nabla \cdot \vec{E} = \frac{1}{N}\left\{ R_p^2(\nabla \cdot \vec{E} - n) + \frac{1}{M^2}\nabla(\nabla \cdot \vec{E} - n) \right\},$$

with the condition

(4.14)
$$\nabla \times \vec{E} = 0.$$

We have omitted the subscripts e and i without introducing ambiguity.

We look for a solution $n, \vec{w}$ and $\vec{E}$ of the system (4.11–14) that describes a perturbation of a given constant solution $(1, 0, 0)$ expressed as an asymptotic series in terms of a small parameter ϵ:

$$\begin{cases} n = 1 + \epsilon n_1 + \epsilon^2 n_2 + \dots \\ \vec{w} = \epsilon \vec{w}_1 + \epsilon^2 \vec{w}_2 + \dots \\ \vec{E} = \epsilon^{1/2}(\epsilon \vec{E}_1 + \epsilon^2 \vec{E}_2 + \dots) \\ N^* = \epsilon^{1/2} N \end{cases}$$

where

(4.15)
$$\vec{E} = \vec{E}(\xi, \eta^\alpha), \quad n_j = n_j(\xi, \eta^\alpha), \quad \vec{w}_j = \vec{w}_j(\xi, \eta^\alpha),$$

$j = 1, 2, \dots, \alpha = 0, 1, 2, 3; x^0 = t$ with

(4.16)
$$\begin{cases} \eta^\alpha = \epsilon^{3/2} x^\alpha \\ \xi = \epsilon^{3/2}\varphi/|\nabla\varphi| \end{cases}.$$

The evolution equation is

(4.17)
$$\frac{\partial \pi}{\partial \sigma} + \pi \frac{\partial \pi}{\partial \xi} + \frac{l_D^2}{2M}\frac{\partial^3 \pi}{\partial \xi^3} + \pi \frac{\partial}{\partial \sigma}\log \tilde{\theta}$$
$$- N^* l_D^2 M \frac{\partial}{\partial \xi}\left(\frac{\partial \pi}{\partial \sigma} + \pi \frac{\partial}{\partial \sigma}\log \tilde{\theta} + \pi \frac{\partial \pi}{\partial \xi} \right) = 0,$$

where π is the variable parameter related to the first order field (Giambò, Palumbo, and Pantano, 1982). Equation (4.17) can be investigated in the particular case when $N^* \simeq \delta \ll 1$ and the wave surfaces are *quasi-plane*, i.e. $\partial/\partial\sigma \log \tilde{\theta} \ll 1$.

Then Equation (4.17) can be written as

(4.18)
$$\frac{\partial W}{\partial \tau} - 6W \frac{\partial W}{\partial \xi} + \frac{\partial^3 W}{\partial \xi^3} = R(W)$$

where we have put

$$\tau = \frac{l_D^2}{2M} \sigma, \qquad W = -\frac{1}{3} \frac{M}{l_D^2} \pi$$

and

(4.19)
$$R(W) = -W \frac{\partial}{\partial \sigma} \log \tilde{\theta} - \nu \frac{\partial^4 W}{\partial \xi^4} + 0(\delta)$$

with $\nu = N^* l_D M$.

Equation (4.18) is a perturbed KdV equation and can be treated by a perturbation method.

By putting

(4.20)
$$W(\xi, \tau) = W_s(\xi, \tau) + \delta W(\xi, \tau)$$

where W_s represents a soliton moving with a changing shape and δW is a shelf appearing during the propagation behind the soliton,

(4.21)
$$W_s = -2\chi^2 \operatorname{sech}^2 z$$

with

(4.22)
$$z = \chi(\tau)[\xi - \overline{\xi}(\tau)],$$

we obtain

(4.23)
$$\frac{d\chi}{d\tau} = -\frac{1}{4z} \int_{-\infty}^{+\infty} R(W_s) \operatorname{sech}^2 z \, dz$$

and

(4.24)
$$\frac{d\overline{\xi}}{d\tau} = 4\chi^2 + 0(\delta).$$

Since $R(W_s)$ contains geometrical and collisional terms, the shape of the soliton and consequently its velocity is determined by the mutual balancing of these two dissipative terms. In this case equation (4.18) can be written as

(4.25)
$$\frac{d\chi}{d\tau} + \frac{2\chi}{3\tilde{\theta}} \frac{d\tilde{\theta}}{d\tau} + \frac{65}{21} \nu \chi^5 = 0.$$

This is just a Bernoulli equation and its solutions are

(4.26)
$$\chi = \tilde{\theta}^{-2/3} \left(\frac{260}{21} \nu \int_0^\tau \tilde{\theta}^{-8/3} d\tau + \chi_0^{-4} \right)^{-1/4}.$$

Also the shelf can be determined by the mutual action of geometrical and collisional terms.

REFERENCES

- N. Asano, *Reductive perturbation method for nonlinear wave propagation in inhomogeneous media*, J. Phys. Soc. Jpn. **29**, 220 (1970).
- M. Bartuccelli, *The Generalized Burgers' equation in Radiative Magnetodynamics*. Lett. Nuovo Cimento **36**, 21 (1983).
- G.W. Bluman, J.D. Cole, *Similarity Methods for differential Equations*. Springer-Verlag (1974).
- T. Brugarino, E. De Pascale, P. Pantano, *Soluzione di similarità per l'equazione cilindrica di Burgers e per l'equazione cilindrica di Korteveg-De Vries*, Proceedings of the VI National Congress AIMETA, Genoa, October 7–9, 1982.
- T. Brugarino, P. Pantano, *Nonlinear acoustic waves in thermoviscous fluids*, Atti Acad. PA., Serie IV, XXXIX, 303 (1979).
- D.G. Crighton, *Model equations of nonlinear acoustics*, Ann. Rev. Fluid. Mech. **17**, 11 (1979).
- S. Giambò, A. Palumbo, P. Pantano, *Pertubated three-dimensonal solitons in a collisional plasma*, Proceedings of the VI National Congress AIMETA, Genoa, October 7–9, 1982.
- J. J. C. Nimmo and D. G. Crighton, *Bäcklund transformations for nonlinear parabolic equations: the general results*, Proc. Roy. Soc. Lond. A. **384**, 381 (1982).
- A. Palumbo, *Nonlinear Three-Dimensional Propagation of Acoustic Waves in Radiative Gasdynamics*. Lett. Nuovo Cimento **37**, 588 (1983).
- V. Rizun, I.K. Engel'brechkt, *application of the Burgers' equation with a variable coefficient to the study of nonplanar wave transient*, Appl. Math. Mech. PMM **39**, 524 (1975).
- P. L. Sachdev, *Nonlinear Diffusive Waves*, Cambridge University Press, Cambridge (1987).
- A. Sedov, G.A. Nariboli, *Burgers' equation for plane and isotropic magnetogasdynamic waves*, J. Phys. Soc. Jpn. **44**, 1380 (1978).
- G.B. Whitham, *Linear and Nonlinear Waves*, Wiley, New York (1974).

INTERACTION OF DISPERSIVE WAVES

Introduction

One of the fundamental features of nonlinear waves is that different waves interact with each other and produce new waves. In this chapter, we derive asymptotic equations for the resonant interaction of small amplitude dispersive waves. In the simplest case, the wave amplitudes satisfy the three wave resonant interaction (TWRI) equations.

Our discussion is not intended to be comprehensive. For additional information on applications of the theory, see Ablowitz and Segur (1981), Craik (1985), and Weiland and Wilhelmsson (1977). For an analysis of the TWRI equations using the inverse scattering transform, see Kaup, Reiman and Bers (1979) and Kaup (1980). Two specific topics we do not discuss are the use of variational formulations (Simmons, 1969), and the role of resonant interactions in random wave-fields (Hasselmann, 1966, Benney and Saffman, 1966, Benney and Newell, 1969, Segur, 1984).

We consider dispersive waves exclusively. However, very similar asymptotic expansions can be used to derive nonconservative TWRI equations and coupled Ginzburg-London equations for bifurcation phenomena in systems which are near a point of marginal instability (Crai, 1985).

1 The tri-resonance condition

Suppose two waves with different frequencies propagate through some medium. For example, suppose we fire two laser beams through a quartz crystal; or we use an oscillating paddle to propagate two internal waves down a tank filled with salt water. If the waves are linear, then their amplitudes superpose. This leads to interference between the waves but to no other effects.

If the waves are nonlinear, then they interact and produce new waves. We have two aims: first, to determine what new waves are produced by this nonlinear interaction; and second, to derive equations that describe the evolution of the amplitudes of the interacting waves. We study the first problem in this section and the second problem in the next section.

Most waves are of rather small amplitude. Linearized theory describes such waves well for a finite time. However, after a long enough interval of time cumulative nonlinear effects lead to a significant change in the wave field. When we consider the propagation of small amplitude waves over these long times, we say that the waves are weakly nonlinear. In this chapter we analyze the interaction of weakly nonlinear dispersive waves. In the next chapter we shall analyze the interaction of weakly nonlinear hyperbolic waves.

The theory for weakly nonlinear wave interactions is well established. There is no comparable theory for the interaction of large amplitude waves. A large amplitude theory is possible for waves that are modelled by completely integrable equations, such as the KdV equation or the NLS equation. However, completely integrable equations are very special because they have an underlying linear structure. It is unlikely that nonintegrable equations can be treated in a similar fashion. We shall consider only the interaction of small amplitude waves.

The problem of weakly nonlinear wave interactions involves two time scales. The "fast" time scale is a typical wave period. The "slow" time scale is a time that is characteristic of the nonlinear interaction. We can classify wave interactions into two types: resonant and nonresonant. In resonant interactions, the average energy exchange between waves is nonzero. Over sufficiently long time intervals, the interaction leads to significant changes in the wave field. In nonresonant ineractions, the average energy exchange is zero. Consequently, the energy of the waves produced by the interaction is always much smaller than the energy of the original waves, and linearized theory provides a uniformly valid first approximation to the wave field. Resonant interactions occur when a certain resonance condition is satisfied. The resonance condition depends upon the linearized dispersion relation of the wave motion, as we now explain.

The simplest example of resonance is the forced simple harmonic oscillator. Suppose that the natural frequency of the oscillator is ω_0 and that the forcing has frequency ω and amplitude a. The nonresonant case is $\omega \neq \omega_0$, and the amplitude of the response of the oscillator is proportional to $(\omega - \omega_0)^{-1}a$; the resonant case is $\omega = \omega_0$, and the amplitude of the response grows linearly in time.

To discuss the resonant interaction of waves, suppose that the linearized dispersion relation is $D(\omega, \vec{k}) = 0$. We assume that the equation modelling the wave motion has real coefficients. Then $D(\omega, \vec{k}) = 0$ implies that $D(-\omega, -\vec{k}) = 0$. We may think of the effect of nonlinearity as introducing small forcing terms on the linearized equations. Resonance occurs when one of these forcing terms has the same frequency and wave number as a solution of the unforced linearized equations.

The quickest possible weakly nonlinear effects are those due to quadratic nonlinearity, and we treat this case first. Let us consider the interaction of two plane waves with frequencies ω_1 and ω_2, and wave numbers $\vec{k}_1$ and $\vec{k}_2$. Quadratically nonlinear combinations of such waves leads to sum and difference forcings on the linearized equations, with frequency $\omega_1 + \omega_2$ and wave number $\vec{k}_1 + \vec{k}_2$, and frequency $\omega_1 - \omega_2$ and wave number $\vec{k}_1 - \vec{k}_2$. Resonance occurs when one of these pairs satisfies the dispersion relation of the wave. Thus, the waves are nonresonant if

$$(1.1) \qquad D(\omega_1 + \omega_2, \vec{k}_1 + \vec{k}_2) \neq 0, \quad \text{and} \quad D(\omega_1 - \omega_2, \vec{k}_1 - \vec{k}_2) \neq 0.$$

The waves are resonant if

$$(1.2) \qquad D(\omega_1 + \omega_2, \vec{k}_1 + \vec{k}_2) = 0,$$

or

(1.3)
$$D(\omega_1 - \omega_2, \vec{k}_1 - \vec{k}_2) = 0.$$

Case (1.2) is a sum resonance. The waves produce a new wave with higher frequency $\omega_3 = \omega_1 + \omega_2$ and the wave number $\vec{k}_3 = \vec{k}_1 + \vec{k}_2$. Case (1.3) is a difference resonance. The waves produce a new wave with frequency $\omega_3 = \omega_1 - \omega_2$ and the wave number $\vec{k}_3 = \vec{k}_1 - \vec{k}_2$. We can write the resonance condition in a symmetrical way by choosing the signs of $(\omega_j, \vec{k}_j)$, appropriately. A wave with negative frequency is the same thing as the wave with positive frequency and oppositely directed wave number vector. The condition for quadratically nonlinear resonance to occur is then

(1.4)
$$D(\omega_j, \vec{k}_j) = 0, \qquad j = 1, 2, 3,$$
$$\omega_1 + \omega_2 + \omega_3 = 0,$$
$$\vec{k}_1 + \vec{k}_2 + \vec{k}_3 = 0.$$

Equation (1.4) is called the *triresonance condition*, and the three interacting waves are called a *resonant triad.*

There are two interesting special cases of the triresonance condition. The first is *second harmonic resonance*, for which

(1.5)
$$D(\omega, \vec{k}) = 0, \qquad D(2\omega, 2\vec{k}) = 0.$$

This corresponds to

$$\omega_1 = \omega_2 = \omega, \qquad \omega_3 = -2\omega,$$
$$\vec{k}_1 = \vec{k}_2 = \vec{k}, \qquad \vec{k}_3 = -2\vec{k}.$$

If (1.5) holds, then the nonlinear self-interaction of a wave with frequency ω and wave number $\vec{k}$, produces a harmonic with twice the frequency. This is called *superharmonic resonance.* Conversely, a wave with frequency 2ω and wave number $2\vec{k}$, which is perturbed by a harmonic of frequency ω and wave number $\vec{k}$, may lose its energy to this harmonic. This is called *subharmonic instability.*

The second special case of resonant three wave interactions is called *long wave-short wave resonance.* It occurs when

(1.6)
$$\omega_1 = \omega + \frac{1}{2}\epsilon, \qquad \omega_2 = \omega - \frac{1}{2}\epsilon, \qquad \omega_3 = \epsilon,$$
$$\vec{k}_1 = \vec{k} + \frac{1}{2}\vec{\delta}, \qquad \vec{k}_2 = \vec{k} - \frac{1}{2}\vec{\delta}, \qquad \vec{k}_3 = \vec{\delta},$$

where ϵ/ω, $|\vec{\delta}|/|\vec{h}| \ll 1$. This corresponds to the interaction of two nearly identical short waves and a long wave. Expanding the dispersion relation up to linear terms in ϵ and δ, shows that

(1.7)
$$\vec{C}(\omega, \vec{k}) \cdot \hat{\delta} = |\delta|^{-1}\epsilon, \qquad \hat{\delta} = |\delta|^{-1}\vec{\delta},$$

where $\vec{C} = \nabla_{\vec{k}}\omega$ is the group velocity of the wave. When (1.6) is satisfied, a significant exchange of energy between waves of widely differing length scales is possible. In one space dimension, the resonance condition states that the group velocity of the short wave is equal to the phase velocity of the long wave. Modulations of the short wave train resonate with the long wave.

The extreme case of (1.6) is $\epsilon, \vec{\delta} = 0$. Then $\omega_1 = \omega_2 = \omega$, $\vec{k}_1 = \vec{k}_2 = \vec{k}$, $\omega_3 = 0$, $\vec{k}_3 = 0$, and one has *mean–field resonance*. The nonlinear self-interaction of a wave induces a mean field, which is of the same order of magnitude as the wave amplitude. Even nonresonant mean-field interactions may be important, since mean-fields lead to new effects, like net transport of mass by the wave. Moreover, the interaction of waves and mean-field often has special features that are not typical of wave-wave interactions (Grimshaw, 1984).

We derive equations for the wave amplitudes of a resonant triad in the next section. To do this, it is helpful to make an order of magnitude estimate of the time scale for the exchange of energy between the waves in a resonant triad. Suppose that the dimensionless wave amplitudes are of the order $\alpha \ll 1$. Let T_0 be a time scale characteristic of the nonlinear effects. We choose T_0 so that it is independent of the wave amplitude. For example, suppose that the wave is modelled by an equation of the form

$$u_t + L(\partial_x)u + c_0 u u_x = 0,$$

where $u(x,t)$ is a nondimensionalised field variable, and $L(\partial_x)$ is a linear differential operator. Then $T_0 = (c_0 k)^{-1}$, where the wave number of the wave is of the order k. For a wave modelled by an equation like

$$u_t + L(\partial_x)u + \Omega u^2 = 0,$$

the characteristic time scale is $T_0 = \Omega^{-1}$. The rate of change of a wave amplitude due to the interaction of two other waves is of the order $\alpha^2 T_0^{-1}$. If the interaction is resonant, then the cumulative effect, after time t produces a wave with amplitude of the order $\alpha^2 t T_0^{-1}$. The amplitude of the new wave will grow to the same order of magnitude α as the original waves after a time $t = O(T_N)$ where

$$T_N = \alpha^{-1} T_0.$$

Thus, the time scale for quadratically nonlinear interactions is inversely proportional to the wave amplitude.

It may be appropriate to treat three waves as resonant even if the triresonance condition (1.4) is not exactly satisfied. For example, suppose that

$$\omega_1 + \omega_2 + \omega_3 = \Omega,$$

$$\vec{k}_1 + \vec{k}_2 + \vec{k}_3 = 0,$$

where $\Omega \ll \omega_j$ is a small parameter which measures the "detuning" of the waves. The forced wave, with frequency $\omega_1 + \omega_2$, drifts out of phase from the free wave, with frequency ω_3, after times $t = O(T_D)$, where

$$T_D = \Omega^{-1}.$$

If the detuning is negligible over the time scale T_N for resonant interactions to occur, then the waves will resonate. This implies that:

$$\alpha \gg \Omega T_0 \qquad \text{resonant}$$

$$\alpha \ll \Omega T_0 \qquad \text{nonresonant.}$$

If α and ΩT_0 are the same order of magnitude as $\alpha \to 0$, then a balance between resonant interaction and detuning is possible. This is discussed further in the next section.

It may happen that the triresonance condition (1.4) has no real solutions. In that case, it is necessary to consider higher order resonances. Nonlinearities of order n lead to forcing terms on the linearized equations which have frequencies and wave numbers of the form

$$\omega_1 \pm \omega_2 \pm \ldots \pm \omega_n \qquad \text{and} \qquad \pm \vec{k}_1 \pm \vec{k}_2 \pm \ldots \pm \vec{k}_n \,.$$

Resonance occurs if one of these pairs—call it ω_{n+1} and $\vec{k}_{n+1}$—satisfies the dispersion relation. Choosing the signs of ω_j and $\vec{k}_j$ appropriately, the condition for the degree n resonant interaction of $n+1$ waves is

$$D(\omega_j, \vec{k}_j)\,, \qquad j = 1,\ldots,n+1\,,$$

(1.8)
$$\sum_{j=1}^{n+1} \omega_j = 0\,, \qquad \sum_{j=1}^{n+1} \vec{k}_j = 0\,.$$

The time scale for the resonance is

$$T_N = \alpha^{1-n} T_0\,.$$

The interaction is slower for higher degrees of nonlinearity.

Just as second harmonic resonance is a special case of (1.2), n^{th} harmonic resonance is a special case of (1.8). It occurs when

$$D(\omega, \vec{k}) = D(n\omega, n\vec{k}) = 0\,.$$

A simple example where four wave interactions occur, is the nonlinear self-interaction of a single dispersive wave. Suppose that the wave has frequency ω, wave number $\vec{k}$, and dispersion relation $\omega = W(\vec{k})$. The only three wave interactions that such a wave can participate in are with the mean field and with its second harmonic. If $W(0) \neq 0$ and $W(2\vec{k}) \neq 0$, these interactions are nonresonant. On the other hand, a single real wave always allows the four-wave interaction,

$$\omega + \omega - \omega - \omega = 0\,, \qquad \vec{k} + \vec{k} - \vec{k} - \vec{k} = 0\,.$$

The dominant nonlinear effect on the wave is therefore cubic and one obtains the nonlinear Schrödinger equation,

$$i A_t + A_{xx} \pm |A|^2 = 0\,.$$

As we shall see in the next chapter, a nondispersive hyperbolic wave always allows second harmonic resonance. Consequently, the dominant nonlinearity is quadratic and one obtains the inviscid Burgers equation

$$u_t + uu_x = 0.$$

The different degrees of nonlinearity in these two asymptotic equations is due to the different properties of the associated linearized dispersion relations.

Two dispersive waves resonate with a third wave only for special wave numbers $\{\vec{k}_1, \vec{k}_2\}$, but they always satisfy the four-wave resonance condition

$$\omega_1 + \omega_2 - \omega_1 - \omega_2 = 0, \qquad \vec{k}_1 + \vec{k}_2 - \vec{k}_1 - \vec{k}_2 = 0.$$

Since four-wave resonance always occurs, it is not usually necessary to consider the slower five-wave resonances.

The fact that a set of waves satisfies one of the above resonance conditions does not, by itself, guarantee that the waves will interact. It is also necessary to have some nonlinear coupling between the waves. In the extreme case of a linear system waves never interact, whether or not they satisfy a resonance condition. The strength of the coupling between waves is measured by an interaction coefficient. The value of this coefficient depends on the nonlinear terms in the equations modelling the wave. The waves only interact if the interaction coefficient is nonzero. In the next section we derive the equations for the amplitudes of a set of resonantly interacting waves. In particular, this procedure gives an explicit expression for the interaction coefficients.

2   Quadratically nonlinear interaction of dispersive waves

In this section we derive the three wave resonant interaction equations that describe the quadratically nonlinear resonant interaction of small-amplitude dispersive waves. We suppose that the waves are modelled by a constant coefficient, first order system,

(2.1) $$A^\alpha(\mathbf{U})\partial_\alpha \mathbf{U} + \mathbf{g}(\mathbf{U}) = 0.$$

In (2.1), $\partial_\alpha = \dfrac{\partial}{\partial x^\alpha}$,

$$x = (x^0, x^1, \ldots, x^m) \in \mathbb{R}^{m+1},$$

$\mathbf{U}$ and $\mathbf{g}(\mathbf{U})$ are real N-vectors, $A^\alpha(\mathbf{U})$ is an $N \times N$ matrix, and summation is implied over α. In this section we do not need to explicitly distinguish the time variable, so it is convenient to write $x^0 = t$. We shall also use the frequency-wave number vector, which we call the wave vector for short, defined by

$$\kappa = (\kappa_0, \kappa_1, \ldots, \kappa_m) \in \mathbb{R}^{m+1},$$

$$\kappa_0 = -\omega, \qquad (\kappa_1, \ldots, \kappa_m) = \vec{k}.$$

We consider waves propagating through a uniform medium $\mathbf{U} = \mathbf{U}_0$, where the constant vector $\mathbf{U}_0$ satisfies

$$\mathbf{g}(\mathbf{U}_0) = 0.$$

The linearization of (2.1) about $\mathbf{U} = \mathbf{U}_0$ is

(2.2) $$\mathcal{L}[\mathbf{U}] = 0,$$

where

(2.3)
$$\mathcal{L}[\mathbf{U}] = \mathcal{A}_0^\alpha \partial_\alpha \mathbf{U} + \mathcal{G}\mathbf{U},$$
$$\mathcal{A}_0^\alpha = \mathcal{A}^\alpha(\mathbf{U}_0),$$
$$\mathcal{G} = \nabla\mathbf{g}(\mathbf{U}_0).$$

The plane wave solutions of (2.2) are

(2.4) $$\mathbf{U} = \exp(i\kappa \cdot x)\mathbf{R}(\kappa),$$

where κ satisfies the dispersion relation

(2.5) $$\Delta(\kappa) := \det(i\kappa_\alpha \mathcal{A}_0^\alpha + \mathcal{G}_0) = 0,$$

and $\mathbf{R}(\kappa)$ is a right null vector

(2.6) $$(i\kappa_\alpha \mathcal{A}_0^\alpha + \mathcal{G})\mathbf{R}(\kappa) = 0.$$

We also denote a left null vector by $\mathbf{L}(\kappa)$, where

(2.7) $$\mathbf{L}^*(\kappa) \cdot (i\kappa_\alpha \mathcal{A}_0^\alpha + \mathcal{G}) = 0.$$

To analyze the interaction of waves like (2.4), we introduce a small parameter ϵ which measures the wave amplitude. We define "slow" time and space scales by

(2.8) $$X^\alpha = \epsilon x^\alpha.$$

As discussed in Section 1, these are the appropriate variables for describing the slow effect of quadratic nonlinearities on the waves. We look for an asymptotic expansion of solutions to (2.1) of the form,

(2.9) $$\mathbf{U}(x; \epsilon) = \mathbf{U}_0 + \epsilon\mathbf{U}_1(x, X) + \epsilon^2\mathbf{U}_2(x, X) + O(\epsilon^3), \quad \text{as} \quad \epsilon \to 0_+,$$

where $\mathbf{U}_1$ is a sum of plane waves,

(2.10) $$\mathbf{U}_1 = \sum_{j=1}^{J} a_j(X)\exp(i\kappa_j \cdot x)\mathbf{R}_j + c.c., \quad \mathbf{R}_j = \mathbf{R}(\kappa_j).$$

In (2.10), each wave vector $\kappa_j = (\kappa_{j0}, \ldots, \kappa_{jm})$ satisfies the dispersion relation (2.5). The scalar $a_j(X)$ is the amplitude of the jth wave.

We shall say that a set of wave vectors $\{\kappa_1, \ldots, \kappa_J\}$ is *nondegenerate* if $\kappa_j \neq 0$, $1 \leq j \leq J$, and $\kappa_p \neq \pm\kappa_q$ for any $1 \leq p \neq q \leq J$. We shall call the set *closed* if it satisfies the following condition:

$$\Delta(\kappa_p \pm \kappa_q) = 0 \quad \text{implies that} \quad \kappa_p \pm \kappa_q = \pm\kappa_j$$
$$\text{for some} \quad 1 \leq j \leq J.$$

That is, the set is closed if it contains all wave vectors which can be produced by resonant interactions between a pair of waves in the set. We assume that the wave vectors in (2.10) form a nondegenerate, closed set.

Our aim is to construct an asymptotic approximation $\epsilon\mathbf{U}_1$ of $\mathbf{U}$ which is valid when $|x| = O(\epsilon^{-1})$ and $|X| = O(1)$. If $\mathbf{U}_2$, in (2.9), contains terms—called *secular terms*—with linear, or super-linear, growth in $|x|$, then

$$\epsilon^2 \mathbf{U}_2(x, X) \geq O(\epsilon) \qquad \text{when} \qquad x = O(\epsilon^{-1}).$$

Then $\epsilon\mathbf{U}_1$ is not a uniformly valid asymptotic approximation for times and propagation distances of the order ϵ^{-1}. We therefore require that $\mathbf{U}_2$ is sublinear in x, meaning that

$$(2.11) \qquad\qquad \lim_{|x| \to +\infty} |x|^{-1} \mathbf{U}_2(x, X) = 0.$$

Imposing (2.11) on our asymptotic expansion leads to a set of equations for the evolution of the wave amplitudes.

We derive the pertubation equations by the method of multiple scales. In this method, the "slow" variables X and the "fast" variables x are treated as independent variables. We expand x-derivatives as

$$(2.12) \qquad\qquad \partial_\alpha = \partial_{x^\alpha} + \epsilon\partial_{X^\alpha}.$$

We also Taylor expand the coefficients of (2.1) about $\mathbf{U} = \mathbf{U}_0$,

$$(2.13) \qquad \begin{aligned} \mathcal{A}^\alpha(\mathbf{U}) &= \mathcal{A}_0^\alpha + \nabla\mathcal{A}_0^\alpha \cdot (\mathbf{U} - \mathbf{U}_0) + O(|\mathbf{U} - \mathbf{U}_0|^2), \\ \mathbf{g}(\mathbf{U}) &= \mathcal{G}(\mathbf{U} - \mathbf{U}_0) + \frac{1}{2}\nabla^2\mathbf{g}_0 \cdot (\mathbf{U} - \mathbf{U}_0)(\mathbf{U} - \mathbf{U}_0) + O(|\mathbf{U} - \mathbf{U}_0|^3), \end{aligned}$$

where subscripts "0" indicate evaluation at $\mathbf{U} = \mathbf{U}_0$. Substituting (2.9), (2.12), and (2.13) into (2.1), and equating coefficients of ϵ and ϵ^2 to zero in the result implies that

$$(2.14) \qquad\qquad \mathcal{L}[\mathbf{U}_1] = 0,$$

$$(2.15) \qquad \mathcal{L}[\mathbf{U}_2] + \mathcal{A}_0^\alpha \partial_{X^\alpha}\mathbf{U}_1 + \nabla\mathcal{A}_0^\alpha \cdot \mathbf{U}_1 \partial_{x^\alpha}\mathbf{U}_1 + \nabla^2\mathbf{g}_0 \cdot \mathbf{U}_1\mathbf{U}_1 = 0.$$

Here $\mathcal{L}$ is the linearized operator in ∂_x, defined in (2.3). Thus, $\mathbf{U}_1$ solves the homogeneous linearized equations, and $\mathbf{U}_2$ solves the linearized equations with a forcing term which is quadratic in $\mathbf{U}_1$. This justifies the informal discussion given in Section 1.

Equation (2.14) is satisfied by any function of the form (2.10). Use of (2.10) in (2.15), implies that

$$(2.16) \quad \mathcal{L}[\mathbf{U}_2] = \sum_{j=1}^{J} \mathbf{F}_j \exp(i\kappa_j \cdot x) + \sum_{p=1}^{J}\sum_{q=1}^{J} \{a_p a_q \mathbf{S}_{pq} \exp i(\kappa_p + \kappa_q) \cdot x$$

$$+ a_p^* a_q \mathbf{D}_{pq} \exp i(-\kappa_p + \kappa_q) \cdot x\} + c.c. = 0\,,$$

where

$$(2.17) \quad \begin{aligned} \mathbf{F}_j &= -\mathcal{A}_0^\alpha \mathbf{R}_j \partial_{X^\alpha} a_j\,, \\ \mathbf{S}_{pq} &\equiv \{i\kappa_{q\alpha} \nabla \mathcal{A}_0^\alpha \cdot \mathbf{R}_p \mathbf{R}_q + \frac{1}{2}\nabla^2 g_0 \cdot \mathbf{R}_p \mathbf{R}_q\}\,, \\ \mathbf{D}_{pq} &= -\{i\kappa_{q\alpha} \nabla \mathcal{A}_0^\alpha \cdot \mathbf{R}_p^* \mathbf{R}_q + \frac{1}{2}\nabla^2 g_0 \cdot \mathbf{R}_p^* \mathbf{R}_q\}\,. \end{aligned}$$

Equation (2.16) is a linear equation for $\mathbf{U}_2$, in which the "slow" variables X occur as parameters. It therefore suffices to consider an equation with a single term on the right hand side, namely

$$(2.18) \quad \mathcal{L}[\mathbf{U}] = \exp(i\kappa \cdot x)\mathbf{F}\,,$$

where $\mathbf{F}$ is an N-vector which is independent of x. The solution of (2.16) may then be obtained by superposition.

There are two cases to consider. The first is when $\Delta(\kappa) \neq 0$. Then a particular solution of (2.16) is

$$(2.19) \quad \mathbf{U} = \exp(i\kappa \cdot x)(i\kappa_\alpha \mathcal{A}_0^\alpha + \mathcal{G})^{-1}\mathbf{F}\,,$$

which is uniformly bounded in x. The second case is when $\Delta(\kappa) = 0$. Then the solution of (2.17) may be unbounded as $|x| \to +\infty$. If $\Delta(\kappa) = 0$, we shall assume that

$$(2.20) \qquad \text{zero is a simple eigenvalue of } (i\kappa_\alpha \mathcal{A}_0^\alpha + \mathcal{G})\,.$$

This condition means that there is only one wave mode with a given wave vector. For example, (2.19) excludes isotropic transverse waves in three space dimensions— which have two modes with different polarizations—and "mode crossing", where the frequencies of two different modes coincide at some wavenumber. The following proposition summarizes when (2.17) has sublinear solutions.

PROPOSITION (2.1). *Suppose that $\Delta(\kappa) = 0$, and (2.19) is satisfied. Then (2.17) has sublinear solutions, satisfying (2.11), if and only if*

$$(2.21) \qquad \mathbf{L}^*(\kappa) \cdot \mathbf{F} = 0\,,$$

where $\mathbf{L}^*(\kappa)$ *is a left null vector defined in (2.7). The sublinear solution may be chosen to be bounded in x.*

PROOF. First suppose that (2.20) holds. Then, because of assumption (2.19), there is a vector $\mathbf{W}$ such that

$$(i\kappa_\alpha A_0^\alpha + \mathcal{G})\mathbf{W} = \mathbf{F}.$$

It follows that a bounded solution of (2.17) is

$$\mathbf{U}(x) = \exp(i\kappa \cdot x)\mathbf{W}.$$

Conversely, suppose that (2.17) has a sublinear solution. If $\kappa = 0$, then taking the scalar product of (2.17) with $\mathbf{L}^*(\kappa)$ gives (2.21), so we may suppose that $\kappa \neq 0$. We make a linear change of coordinates to

$$\bar{x} = (\bar{x}^0, \bar{x}^1, \ldots, \bar{x}^m),$$

where $\bar{x}^\alpha = l_\beta^\alpha x^\beta$ and $\bar{x}^0 = \kappa \cdot x$. Equation (2.17) becomes

$$(2.22) \qquad \overline{A}^\alpha \partial_{\bar{x}^\alpha} \mathbf{U} + \mathcal{G}\mathbf{U} = \exp(i\bar{x}^0)\mathbf{F},$$

where

$$(2.23) \qquad \overline{A}^\alpha = l_\beta^\alpha A_0^\beta, \qquad \overline{A}^0 = \kappa_\beta A_0^\beta.$$

We take the scalar product of (2.21) with $\exp(-i\bar{x}^0)\mathbf{L}^*(\kappa)$, average the result over

$$\{\bar{x} \in \mathbb{R}^{m+1} \mid |\bar{x}| \leq T\},$$

and let $T \to +\infty$. The average of the right hand side is $\mathbf{L}^*(\kappa) \cdot \mathbf{F}$, and (2.20) follows from the fact that the average of the left hand side is zero when $\mathbf{U}$ is sublinear. To show this fact, we note that, for $\alpha = 1, \ldots, m$

$$\frac{1}{2T}\int_{-T}^{T} \overline{A}^\alpha \partial_{\bar{x}^\alpha} \mathbf{U} \, dx^\alpha = \frac{1}{2T} \overline{A}^\alpha [\mathbf{U}]_{x^\alpha = -T}^{x^\alpha = T},$$

which approaches zero as $T \to +\infty$ because $\mathbf{U}$ is sublinear. Also, integration by parts shows that

$$\frac{1}{2T}\int_{-T}^{T} \exp(i\bar{x}^0)\mathbf{L}^*(\kappa) \cdot (\overline{A}^0 \partial_{\bar{x}^0}\mathbf{U} + \mathcal{G}\mathbf{U}) \, dx^0 =$$

$$= \frac{1}{2T}\overline{A}^0[\exp(-i\bar{x}^0)\mathbf{U}]_{x^0 = -T}^{x^0 = T} + \frac{1}{2T}\int_{-T}^{T} \exp(-ix^0)\mathbf{L}^*(\kappa) \cdot (i\overline{A}^0 + \mathcal{G})\mathbf{U} \, dx^0.$$

The first term on the right hand side approaches zero, because $\mathbf{U}$ is sublinear, and the second term is zero from (2.22) and (2.7). □

Now we apply this proposition to (2.15). Since $\{\kappa_1, \ldots, \kappa_j\}$ is closed, the only secular producing terms on the right hand side of (2.16) are those proportional to $\exp(\pm i\kappa_j \cdot x)$. Since the right hand side is real, (2.15) has a sublinear solution for U_2 if and only if the coefficient of $\exp(\pm i\kappa_j \cdot x)$ is orthogonal to L_j for $j = 1, \ldots, J$. This implies that the wave amplitudes solve the following system of equations,

$$
(2.24) \qquad C_j^\alpha \partial_{X^\alpha} a_j + \sum_{p,q}{}^{(j,s)} i\Gamma_{jpq}^{(s)} a_p a_q + \sum_{p,q}{}^{(j,s*)} i\Gamma_{jpq}^{(s*)} a_p^* a_q^*
$$

$$
+ \sum_{p,q}{}^{(j,d)} i\Gamma_{jpq}^{(d)} a_p^* a_q = 0.
$$

Here,

$$
C_j^\alpha = L_j^* \cdot A_0^\alpha R_j.
$$

The vector $C_j = (C_j^0, \ldots, C_j^m)$ is directed along the rays of the jth wave. The summations in (2.24) are defined as follows:

$$
\sum_{p,q}{}^{(j,s)} \text{ is the sum over all } p, q \text{ such that } \kappa_j = \kappa_p + \kappa_q;
$$

$$
\sum_{p,q}{}^{(j,s*)} \text{ is the sum over all } p, q \text{ such that } \kappa_j = -\kappa_p - \kappa_q;
$$

$$
\sum_{p,q}{}^{(j,d)} \text{ is the sum over all } p, q \text{ such that } \kappa_j = -\kappa_p + \kappa_q.
$$

The interaction coefficients are

$$
i\Gamma_{jpq}^{(s)} = L_j^* \cdot S_{pq},
$$

$$
i\Gamma_{jpq}^{(s*)} = L_j^* \cdot S_{pq}^*,
$$

$$
i\Gamma_{jpq}^{(d)} = L_j^* \cdot (D_{pq} + D_{qp}^*),
$$

where S_{pq} and D_{pq} are given below (2.16).

The simplest, and commonest case, is that of a single resonant triad, which consists of three waves such that

$$
\kappa_1 + \kappa_2 + \kappa_3 = 0,
$$

$$
\Delta(\kappa_j) = 0, \qquad j = 1, 2, 3.
$$

We assume that there are no difference, second harmonic, or mean-field resonances i.e.

$$
\Delta(\kappa_p - \kappa_q) \neq 0, \quad 1 \le p, q \le 3, \quad \Delta(2\kappa_j) \neq 0, \quad \Delta(0) \neq 0.
$$

Then

$$
U(x; \epsilon) = \epsilon \sum_{j=1}^{3} a_j(\epsilon x) \exp(i\kappa_j \cdot x) + O(\epsilon^2),
$$

$$
\text{as} \quad \epsilon \to 0 \quad \text{with} \quad \epsilon x = O(1).
$$

From (2.23), the wave amplitudes $a_j(X)$ satisfy

(2.25)
$$C_1^\alpha \partial_{X^\alpha} a_1 + i\Gamma_1 a_2^* a_3^* = 0,$$
$$C_2^\alpha \partial_{X^\alpha} a_2 + i\Gamma_2 a_3^* a_1^* = 0,$$
$$C_3^\alpha \partial_{X^\alpha} a_3 + i\Gamma_3 a_1^* a_2^* = 0.$$

In (2.24), the interaction coefficients Γ_j are

$$\Gamma_j = -\mathbf{L}_j^* \cdot \left(\kappa_{q\alpha} \nabla \mathcal{A}_0^\alpha \cdot \mathbf{R}_p^* \mathbf{R}_q^* + \kappa_{p\alpha} \nabla \mathcal{A}_0^\alpha \cdot \mathbf{R}_q^* \mathbf{R}_p^* - i\nabla^2 \mathbf{g}_0 \cdot \mathbf{R}_p^* \mathbf{R}_q^* \right),$$

where (j, p, q) is a cyclic permutation of $(1, 2, 3)$. Equations (2.24) are usually called the *three wave resonant interaction* (TWRI) *equations*.

A variation of these equations arises if the three waves are not in perfect resonance. Instead, suppose that

$$\kappa_1 + \kappa_2 + \kappa_3 = \epsilon \delta,$$

where $\delta \in \mathbb{R}^{m+1}$ describes the detuning of the waves, and is assumed to be of the order one as $\epsilon \to 0$. Then (2.24) becomes

$$C_1^\alpha \partial_{X^\alpha} a_1 + i\Gamma_1 a_2^* a_3^* e^{i\delta \cdot X} = 0,$$
$$C_2^\alpha \partial_{X^\alpha} a_2 + i\Gamma_2 a_3^* a_1^* e^{i\delta \cdot X} = 0,$$
$$C_3^\alpha \partial_{X^\alpha} a_3 + i\Gamma_3 a_1^* a_2^* e^{i\delta \cdot X} = 0.$$

If $|\delta| \ll 1$, then these equations reduce to the TWRI equations. If $|\delta| \gg 1$, then the integral of the rapidly varying interaction terms along a ray is $O(|\delta|^{-1})$ and the waves decouple.

Equations for second harmonic resonance also follow from (2.24). We take $\kappa_1 = \kappa_3 = \kappa$, $\kappa_2 = 2\kappa$, where $\Delta(\kappa) = \Delta(2\kappa) = 0$, and $\Delta(0)$, $\Delta(3\kappa) \neq 0$. The asymptotic solution is

$$\mathbf{U} = \epsilon [a_1(\epsilon x) \exp(i\kappa \cdot x)\mathbf{R}_1 + a_2(\epsilon x) \exp(i2\kappa \cdot x)\mathbf{R}_2] + O(\epsilon^2),$$

$\epsilon \to 0$, where

(2.26)
$$C_1^\alpha \partial_{X^\alpha} a_1 + i\Gamma_1 a_1^* a_2 = 0,$$
$$C_2^\alpha \partial_{X^\alpha} a_2 + i\Gamma_2 a_1^2 = 0.$$

In (2.25),

$$\Gamma_1 = \mathbf{L}_1^* \cdot \left[\kappa_\alpha \nabla \mathcal{A}_0^\alpha \cdot \mathbf{R}_2 \mathbf{R}_1^* + 2\kappa_\alpha \nabla \mathcal{A}_0^\alpha \cdot \mathbf{R}_1^* \mathbf{R}_2 - i\nabla^2 \mathbf{g}_0 \cdot \mathbf{R}_1^* \mathbf{R}_2 \right],$$
$$\Gamma_2 = \mathbf{L}_2^* \cdot \left[2\kappa_\alpha \nabla \mathcal{A}_0^\alpha \cdot \mathbf{R}_1 \mathbf{R}_1 - i\nabla^2 \mathbf{g}_0 \cdot \mathbf{R}_1^* \mathbf{R}_1 \right].$$

Amplitude equations for long wave-short wave resonance can be obtained in a similar way, but it is preferable to treat this case separately. The usual equations are of the form

$$ia_t + \mu a_{xx} = \Gamma a u,$$
$$u_t + \lambda(|a|^2)_x = 0,$$

where $a(x,t)$ is the complex amplitude of the short wave envelope and $u(x,t)$ is the real amplitude of the long wave. These equations, like the nonlinear Schrödinger equation and the three wave resonant interaction equations, are solvable by the inverse scattering transform (Dodd, Eilbeck, Gibbon and Morris, 1982). We derive them for a specific example at the end of Section 3.

Finally, we derive equations for the cubically nonlinear interaction of four-waves. We consider four wave interactions among waves with wave vectors $\{\kappa_1, \ldots, \kappa_j\}$. We assume that there are no three-wave resonances, meaning that

$$(2.27) \qquad \Delta(\kappa_p \pm \kappa_q) \neq 0 \quad \text{for all} \quad 1 \leq p, q \leq J.$$

We also assume that the waves are closed in the sense that if

$$\Delta(\pm\kappa_p \pm \kappa_q \pm \kappa_r) = 0,$$

then $\kappa_p \pm \kappa_q \pm \kappa_r = \kappa_j$ or $-\kappa_j$ for some $1 \leq j \leq J$.

The appropriate "slow" variable for cubically nonlinear interactions is $X = \epsilon^2 x$. It is also interesting to include dispersive effects by allowing dependence on the "intermediate" variables, $y = \epsilon x$. We expand solutions of (2.1) as

$$\mathbf{U} = \epsilon \mathbf{U}_1(x, y, X) + \epsilon^2 \mathbf{U}_2(x, y, X) + \epsilon^3 \mathbf{U}_3(x, y, X) + O(\epsilon^4),$$

where

$$(2.28) \qquad \mathbf{U}_1 = \sum_{j=1}^{J} a_j(y, X) \exp(i\kappa_j \cdot x) \mathbf{R}_j + c.c.$$

The equations for $\mathbf{U}_2$ and $\mathbf{U}_3$ are

$$(2.29) \qquad \mathcal{L}[\mathbf{U}_2] + \mathcal{A}_0^\alpha \partial_{y^\alpha} \mathbf{U}_1 + \nabla \mathcal{A}_0^\alpha \cdot \mathbf{U}_1 \partial_{x^\alpha} \mathbf{U}_1 + \frac{1}{2}\nabla^2 \mathbf{g}_0 \cdot \mathbf{U}_1 \mathbf{U}_1 = 0,$$

$$(2.30) \quad \mathcal{L}[\mathbf{U}_3] + \mathcal{A}_0^\alpha \partial_{y^\alpha} \mathbf{U}_2 + \nabla \mathcal{A}_0^\alpha \cdot \mathbf{U}_1 \mathbf{U}_{1_y^\alpha} + \mathcal{A}_0^\alpha \partial_{X^\alpha} \mathbf{U}_1 + \nabla \mathcal{A}_0^\alpha \cdot \mathbf{U}_1 \partial_{x^\alpha} \mathbf{U}_2$$

$$+ \nabla \mathcal{A}_0^\alpha \cdot \mathbf{U}_2 \partial_{x^\alpha} \mathbf{U}_1 + \frac{1}{2}\nabla^2 \mathcal{A}_0^\alpha \cdot \mathbf{U}_1 \mathbf{U}_1 \partial_{x^\alpha} \mathbf{U}_1 + \nabla^2 \mathbf{g}_0 \cdot \mathbf{U}_1 \mathbf{U}_2$$

$$+ \frac{1}{6}\nabla^3 \mathbf{g}_0 \cdot \mathbf{U}_1 \mathbf{U}_1 \mathbf{U}_1 = 0.$$

The solvability condition for (2.29) implies that

$$C_j^\alpha \partial_{y^\alpha} a_j = 0.$$

For simplicity, we assume that a_j depends on y through $\eta = (\eta_1, \ldots, \eta_K)$ where $\eta_k = \mu_k \cdot y$ is constant along the rays associated with each of the phases $\kappa_j \cdot x$. That is,

$$(2.31) \qquad C_j \cdot \mu_k = 0 \quad \text{for all} \quad 1 \leq j \leq J \quad \text{and all} \quad 1 \leq k \leq K.$$

If (2.31) is not satisfied for some $i \neq k$, then averages of $| a_k |^2$ with respect to n_k appear in equation (2.32) below, (Knobloch and De Luca, 1990).

A solution of (2.29) for $\mathbf{U}_2$ is then

$$(2.32)\ \mathbf{U}_2 = \sum_{p=1}^{J}\sum_{q=1}^{J} a_p a_q \exp[i(\kappa_p + \kappa_q) \cdot x]\mathbf{R}_{pq}^{(s)} + a_p^* a_q \exp[i(-\kappa_p + \kappa_q) \cdot x] \cdot \mathbf{R}_{pq}^{(d)}$$

$$+ \sum_{j=1}^{J}\sum_{k=1}^{K} \partial_{\eta_k} a_j \exp(i\kappa_j \cdot x)\mathbf{T}_{jk} + c.c.$$

where

$$[i(\kappa_{p\alpha} + \kappa_{q\alpha})\mathcal{A}_0^\alpha + \mathcal{G}]\mathbf{R}_{pq}^{(s)} + \mathbf{S}_{pq} = 0,$$

$$[i(-\kappa_{p\alpha} + \kappa_{q\alpha})\mathcal{A}_0^\alpha + \mathcal{G}]\mathbf{R}_{pq}^{(d)} + \mathbf{D}_{pq} = 0,$$

$$[i\kappa_j \mathcal{A}_0^\alpha + \mathcal{G}]\mathbf{T}_{jk} + \mu_{ka}\mathcal{A}_0^\alpha \mathcal{R}_j = 0.$$

These equations are uniquely solvable for $\mathbf{R}_{pq}^{(s)}$ and $\mathbf{R}_{pq}^{(d)}$ because of assumption (2.27). The condition in (2.31) implies that we can solve for $\mathbf{T}_{jk}$.

We use (2.28) and (2.32) in (2.30). This gives that

$$(2.33)\ \mathcal{L}[\mathbf{U}_3] + \sum_{j=1}^{J}\left\{ \mathcal{A}_0^\alpha \mathbf{R}_j \partial_{X^\alpha} a_j + \sum_{p=1}^{K}\sum_{q=1}^{K} \mu_{pa}\mathcal{A}_0^\alpha \mathbf{T}_{jq}\partial_{\eta_p \eta_q}^2 a_j \right\} \exp(i\kappa_j \cdot x)$$

$$+ \sum_{p=1}^{J}\sum_{q=1}^{J}\sum_{r=1}^{J}\mathbf{S}_{pqr} a_p a_q a_r \exp[i(\kappa_p + \kappa_q + \kappa_r) \cdot x]$$

$$+ \sum_{p=1}^{J}\sum_{q=1}^{J}\sum_{r=1}^{J}\mathbf{D}_{pqr} a_p a_q a_r^* \exp[i(\kappa_p + \kappa_q - \kappa_r) \cdot x] + c.c. + \text{N.S.T.} = 0.$$

In (2.33), N.S.T. stands for non-secular producing terms, proportional to $\exp[i(\pm\kappa_p \pm\kappa_q) \cdot x]$, and

$$\mathbf{S}_{pqr} = i(\kappa_{p\alpha} + \kappa_{q\alpha})\nabla \mathcal{A}_0^\alpha \cdot \mathbf{R}_r \mathbf{R}_{pq}^{(s)} + i\kappa_{r\alpha}\nabla \mathcal{A}_0^\alpha \cdot \mathbf{R}_{pq}^{(s)}\mathbf{R}_r$$

$$+ \nabla^2 g_0 \cdot \mathbf{R}_{pq}^{(s)}\mathbf{R}_r + \frac{1}{2}i\kappa_{r\alpha}\nabla^2 \mathcal{A}_0^\alpha \cdot \mathbf{R}_p \mathbf{R}_q \mathbf{R}_r + \frac{1}{6}\nabla^3 g_0 \cdot \mathbf{R}_p \mathbf{R}_q \mathbf{R}_r,$$

$$\mathbf{D}_{pqr} = i(\kappa_{p\alpha} + \kappa_{q\alpha})\nabla \mathcal{A}_0^\alpha \cdot \mathbf{R}_r^* \mathbf{R}_{pq}^{(s)} - i\kappa_{r\alpha}\nabla \mathcal{A}_0^\alpha \cdot \mathbf{R}_{pq}^{(s)}\mathbf{R}_r^*$$

$$+ \nabla^2 g_0 \cdot \mathbf{R}_{pq}^{(s)}\mathbf{R}_r^* + i(\kappa_{q\alpha} - \kappa_{r\alpha})\nabla \mathcal{A}_0^\alpha \cdot \mathbf{R}_p[\mathbf{R}_{rq}^{(d)} + \mathbf{R}_{qr}^{(d)*}]$$

$$+ i\kappa_{p\alpha}\nabla \mathcal{A}_0^\alpha \cdot [\mathbf{R}_{rq}^{(d)} + \mathbf{R}_{qr}^{(d)*}]\mathbf{R}_p + \nabla^2 g_0 \cdot [\mathbf{R}_{rq}^{(d)} + \mathbf{R}_{qr}^{(d)*}]\mathbf{R}_p$$

$$+ i\kappa_{p\alpha}\nabla^2 \mathcal{A}_0 \cdot \mathbf{R}_q \mathbf{R}_r^* \mathbf{R}_p - \frac{1}{2}i\kappa_{r\alpha}\nabla^2 \mathcal{A}_0^\alpha \cdot \mathbf{R}_p \mathbf{R}_q \mathbf{R}_r^*.$$

Eliminating secular terms from (2.33) gives a system of equations for the wave amplitudes

$$(2.34)\qquad C_j^\alpha \partial_{X^\alpha} a_j + \sum_{p=1}^{K}\sum_{q=1}^{K} i\Lambda_{jpq}\partial_{\eta_p \eta_q}^2 a_j + \sum_{p,q,r}{}^{(j,s)}i\Gamma_{jpqr}^{(s)} a_p a_q a_r$$

$$+ \sum_{p,q,r}^{(j,s*)} i\Gamma_{jpqr}^{(s*)} a_p^* a_q^* a_r^* + \sum_{p,q,r}^{(j,d)} i\Gamma_{jpqr}^{(d)} a_p a_q a_r^*$$

$$+ \sum_{p,q,r}^{(j,d*)} i\Gamma_{jpqr}^{(d*)} a_p^* a_q^* a_r = 0 \,.$$

In (2.34), the summations are defined as follows,

$\sum_{p,q,r}^{(j,s)}$ is the sum over all p, q, r such that

$$\kappa_j = \kappa_p + \kappa_q + \kappa_r \,,$$

$\sum_{p,q,r}^{(j,s*)}$ is the sum over all p, q, r such that

$$\kappa_j = -\kappa_p - \kappa_q - \kappa_r \,,$$

$\sum_{p,q,r}^{(j,d)}$ is the sum over all p, q, r such that

$$\kappa_j = \kappa_p + \kappa_q - \kappa_r \,,$$

$\sum_{p,q,r}^{(j,d*)}$ is the sum over all p, q, r such that

$$\kappa_j = -\kappa_p - \kappa_q + \kappa_r \,.$$

The interaction coefficients are

$$i\Gamma_{jpqr}^{(s)} = \mathbf{L}_j^* \cdot \mathbf{S}_{pqr} \,,$$
$$i\Gamma_{jpqr}^{(s*)} = \mathbf{L}_j^* \cdot \mathbf{S}_{pqr}^* \,,$$
$$i\Gamma_{jpqr}^{(d)} = \mathbf{L}_j^* \cdot \mathbf{D}_{pqr} \,,$$
$$i\Gamma_{jpqr}^{(d*)} = \mathbf{L}_j^* \cdot \mathbf{D}_{pqr}^* \,.$$

The coefficient of the dispersive term is

$$i\Lambda_{jpq} = \mathbf{L}_j^* \cdot \mu_{p\alpha} A_0^\alpha \mathbf{T}_{jq} \,.$$

As a special case, we consider the four-wave interaction of two waves with wave vectors κ_1 and κ_2. We assume that the only resonances are the automatic ones,

$$\kappa_1 + \kappa_2 - \kappa_1 - \kappa_2 = 0 \,,$$

$$\kappa_j + \kappa_j - \kappa_j - \kappa_j = 0 \,, \qquad j = 1, 2 \,.$$

We also suppose that the wave amplitudes depend on a single transverse variable, $\eta = \epsilon\mu \cdot x$. The above solution becomes

$$\mathbf{U}(x; \epsilon) = \epsilon a_1 (\epsilon\mu \cdot x, \epsilon^2 x) \exp(i\kappa_1 \cdot x) \mathbf{R}_1$$

$$+ \epsilon a_2 (\epsilon\mu \cdot x, \epsilon^2 x) \exp(i\kappa_2 \cdot x) \mathbf{R}_2 + O(\epsilon^2) \,,$$

as $\epsilon \to 0+$ with $\epsilon^2 x = 0(1)$. The wave amplitudes $\{a_j(\eta, X)\}$ satisfy equations of the form

$$C_1^\alpha \partial_{X^\alpha} a_1 + i\Lambda_1 \partial_{\eta\eta}^2 a_1 + i\Gamma_{11} |a_1|^2 a_1 + i\Gamma_{12} |a_2|^2 a_1 = 0 \,,$$

$$C_2^\alpha \partial_{X^\alpha} a_2 + i\Lambda_2 \partial_{\eta\eta}^2 a_2 + i\Gamma_{21} |a_1|^2 a_2 + i\Gamma_{22} |a_2|^2 a_2 = 0 \,.$$

If $a_2 = 0$, these equations reduce to the nonlinear Schrödinger equation for a_1.

3   A long wave equation

The following equation provides a simple example of three wave interactions,

$$(3.1) \qquad\qquad u_t + u u_x + \alpha u_{xxx} + \beta \partial_x^5 u = 0 \,,$$

where $x, t, u(x,t) \in \mathbb{R}$. The tri-resonance condition for (3.1) can be solved exactly.

First, we explain how equation (3.1) arises as an asymptotic description of long waves.

Suppose that the linearized dispersion relation of a wave motion is

$$(3.2) \qquad\qquad \omega = W(k; \mu), \quad k \in \mathbb{R} \,,$$

In (3.2), μ is a dimensionless parameter. We assume that $W(-k; \mu) = -W(k; \mu)$, and that W is analytic at $k = 0$. For long waves, (3.2) may be approximated by its Taylor expansion

$$(3.3) \qquad W(k; \mu) = c_0(\mu)k - a(\mu)k^3 + b(\mu)k^5 + O(k^7) \,, \text{ as } k \to 0 \,.$$

Using a Galilean transformation, we can set $c_0 = 0$. If a is nonzero, then bk^5 is negligible compared to ak^3 when $k \ll 1$. Weakly nonlinear long waves are then described asympotically by a KdV equation, namely (3.1) with $\alpha = a$ and $\beta = 0$. The linearized dispersion relation of the KdV equation agrees with (3.3) to $O(k^3)$. It may happen that a vanishes at a particular value $\mu = \mu_0$. In that case, it is not consistent to neglect bk^5 in comparison with ak^3 when μ is close to μ_0. Instead, weakly nonlinear long waves are described by (3.1) with

$$\alpha = \frac{da}{d\mu}(\mu_0)(\mu - \mu_0) \,, \quad \beta = b(\mu_0) \,.$$

Let us give two example. The first is gravity-capillary waves on the free surface of a fluid ("water waves"). We nondimensionalize by the mean depth h, the gravitational acceleration g, and the fluid density ρ. The linearized dispersion relation for waves propagating in the positive x-direction is (Whitham, 1974)

$$(3.4) \qquad\qquad \omega = [k(1 + \tau k^2) \tan k]^{1/2} \,.$$

In (3.4), τ is the Bond number,

$$\tau = \frac{T}{\rho g h^2} \,,$$

where T is the surface tension of the fluid. Expanding (3.4) as $k \to 0$ implies that

$$(3.5) \qquad \omega = k + \frac{1}{2}\left(\tau - \frac{1}{3}\right)k^3 + \frac{1}{4}\left(\frac{14}{90} - \tau^2\right)k^5 + O(k^7) \,.$$

The coefficient of k^3 vanishes when $\tau = 1/3$. In the limit $\tau \to 1/3$, small amplitude, long water waves are described by (3.1), with

$$\alpha = \frac{1}{2}\left(\frac{1}{3} - \tau\right) \text{ and } \beta = \frac{1}{90},$$

A formal derivation of (3.1) for water waves is given in Hasimoto (1970) and Hunter and Scheurle (1988).

For a layer of water on the earth's surface, a Bond number of one third corresponds to a depth of 5mm. When ripple tanks are used to simulate nondispersive wave phenomena, they are filled to this depth in order to minimize dispersive effects. However, a typical Reynolds number of the flow associated with such a shallow water wave is ten. Thus, it is not realistic to neglect viscous damping. It should be possible to carry out experiments in space with smaller viscous effects. For example, when $g = 10^{-2}\text{m/s}^2$, the critical depth is 15 cm and the Reynolds number is 60 (Zufiria, 1987).

The second physical example is magneto-acoustic waves in a plasma. We nondimensionalize by the Alfvén speed, and by $(\omega_i\omega_e)^{1/2}$, where ω_i is the ion plasma frequency and ω_e is the electron plasma frequency. The linearized dispersion relation for a magneto-acoustic wave is (Kakutani and Ono, 1969)

$$(3.6)\, \omega = \frac{1}{2}k(1 + k^2)^{-1}\left\{\left[(1 + \cos\phi)^2 + \left[\left(\rho + \frac{1}{\rho}\right)\cos^2\phi + \sin^2\phi + 2\cos\phi\right]k^2\right]^{1/2} \right.$$

$$\left. + \left[(1 - \cos\phi)^2 + \left[\left(\rho + \frac{1}{\rho}\right)\cos^2\phi + \sin^2\phi - 2\cos\phi\right]k^2\right]^{1/2}\right\}.$$

In (3.6), ϕ is the angle between the wave number vector of the magneto-acoustic wave and the undisturbed magnetic field and

$$\rho = \frac{m_i}{m_e},$$

where m_i is the mass of the ions and m_e is the mass of the electrons. The power series expansion of ω as $k \to 0$ is

$$(3.7) \qquad \omega = k - \frac{1}{2}\left[1 - (\rho^{1/2} - \rho^{-1/2})^2 \cot^2\phi\right] k^3 + O(k^5).$$

The coefficient of k^3 vanishes when the magneto–acoustic wave is at a critical angle ϕ_0 to the magnetic field, where

$$\phi_0 = \tan^{-1}\left[\rho^{1/2} - \rho^{-1/2}\right].$$

For long waves with ϕ close to ϕ_0, ω has the expansion

$$\omega \sim k - \frac{\phi - \phi_0}{\cos\phi_0 \sin\phi_0}k^3 - \frac{1}{2\sin^2\phi_0}k^5.$$

Thus, magneto-acoustic waves are described asymptotically by (3.1) with

$$\alpha = \frac{\phi - \phi_0}{\cos \phi_0 \sin \phi_0} \qquad \beta = -\frac{1}{2 \sin^2 \phi_0} \, .$$

This equation is derived by Kakutani and Ono (1969). However, they consider waves which propagate exactly at the critical angle ϕ_0. Therefore they obtain (3.1) with $\alpha = 0$.

In the expansion leading to (3.1), the parameter ρ is assumed to be of order one with respect to the amplitude and wave number parameter ϵ. In fact, ρ is numerically large because ions are much more massive than electrons. For example, in a proton-electron plasma, $\rho = 1837$, which implies that $\phi_0 = 88.70$. The coefficient of k^3 is positive for waves whose wave number vectors are in a narrow wedge of width $2.6°$. As a result, it is likely that (3.1) gives a good quantitative approximation only for very small values of ϵ.

We note in passing that although unidirectional water and magneto-acoustic waves are described by the same equation (3.1), the corresponding KP-type equations are different. The reason is that water waves are isotropic whereas magneto-acoustic waves are anisotropic. Let $\mathbf{k} = (k, l)$ be the wave number vector. The dispersion relation for water waves is

$$(3.8) \qquad \omega = |\mathbf{k}| + \frac{1}{2}\left(\tau - \frac{1}{3}\right)|\mathbf{k}|^3 + \frac{1}{4}\left(\frac{14}{90} - \tau^2\right)|\mathbf{k}|^5 + O(|\mathbf{k}|^7) \, .$$

Expanding (3.8) as $\tau \to \frac{1}{3}$ and $k, l \to 0$ with $l \ll k$ gives

$$(3.9) \qquad \omega \sim k + \frac{l^2}{2k} + \frac{1}{2}\left(\tau - \frac{1}{3}\right)k^3 + \frac{1}{90}k^5 \, .$$

The KP equation with linearized dispersion relation (3.9) is

$$\partial_x\left[u_t + u_x + u u_x + \frac{1}{2}\left(\tau - \frac{1}{3}\right)u_{xxx} + \frac{1}{90}\partial_x^5 u\right] + \frac{1}{2}u_{yy} = 0 \, .$$

For magneto-acoustic waves, let k be the component of the wave number vector along the critical angle ϕ_0 and l the orthogonal component. Then

$$\phi - \phi_0 \sim \frac{l}{k} \, ,$$

and the dispersion relation for $l \sim k^3 \ll 1$ is approximately

$$\omega \sim k + \frac{l^2}{2k} - \frac{1}{\cos \phi_0 \sin \phi_0}k^2 l - \frac{1}{2 \sin^2 \phi_0}k^5 \, .$$

The KP equation is therefore

$$\partial_x\left[u_t + u_x + u u_x + \frac{1}{\cos \phi_0 \sin \phi_0}u_{xxy} - \frac{1}{2 \sin^2 \phi_0}\partial_x^5 u\right] + \frac{1}{2}u_{yy} = 0 \, .$$

Now we calculate the resonant triads for (3.1). Linearizing about $u = 0$, its dispersion relation is

(3.10)
$$\omega = -\alpha k^3 + \beta k^5 .$$

The tri-resonance condition is

(3.11)
$$k_1 + k_2 + k_3 = 0 ,$$
$$\beta(k_1^5 + k_2^5 + k_3^5) - \alpha(k_1^3 + k_2^3 + k_3^3) = 0 .$$

We assume that $k_j \neq 0$. Eliminating k_3 from (3.11) gives

(3.12)
$$(k_1 + k_2)^2 + \frac{1}{3}(k_1 - k_2)^2 = \gamma^2 ,$$

where $\gamma = [4\alpha/5\beta]^{1/2}$. A parametric solution of (3.12) is

$$k_1 = \gamma \cos\left(t - \frac{\pi}{3}\right) ,$$

$$k_2 = \gamma \cos\left(t + \frac{\pi}{3}\right) ,$$

$$k_3 = -\gamma \cos t .$$

The dispersion relation (3.10) has resonant triads only when $\alpha\beta > 0$. For water waves, this occurs when $\tau < 1/3$; for magneto-acoustic waves it occurs when $\phi < \phi_0$. The KdV equation does not have resonant triads. Two wave numbers participate in a triad if they lie on the ellipse (3.12). This example shows clearly that two waves do not resonate unless they are specially "tuned". A wave with wave number k participates in exactly one resonant triad if $|k| \leq \gamma$. The other two wave numbers are given by the k_2 coordinates of the points of intersection of the line $k_1 = k$ with the ellipse (3.12). Second harmonic resonance occurs when $k_1 = \gamma, k_2 = k_3 = \gamma/2$.

To derive the TWRI equations for (3.1), we expand u as

(3.13)
$$u = \epsilon u_1(x, t, X, T) + \epsilon^2 u_2(x, t, X, T) + O(\epsilon^3) ,$$

where $X = \epsilon x, T = \epsilon t$. We expand derivatives as

(3.14)
$$\partial_t \rightarrow \partial_t + \epsilon \partial_T ,$$
$$\partial_x \rightarrow \partial_x + \epsilon \partial_X ,$$
$$\partial_x^3 \rightarrow \partial_x^3 + 3\epsilon \partial_x^2 \partial_X + O(\epsilon^2) ,$$
$$\partial_x^5 \rightarrow \partial_x^5 + 5\epsilon \partial_x^4 \partial_X + O(\epsilon^2) .$$

We use (3.13) and (3.14) in (3.1) and equate coefficients of ϵ and ϵ^2 to zero. The result is that

(3.15)
$$u_{1t} + \alpha u_{1xxx} + \beta \partial_x^5 u_1 = 0 ,$$

(3.16)　　　$u_{2t} + \alpha u_{2xxx} + \beta \partial_x^5 u_2 + u_{1T} + (3\alpha \partial_x^2 + 5\beta \partial_x^4)u_{1X} + u_1 u_{1x} = 0.$

A solution of (3.15) is

(3.17)　　　　　　　　$u_1 = \sum_{j=1}^{3} a_j(X,T) \exp i(k_j x - \omega_j t) + c.c.,$

where (ω_j, k_j) satisfies the dispersion relation (3.10) and the tri-resonance condition (3.11). Eliminating secular terms from (3.16) and using (3.11) implies that

(3.18)　　　　　　　　$a_{jT} + C_j a_{jX} + i k_j a_p^* a_q^* = 0.$

Here, (j,p,q) runs through cyclic permutations of (1,2,3), and C_j is the group velocity of the jth wave,

$$C_j = 3\alpha k_j^2 + 5\beta k_j^4.$$

Since each pair of waves participates in at most one resonant triad, there are no multiple triads for this equation.

　　Equation (3.1) has a conservation law with a quadratic flux, namely

(3.19)　　　　$\partial_t \left\{ \frac{1}{2}u^2 \right\} + \partial_x \left\{ \frac{1}{3}u^3 + \alpha \left[uu_{xx} - \frac{1}{2}u_x^2 \right] + \right.$

$$\left. + \beta \left[uu_{xxxx} - u_x u_{xxx} + \frac{1}{2}u_{xx}^2 \right] \right\} = 0.$$

Equation (3.18) has the corresponding averaged conservation law,

$$\partial_T \left\{ \sum_{j=1}^{3} |a_j|^2 \right\} + \partial_X \left\{ \sum_{j=1}^{3} C_j |a_j|^2 \right\} = 0.$$

　　Finally we derive the equations for long wave-short wave resonance. Without loss of generality we assume that $\alpha = \beta = 1$ in (3.1). From (3.10), the phase velocity of a long wave is zero. Resonance occurs with a short wave whose group velocity is zero. This implies that

(3.20)　　　　　　　$k = \left[\frac{5}{3} \right]^{1/2}, \quad \omega = \frac{10k}{9}.$

The appropriate expansion is

(3.21)　　　$u = \epsilon u_1 + \epsilon^{4/3} u_2 + \epsilon^{5/3} u_3 + \epsilon^2 u_4 + \epsilon^{7/3} u_5 + \epsilon^{8/3} u_6 + O(\epsilon^3),$

where

$$u_j = u_j(\xi, \tau, \theta),$$

and the multiple scale variables are evaluated at

$$\xi = \epsilon^{2/3} x, \quad \tau = \epsilon^{4/3} t, \quad \theta = kx - \omega t.$$

The linearized operator in (3.1) has the expansion

$$\partial_t + \partial_x^3 + \partial_x^5 = L_0 + \epsilon^{2/3}L_1\partial_\xi + \epsilon^{4/3}L_2\partial_{\xi\xi} + \epsilon^{4/3}\partial_\tau + O(\epsilon^2)$$

where the L_j are differential operators in ∂_θ,

$$L_0 = -\omega\partial_\theta + k^3\partial_\theta^3 + k^5\partial_\theta^5,$$
$$L_1 = 3k^2\partial_\theta^2 + 5k^4\partial_\theta^4,$$
$$L_2 = 3k\partial_\theta + 10k^3\partial_\theta^3.$$

Using (3.21) in (3.1), expanding derivatives, and equating coefficients of $\epsilon^{1/3}$ to zero implies that

(3.22) $$L_0[u_1] = 0,$$
(3.23) $$L_0[u_2] = 0,$$
(3.24) $$L_0[u_3] + L_1[u_{1\xi}] = 0,$$
(3.25) $$L_0[u_4] + L_1[u_{2\xi}] + ku_1u_{1\theta} = 0,$$
(3.26) $$L_0[u_5] + L_1[u_{2\xi}] + L_2[u_{1\xi\xi}] + u_{1\tau} + k[u_1u_2]_\theta = 0,$$
(3.27) $$L_0[u_6] + L_1[u_{3\xi}] + L_2[u_{2\xi\xi}] + u_{2\tau} + [u_1u_{1\xi}] + k[u_1u_3]_\theta + ku_2u_{2\theta} = 0.$$

A solution of (3.22) and (3.23) is

$$u_1 = a(\xi,\tau)e^{i\theta} + c.c.$$
$$u_2 = b(\xi,\tau),$$

where $\omega = k^5 - k^3$. Here, a is the complex amplitude of the short wave and b is the real amplitude of the long wave. To eliminate secular producing terms, which are proportional to $e^{i\theta}$, from (3.24) we require that $L_1[e^{i\theta}] = 0$. This is the long wave-short wave resonance condition. It implies that the group velocity $5k^4 - 3k^2$ vanishes, so that k is given by (3.20). There are no secular producing terms in (3.25). Eliminating secular producing terms from (3.26) and (3.27) leads to the long wave-short wave equations

(3.28)
$$ia_\tau + (10k^3 - 3k)a_{\xi\xi} = kab,$$
$$b_\tau + (|a|^2)_\xi = 0.$$

4   Internal waves

In this section we study three-wave interactions for equations (4.9) below. This system describes short internal waves. Internal waves are gravity waves in the interior of a stably stratified fluid. They are common in the atmosphere and the ocean. By a short wave, we mean a wave with wavelength much less than the scale height of the stratification. The effect of the compressibility on short internal waves is small (Lighthill, 1978). We therefore assume that the flud is incompressible. The equations of motion are

$$\frac{D\rho}{Dt} = 0 \,,$$

(4.1) $$\operatorname{div} \vec{u} = 0 \,,$$

$$\rho \frac{D\vec{u}}{Dt} + \nabla P + \rho g \hat{e} = 0 \,.$$

In (4.1), $D/Dt = \partial_t + \vec{u} \cdot \nabla$ is the material derivative, g is the acceleration due to gravity, and $\hat{e}$ is the unit vector in the vertical z-direction. The dependent variables are the density ρ, the pressure P, and the velocity $\vec{u}$. Equation $(4.1)_2$ is the incompressibility condition. Equations $(4.1)_1$ and $(4.1)_3$ are balance of mass and momentum.

We suppose that the unperturbed fluid is in hydrostatic equilibrium,

$$\rho = \rho_0(z) \,, \quad P = P_0(z) \,, \quad \vec{u} = 0 \,,$$

where $dP_0/dz + g\rho_0 = 0$. We denote the density and pressure perturbations by ρ' and P'. Using

$$\rho = \rho_0 + \rho' \,, \quad P = P_0 + P' \,,$$

in (4.1) gives

$$\frac{D\rho'}{Dt} + \hat{e} \cdot \vec{u} \frac{d\rho_0}{dz} = 0 \,,$$

(4.2) $$\operatorname{div} \vec{u} = 0 \,,$$

$$(\rho_0 + \rho') \frac{D\vec{u}}{Dt} + \nabla P' + \rho' g \hat{e} = 0 \,.$$

In order to simplify (4.2) and also to estimate typical length and time scales for three-wave interactions of internal waves, we introduce several parameters. The stratification is characterized by the scale height and the Brunt-Vaisala frequency. The scale height H is a typical value of $\rho_0[d\rho_0/dz]^{-1}$, and is the length scale of variations in the stratification. The Brunt-Vaisala or buoyancy frequency is $[-g\rho^{-1}d\rho_0/dz]^{1/2}$. It is the frequency of vertical oscillations in the stratified fluid and is the maximum frequency of internal waves. We denote a typical value of the Brunt-Vaisala frequency by N.

The wave is characterized by a typical wave number k and a dimensionless amplitude ϵ. We define ϵ by

$$\frac{\text{wave vorticity}}{\text{Brunt-Vaisala frequency}} = \frac{|\operatorname{curl} \vec{u}|}{N} = O(\epsilon) \,.$$

We define a second dimesionless parameter δ by

$$\delta = \frac{1}{kH} \,.$$

We assume that

(4.3)
$$\delta \ll 1 \quad \text{short wave},$$
$$\epsilon \ll 1 \quad \text{weak wave}.$$

To estimate when nonlinear effects are important, we consider the rate of change of $\vec{u}$ due to the convective nonlinear terms $\vec{u} \cdot \nabla \vec{u}$. These terms are of the order $N^2 \epsilon^2 k^{-1}$. The cumulative effect of nonlinearity over time T_N is important at leading order in the wave amplitude when $N^2 \epsilon^2 k^{-1} T_N = N \epsilon k^{-1}$. Thus

(4.4)
$$T_N = \frac{1}{\epsilon N}.$$

The group velocity is of the order $N k^{-1}$, so in time T_N a wave packet propagates a distance L_N where

(4.5)
$$L_N = \frac{1}{\epsilon k}.$$

The estimate in (4.4) is correct only if the Brunt-Vaisala frequency does not change appreciably during resonance. The Brunt-Vaisala frequency varies over distances of the order H (except in the special case of an exponentially stratified fluid when the Brunt-Vaisala frequency is identically constant). Therefore, resonant interactions occur provided that $L_N \ll H$, which implies that $\delta \ll \epsilon$. In fact, we shall see below that variations in ρ_0 can be completely neglected only if

(4.6)
$$\delta \ll \epsilon^2.$$

To substantiate the rough estimates given above, we nondimensionalize (4.2). We define dimensionless variables, which are starred, by

(4.7)
$$\rho_0(z) = R \rho_0^* \left(\frac{z}{H} \right),$$
$$\rho' = \delta R \rho^*, \quad \vec{u} = N k^{-1} \vec{u}^* \quad P = R N^2 k^{-2} P^*,$$
$$\vec{x} = k^{-1} \vec{x}^*, \quad T = N^{-1} t^*.$$

In (4.7), R is any convenient reference density. Using (4.7) in (4.2) gives

(4.8)
$$\rho_{t^*}^* + \hat{u}^* \cdot \vec{e} \frac{d}{dZ} \rho_0^* (\delta z^*) + \vec{u}^* \cdot \nabla_* \rho^* = 0,$$
$$\operatorname{div}_* \vec{u}^* = 0,$$
$$[\rho_0^* (\delta z^*) + \delta \rho^*][\vec{u}_{t^*}^* + \vec{u}^* \cdot \nabla_* \vec{u}^*] + \nabla_* P^* + \rho^* \hat{e} = 0.$$

Quadratically nonlinear interactions of waves of amplitude ϵ take place over distances $z^* = O(\epsilon^{-1})$. When (4.6) is satisfied, $Z = \delta z^* \ll \epsilon$. Freezing the coefficients of (4.8) at $\delta z^* = 0$ leads to errors which are $o(\epsilon^2)$ when $\vec{u}^*$ and ρ^* are of the order ϵ.

Such errors are negligible because we shall only expand up to terms of the order ϵ^2. We use

$$R = \rho_0(0), \quad H = \rho_0(0)\left[\frac{d}{dz}\rho_0(0)\right]^{-1}$$

to nondimensionalize, which implies that

$$\rho_0^*(0) = \frac{d}{dZ}\rho_0^*(0) = 1.$$

We also neglect the term $\delta\rho^*$ in $(4.8)_3$ because it is $o(\epsilon^2)$. Replacing the density in the acceleration term of $(4.8)_3$ by a constant reference density is called the Boussinesq approximation. Here, it follows from the short wave assumption (4.6). The resulting simplified equations are

(4.9)
$$\rho_t + \vec{u} \cdot \hat{e} + \vec{u} \cdot \nabla\rho = 0,$$
$$\text{div } \vec{u} = 0,$$
$$\vec{u}_t + \vec{u} \cdot \nabla\vec{u} + \nabla P + \rho\hat{e} = 0.$$

In (4.9) and below, we drop the stars on the nondimensional variables.

Equation (4.9) is a first order system for $\mathbf{U} = (\rho, p, \vec{u})^t$,

(4.10)
$$\mathcal{L}[\mathbf{U}] + \mathcal{Q}[\mathbf{U}] = 0,$$

where

(4.11)
$$\mathcal{L} = \begin{pmatrix} 1 & 0 & 0 \\ 0 & 0 & 0 \\ 0 & 0 & 1 \end{pmatrix}\partial_t + \begin{pmatrix} 0 & 0 & 0 \\ 0 & 0 & \nabla^t \\ 0 & \nabla & 0 \end{pmatrix} + \begin{pmatrix} 0 & 0 & \hat{e}^t \\ 0 & 0 & 0 \\ \hat{e} & 0 & 0 \end{pmatrix},$$
$$\mathcal{Q}[\mathbf{U}] = \begin{pmatrix} \vec{u} \cdot \nabla & 0 & 0 \\ 0 & 0 & 0 \\ 0 & 0 & \vec{u} \cdot \nabla \end{pmatrix}\mathbf{U}.$$

First, let us consider the linearized system

(4.12)
$$\mathcal{L}[\mathbf{U}] = 0.$$

The plane wave solutions of (4.12) are

(4.13)
$$\begin{bmatrix} \rho \\ P \\ \vec{u} \end{bmatrix} = \exp\{i\vec{k} \cdot \vec{x} - i\omega t\}\begin{bmatrix} -i \\ \vec{k} \cdot \hat{e}/k^2 \\ \vec{q} \end{bmatrix},$$

where

(4.14)
$$\omega^2 = 1 - k^{-2}(\vec{k} \cdot \hat{e})^2,$$

(4.15)
$$\vec{q} = \omega^{-1}[k^{-2}(\vec{k} \cdot \hat{e})\vec{k} - \hat{e}].$$

Equation (4.15) is the dispersion relation of internal waves. It can also be written as

$$\omega^2 = \cos^2 \theta \, ,$$

where θ is the angle of the wave number vector to the horizontal. The frequency depends on the direction of propagation of the internal wave and is independent of the wavelength. The maximum dimensionless frequency is one, which corresponds to the Brunt-Vaisala frequency. This occurs for horizontal $\vec{k}$ and vertical oscillations of the fluid particles. When the wave number vector is vertical the particle oscillations are horizontal and the frequency is zero. The velocity varies in phase with the pressure and out of phase with the density.

The vector $\vec{q}$ is in the vertical plane through $\vec{k}$ and is orthogonal to $\vec{k}$. Therefore, from (4.13), internal waves are transverse. Also, the waves are rotational and

$$|\text{curl } \vec{u}| = |\vec{k}| \, ,$$

so that the vorticity gives a measure of the wave amplitude.

The group velocity $\vec{C} = \nabla_{\vec{k}} \omega$ is

(4.16) $$\vec{C} = |\vec{k}|^{-2} (\vec{k} \cdot \vec{e}) \vec{q}.$$

The group velocity of an internal wave is orthogonal to the phase velocity and lies along wavefront. This is in contrast to isotropic waves which have parallel group and phase velocities. A striking demonstration of this anisotropy is the "St. Andrews Cross" (see Whitham, 1974, or Lighthill, 1978, for a photograph). A source of cylindrical internal waves generates waves in four radial directions. These directions are parallel to the group velocity of waves with frequency equal to the source frequency. The wavefronts of the internal waves propagate across each radial "arm" of the cross.

The solution (4.13–15) is invalid when $\omega = 0$. In that case if $\vec{k}$ is not parallel to $\hat{e}$, the solution of (4.12) is

$$\begin{pmatrix} \rho \\ p \\ \vec{u} \end{pmatrix} = \exp(i\vec{k} \cdot \vec{x}) \begin{pmatrix} 0 \\ 0 \\ \vec{q} \end{pmatrix} ,$$

where $\vec{q} = |k|^{-1}(\vec{k} \times \hat{e})$. This solution is a spatially periodic horizontal shear flow. If $\vec{k}$ is parallel to $\hat{e}$ then the solution is

$$\begin{pmatrix} \rho \\ P \\ \vec{u} \end{pmatrix} = \exp(i\vec{k} \cdot \vec{x}) \left\{ \alpha \begin{pmatrix} -i \\ 1/\vec{k} \cdot \hat{e} \\ 0 \end{pmatrix} + \beta \begin{pmatrix} 0 \\ 0 \\ \vec{q} \end{pmatrix} \right\}$$

where α, β are arbitrary complex constants and $\vec{q}$ is a unit horizontal vector. If $\alpha = 0$, this is a horizontal flow. If $\beta = 0$, it is a small hydrostatic pertubation in the density and pressure. The general solution is a linear combination of the two. The only solutions which are the limit of internal wave solutions (4.13) as $\omega \to 0$ are those with $\beta = \pm \alpha$.

The dispersion relation (4.14) has many resonant triads. McComas and Bretherton (1977) identify three types of triads that contribute most to the transfer of energy between internal waves in the ocean. They are:

(a) parametric instability – consisting of two short waves with nearly opposite wave number vectors and equal frequencies and a long wave with twice the frequency,

$$\vec{k}_1 \simeq \vec{k}, \quad \vec{k}_2 \simeq -\vec{k}, \quad |\vec{k}_3| \ll |\vec{k}|\,;$$
$$\omega_1 \simeq \omega, \quad \omega_2 \simeq \omega, \quad \omega_3 \simeq -2\omega\,;$$

(b) induced diffusion – consisting of one long, low frequency wave and two nearly identical short, high-frequency waves,

$$\vec{k}_1 \simeq \vec{k}, \quad \vec{k}_2 \simeq -\vec{k}, \quad |\vec{k}_3| \ll |\vec{k}|\,;$$
$$\omega_1 \simeq \omega, \quad \omega_2 \simeq -\omega, \quad \omega_3 \ll \omega\,;$$

(c) elastic scattering – consisting of two waves with nearly the same horizontal wavenumber and frequency and a nearly vertical low frequency wave,

$$\vec{k}_1 \simeq \vec{k}_H - m\hat{e}, \quad \vec{k}_2 \simeq -\vec{k}_H - m\hat{e}, \quad \vec{k}_3 \simeq 2m\hat{e}\,;$$
$$\omega_1 \simeq \omega, \quad \omega_2 \simeq -\omega, \quad \omega_3 \ll \omega\,;$$

Here $\vec{k}_H$ is a horizontal vector.

We can solve the tri-resonance condition explicitly if the wave number vectors of a triad lie in the same vertical plane. We let $\vec{k} = (k, m)$ where k is the horizontal component and m is the vertical component of $\vec{k}$. Pick any number $k \in \mathbb{R}$, and choose signs $\sigma, \sigma_1, \sigma_2, \sigma_3 \in \{-1, +1\}$, and angles $\theta_1, \theta_2, \theta_3 \in S^1$ such that

$$\sigma_1 \cos \theta_1 + \sigma_2 \cos \theta_2 + \sigma_3 \cos \theta_3 = 0\,.$$

Let

(4.17) $$k_1 = k \sin(\theta_3 - \theta_2), \quad k_2 = k \sin(\theta_1 - \theta_3), \quad k_3 = k \sin(\theta_2 - \theta_1)\,.$$

Then the following waves form a resonant triad:

(4.18) $$\omega_j = \sigma_j \cos \theta_j\,, \quad \vec{k}_j = k_j[\cos \theta_j, \sigma \sin \theta_j]\,, \quad j = 1, 2, 3\,.$$

One can check that

$$\omega_j^2 = \cos^2 \theta_j\,, \quad \sum_{j=1}^{3} \omega_j = 0\,, \quad \sum_{j=1}^{3} \vec{k}_j = 0\,.$$

By choosing k and θ_j appropriately we may take $\sigma = \sigma_1 = \sigma_2 = \sigma_3 = 1$ without loss of generality.

To derive the TWRI equations for (4.9) we use the expansion described in Section 2. For a single triad, the result is that

$$(4.19) \quad \begin{pmatrix} \rho \\ P \\ \vec{u} \end{pmatrix} = \epsilon \sum_{j=1}^{3} a_j(\epsilon \vec{x}, \epsilon t) \exp\{i \vec{k}_j \cdot \vec{x} - i\omega_j t\} \begin{pmatrix} -1 \\ \vec{k}_j \cdot \hat{e}/k_j^2 \\ \vec{q}_j \end{pmatrix} + c.c. + O(\epsilon^2),$$

where q_j is defined us (4.15) and

$$(4.20) \quad \begin{aligned} a_{jT} + \vec{C}_j \cdot \nabla a_j + i\Gamma_{jrs} a_r^* a_s^* &= 0, \\ (jrs) &= (123), (231), (312). \end{aligned}$$

Here, $\vec{C}_j$ is the group velocity of the jth wave, and the interaction coefficient is

$$(4.21) \quad \Gamma_{jrs} = -\frac{1}{2}[(\vec{q}_r \cdot \vec{k}_s)(\vec{q}_j \cdot \vec{q}_s - 1) + (\vec{q}_s \cdot \vec{k}_r)(\vec{q}_j \cdot \vec{q}_r - 1)].$$

If the wave number vectors are in the same vertical plane and given by (4.17) and (4.18) with $\sigma = \sigma_j = 1$, then (4.21) can be written

$$(4.22) \quad \begin{aligned} \Gamma_{jrs} &= (k_s - k_r)\Gamma, \\ \Gamma &= 2\sin\left[\frac{1}{2}(\theta_3 - \theta_2)\right] \sin\left[\frac{1}{2}(\theta_1 - \theta_3)\right] \sin\left[\frac{1}{2}(\theta_2 - \theta_1)\right]. \end{aligned}$$

Equation (4.10) has two conservation laws with quadratic densities which provide a check on these results (Ripa, 1981). The energy equation is

$$e_t + \operatorname{div}[(e + P)\vec{u}] = 0,$$

where e is the energy density,

$$e = \frac{1}{2}\rho^2 + \frac{1}{2}\vec{u} \cdot \vec{u}.$$

Here, $(1/2)\rho^2$ is the gravitational potential energy and $(1/2)\vec{u} \cdot \vec{u}$ is the kinetic energy of the flow. From (4.19) the averaged energy density is

$$< e >= \epsilon^2 \sum_{j=1}^{3} |a_j|^2 + O(\epsilon^3).$$

Conservation of energy for (4.20) implies that

$$(4.23) \quad \Gamma_{123} + \Gamma_{231} + \Gamma_{312} = 0.$$

For flows in a vertical plane, (4.9) has a second conservation law with a quadratic density, namely

$$(\rho\varsigma)_t + \left(\rho u\varsigma - \frac{1}{2}\rho^2 - \frac{1}{2}u^2 + \frac{1}{2}w^2\right)_x + (\rho w\varsigma - uw)_z = 0,$$

where $\vec{u} = (u, w)^t$ and $\varsigma = u_z - w_x$ is the vorticity. From (4.18),

$$< \rho \varsigma > = \epsilon^2 \sum_{j=1}^{3} k_j |a_j|^2 + O(\epsilon^3).$$

Therefore, this conservation law implies that

(4.24) $$\qquad\qquad k_1 \Gamma_{123} + k_2 \Gamma_{231} + k_3 \Gamma_{312} = 0.$$

Together (4.23) and (4.24) imply that

$$\Gamma_{jrs} = (k_s - k_r)\Gamma,$$

which is consistent with (4.22). The factor Γ can only be found by direct computation.

Let us consider the TWRI equations for each of the limiting triads mentioned above. For simplicity we assume that the wave number vectors lie in a vertical plane. Parametric instability corresponds to (4.18) with

$$\theta_1 \simeq \theta, \quad \theta_2 \simeq \theta, \quad \theta_3 \simeq \phi,$$

where $\cos \phi = -2 \cos \theta$. To leading order in $\delta = \theta_2 - \theta_1$, the frequencies and the wave number vectors are

$$\omega_1 \simeq \cos \theta, \quad \omega_2 \simeq \cos \theta, \quad \omega_3 \simeq \cos \phi,$$
$$\vec{k}_1 \simeq k \sin(\phi - \theta)[\cos \theta, \sin \theta],$$
$$\vec{k}_2 \simeq -k \sin(\phi - \theta)[\cos \theta, \sin \theta],$$
$$\vec{k}_3 \simeq \delta k[\cos \phi, \sin \phi],$$

where $k \in \mathbb{R}$ is arbitrary. The TWRI equations are

$$a_{1t} + \vec{C}_1 \cdot \nabla a_1 + i\Gamma a_2^* a_3^* = 0,$$
(4.25) $$\qquad a_{2t} + \vec{C}_2 \cdot \nabla a_2 + i\Gamma a_3^* a_1^* = 0,$$
$$a_{3t} + \vec{C}_3 \cdot \nabla a_3 - 2i\Gamma a_1^* a_2^* = 0,$$

where, to leading order in δ,

$$\Gamma = -\delta k \sin(\phi - \theta) \sin^2 \left[\frac{1}{2}(\phi - \theta) \right].$$

To see one consequence of these equations, we linearize the spatially independent equations about $a_1 = a_2 = 0, a_3 = a_0$. This gives

$$a_{1t} + i\Gamma a_0^* a_2^* = 0,$$
$$a_{2t} + i\Gamma a_0^* a_1^* = 0,$$
$$a_{3t} = 0.$$

Eliminating a_2 implies that

$$a_{1t} - \Gamma^2 |a_0|^2 a_1 = 0.$$

The solution for a_1 is

$$a_1(t) = K_1 \exp\{\Gamma |a_0| t\} + K_2 \exp\{-\Gamma |a_0| t\}.$$

Thus the low-frequency wave is unstable. Perturbing it by one of the high-frequency waves leads to exponential growth in the high-frequency waves. This phenomenon is called parametric instability.

The induced diffusion triads correspond to

$$\theta_1 \simeq \theta, \quad \theta_2 \simeq \theta + \pi, \quad \theta_3 \simeq \frac{\pi}{2}.$$

We let $\delta = \theta_2 - \theta_1 - \pi$. To leading order in δ, the frequencies and wave numbers are

$$\omega_1 \simeq \cos\theta, \quad \omega_2 \simeq -\cos\theta, \quad \omega_3 \simeq -\delta\sin\theta,$$

$$k_1 \simeq k\cos\theta[\cos\theta, \sin\theta], \quad k_2 \simeq -k\cos\theta[\cos\theta, \sin\theta], \quad k_3 \simeq \delta k[0,1].$$

The TWRI equations are,

(4.26)
$$\begin{aligned}
a_{1t} + \vec{C}_1 \cdot \nabla a_1 + ik\cos^2\theta a_2^* a_3^* &= 0, \\
a_{2t} + \vec{C}_2 \cdot \nabla a_2 - ik\cos^2\theta a_3^* a_1^* &= 0, \\
a_{3t} + \vec{C}_3 \cdot \nabla a_3 - i\delta k\sin^2\theta\cos\theta a_1^* a_2^* &= 0.
\end{aligned}$$

For small δ, the amplitude of the long low-frequency wave changes slowly. However, the presence of the long wave allows the short waves to exchange energy; more correctly, the waves exchange action = energy/frequency. In a random ensemble of waves, these triads result in the "diffusion" of action between waves with nearby frequencies and wave numbers.

Elastic scattering corresponds to

$$\theta_1 \simeq \theta, \quad \theta_2 \simeq \pi - \theta, \quad \theta_3 \simeq \frac{\pi}{2}.$$

We let $\delta = \theta_1 + \theta_2 - \pi$. To leading order in δ, the frequencies and wave numbers are

$$\omega_1 \simeq \cos\theta, \quad \omega_2 \simeq -\cos\theta, \quad \omega_3 \simeq \delta\sin\theta,$$

$$k_1 \simeq -k\cos\theta[\cos\theta, \sin\theta],$$

$$k_2 \simeq k\cos\theta[\cos\theta, -\sin\theta],$$

$$k_3 \simeq k[0, \sin 2\theta].$$

The TWRI equations are

(4.27)
$$\begin{aligned}
a_{1t} + \vec{C}_1 \cdot \nabla a_1 - \frac{1}{2}ik\cos\theta(\cos\theta + \sin 2\theta + \cos 3\theta)a_2^* a_3^* &= 0, \\
a_{2t} + \vec{C}_2 \cdot \nabla a_2 + \frac{1}{2}ik\cos\theta(\cos\theta + \sin 2\theta + \cos 3\theta)a_3^* a_1^* &= 0, \\
a_{3t} + \vec{C}_3 \cdot \nabla a_3 - i\delta k\cos\theta\sin\theta(1 - \sin\theta)(1 + 2\sin\theta)a_1^* a_2^* &= 0.
\end{aligned}$$

For small δ, the high-frequency waves reflect almost elastically off the low-frequency vertical wave.

The interaction between internal waves and a steady periodic disturbance has some interesting points. For waves in the same vertical plane, the solution is

$$
(4.28) \qquad
\begin{pmatrix} \rho \\ P \\ \vec{u} \end{pmatrix}
= \epsilon a_1(\epsilon \vec{x}, \epsilon t)\exp\{i\vec{k}_1 \cdot \vec{x} - i w_1 t\}
\begin{pmatrix} -i \\ \vec{k}_1 \cdot \hat{e}/k_1^2 \\ \vec{q}_1 \end{pmatrix}
$$

$$
+ \epsilon a_2(\epsilon \vec{x}, \epsilon t)\exp\{i\vec{k}_2 \cdot \vec{x} - i w_2 t\}
\begin{pmatrix} -i \\ \vec{k}_2 \cdot \hat{e}/k_2^2 \\ \vec{q}_2 \end{pmatrix}
$$

$$
+ \epsilon a_3(\epsilon \vec{x}, \epsilon t)\exp\{i\vec{k}_3 \cdot \vec{x}\}
\left\{ \alpha \begin{pmatrix} -i \\ 1/\vec{k}_3 \cdot \hat{e} \\ 0 \end{pmatrix} \right.
$$

$$
\left. + \beta \begin{pmatrix} 0 \\ 0 \\ \vec{q}_3 \end{pmatrix} \right\} + c.c. + O(\epsilon^2).
$$

Here, the wave number vectors and the $\vec{q}_j$ are

$$
\begin{aligned}
\vec{k}_1 &= k\cos\theta[\cos\theta, \sin\theta], \\
\vec{k}_2 &= k\cos\theta[\cos\theta, -\sin\theta], \\
\vec{k}_3 &= k[0, \sin 2\theta], \\
\vec{q}_1 &= [-\sin\theta, \cos\theta], \\
\vec{q}_2 &= [-\sin\theta, \cos\theta], \\
\vec{q}_3 &= [1, 0].
\end{aligned}
$$

The TWRI equations are

$$
(4.29) \qquad
\begin{aligned}
a_{1t} + \vec{C}_1 \cdot \nabla a_1 + i\Gamma a_2^* a_3^* &= 0, \\
a_{2t} + \vec{C}_2 \cdot \nabla a_2 - i\Gamma a_1^* a_3^* &= 0, \\
a_{3t} &= 0,
\end{aligned}
$$

where

$$
(4.30) \qquad \Gamma = \frac{1}{2}k\cos\theta[\alpha^*(\cos\theta + \cos 3\theta) - \beta^* \sin 2\theta].
$$

The internal waves reflect off the steady disturbance. The steady disturbance is not affected at all by the interaction.

The choice $\alpha = 1$, $\beta = 0$ corresponds to a steady shear flow. In that case

$$
\Gamma = \frac{1}{2}k\cos\theta(\cos\theta + \cos 3\theta).
$$

The interaction coefficient vanishes when the angle of incidence of the internal wave of the shear flow is $\theta = \pi/4$. The shear flow is completely transparent to an internal wave at that angle of incidence.

The choice $\alpha = 0$, $\beta = -1$ corresponds to a periodic hydrostatic density perturbation. In that case, $\Gamma = (1/2)k \cos \theta \sin 2\theta$, which is nonzero for $0 < \theta < \pi/2$.

A steady disturbance that is the limit of low frequency internal waves corresponds to $\alpha = \pm 1$ and $\beta = 1$. For $\alpha = -1$ the interaction coefficient is

$$\Gamma = -\frac{1}{2}k \cos \theta (\cos \theta + \sin 2\theta + \cos 3\theta).$$

Equation (4.27) for the elastic scattering triads reduces to (4.29) with this interaction coefficient when $\delta = 0$.

Phillips (1968) observes that this three wave interaction gives a mechanism for the trapping of internal waves near the surface of the ocean. Suppose that the ocean surface is at $z = 0$ and assume that there is a steady disturbance below the ocean surface with constant amplitude a_3. We consider incident and reflected waves with amplitudes depending only on the depth z.

The TWRI equations (4.29) imply that

$$- k^{-1} \sin \theta \, a_{1z} + i\Gamma a_3^* a_2^* = 0,$$
$$- k^{-1} \sin \theta \, a_{2z} - i\Gamma a_3^* a_1^* = 0.$$

This equation has a solution which decays to zero as $z \to -\infty$,

$$a_1 = K_1 e^{\gamma z}, \quad a_2 = K_2 e^{\gamma z}.$$

Here K_1 and K_2 are arbitrary complex constants and

$$\gamma = |ka_3\Gamma \csc \theta|.$$

In this solution the upward propagating internal wave reflects off the surface at $z = 0$ and produces a downward propagating wave. The downward propagating wave is then scattered back towards the surface by reflection off the steady disturbance.

5 Passage through resonance

Weak and strong resonance

The frequency and wave number of a nonplanar high frequency wave vary slowly in space and time. The equation of the wavefronts is

$$\phi(\vec{X}, T) = \text{constant},$$

where $\vec{X}, T$ are "slow" space and time variables, and ϕ is the phase of the wave. The local frequency $\omega(\vec{X}, T)$ and the local wavenumber $\vec{k}(\vec{X}, T)$ of the wave are defined by

$$\omega = -\phi_T, \quad \vec{k} = \nabla \phi.$$

They satisfy the local, linearized, dispersion relation of the wave motion,

$$D(\omega, \vec{k}; \ \vec{X}, T) = 0.$$

This equation implies that the phase ϕ satisfies an eikonal equation

$$D(-\phi_T, \nabla\phi; \ \vec{X}, T) = 0.$$

For a nonuniform medium, D depends explicity on $\vec{X}$ and T.

Two waves resonate when

(5.1) $$D[\omega_1(\vec{X}, T) + \omega_2(\vec{X}, T), \ \vec{k}_1(\vec{X}, T) + \vec{k}_2(\vec{X}, T); \ \vec{X}, T] := S(\vec{X}, T) = 0.$$

We distinguish two extreme cases:

(5.2)
$$\begin{aligned}
(a) \quad & S = \text{constant} \\
(b) \quad & S_T^2 + |\nabla S|^2 \neq 0.
\end{aligned}$$

In case (5.2a), the waves are either never in resonance (if $S \neq 0$), or they are in resonance everywhere (if $S = 0$). In case (5.2b), the waves pass through resonance as they propagate across the hypersurface $S(\vec{X}, T) = 0$. We call (5.2a), with $S = 0$, *strong resonance* and (5.2b) *weak resonance*. Grimshaw (1988) gives a general discussion of weak and strong resonance of dispersive waves. He also considers a specific application to wave interactions in a stratified shear flow.

In Sections 1–5, we considered plane waves (i.e. $\vec{k}$ and ω constant) in uniform media (i.e. D independent of $\vec{X}, T$). This implies that S is constant, so that the waves are strongly resonant. In this section, we analyze weakly resonant dispersive wave interactions. Weak resonance is the typical case for nonplanar waves.

We begin with an example which shows that weak resonance can lead to complicated wave patterns. Consider an isotropic wave motion, with a linearized dispersion relation of the form

$$D(\omega, |\vec{k}|) = 0, \qquad \vec{k} \in \mathbb{R}^2.$$

Suppose that there are positive constants $\omega_j, k_j \in \mathbb{R}$, $j = 1, 2, 3$, such that

$$\omega_3 = \omega_1 - \omega_2, \qquad D(\omega_1, k_1) = D(\omega_2, k_2) = D(\omega_3, k_3) = 0.$$

We do not require that $k_3 = k_1 - k_2$. One physical example of such a system is surface and internal waves in a two layer fluid. Higher frequency waves on the free surface of the lighter fluid participate in a difference resonance with lower frequency internal waves on the interface between the fluids (Ablowitz and Segur, 1981).

Suppose that two point sources, with frequencies ω_j and wave numbers k_j, are located at $\vec{x} = \vec{a}$ and $\vec{x} = -\vec{a}$. The phases and wave number vectors are

(5.3)
$$\phi_1 = k_1|\vec{x} - \vec{a}| - \omega_1 t, \qquad \phi_2 = k_2|\vec{x} + \vec{a}| - \omega_2 t,$$

$$\vec{k}_1 = k_1 \frac{\vec{x} - \vec{a}}{|\vec{x} - \vec{a}|}, \qquad \vec{k}_2 = k_2 \frac{\vec{x} + \vec{a}}{|\vec{x} + \vec{a}|}.$$

These waves satisfy the tri-resonance condition when

(5.4) $|\vec{k}_1 - \vec{k}_2| = k_3 \, .$

Using $(5.3)_2$ in (5.4) implies that

(5.5) $\cos \delta(\vec{x}) = \dfrac{k_1^2 + k_2^2 - k_3^2}{2k_1 k_2} \, ,$

where δ is the angle between $\vec{x} - \vec{a}$ and $\vec{x} + \vec{a}$. The locus defined by (5.5) consists of two arcs of circles passing through $\vec{a}$ and $-\vec{a}$, provided that the modulus of the right hand side of (5.5) is less that one. This is true when

$$|k_1 - k_2| < k_3 < k_1 + k_2 \, .$$

We shall consider the upper arc; the behavior of the waves on the lower arc follows by symmetry.

The phase of the wave generated by this interaction satisfies

$$\phi_3(\vec{x}, t) = \Phi_3(\vec{x}) - \omega_3 t \, ,$$

where Φ_3 solves the initial value problem

$$D(\omega_3, |\nabla \Phi_3|) = 0 \, ,$$
(5.6) $$\Phi_3 = k_1|\vec{x} - \vec{a}| - k_2|\vec{x} + \vec{a}| \, ,$$
$$\nabla \Phi_3 = \vec{k}_1 - \vec{k}_2 \quad \text{on } (5.5) \, .$$

The solution of (5.6) is

$\Phi_3 = k_3|\vec{x} - \vec{b}|$ on rays intersecting (5.5) with

$$k_1|\vec{x} - \vec{a}| > k_2|\vec{x} + \vec{a}| \, ,$$

$\Phi_3 = -k_3|\vec{x} - \vec{b}|$ on rays intersecting (5.5) with

$$k_1|\vec{x} - \vec{a}| > k_2|\vec{x} + \vec{a}| \, .$$

Here, $\vec{b}$ is the point on the upper half arc (5.5) such that

$$k_1|\vec{b} - \vec{a}| = k_2|\vec{b} + \vec{a}| \, .$$

The 3-wave is another cylindrical wave whose center is one on the circle (5.5). Part of the wave is an incoming cylindrical wave, which focuses at $\vec{x} = \vec{b}$. The other part is an outgoing cylindrical wave. The ray system is singular at $\vec{x} = \pm \vec{a}$ and $\vec{x} = \vec{b}$ where the waves focus and where one set of rays is tangent to the resonance surface (5.5).

The Ablowitz-Funk-Newell equations

We consider a dispersive wave motion governed by a first order system of PDEs with slowly varying coefficients,

$$(5.7) \qquad \mathcal{A}^{\alpha}(\epsilon x, \mathbf{U})\mathbf{U}_{x^{\alpha}} + \mathbf{g}(\epsilon x, \mathbf{U}) = 0,$$

where

$$x = (x^0, x^1, \cdots, x^m) \in \mathbb{R}^{m+1}.$$

We suppose that (5.7) has real coefficients and $\mathbf{U} \in \mathbb{R}^N$. We do not need to explicity distinguish the time variable, so we let $t = x^0$. We assume without loss of generality that $\mathbf{g}(\epsilon x, 0) = 0$.

We consider the interaction of three waves with phases $\phi_1(\epsilon x)$, $\phi_2(\epsilon x)$, $\phi_3(\epsilon x)$. We denote the frequency-wave number vectors by

$$\kappa_j(X) = D\phi_j(X), \qquad D := (\partial_{X^0}, \ldots, \partial_{X^m}).$$

Each frequency-wave number vector $\kappa_j = (\kappa_{j0}, \ldots, \kappa_{jm})$ satisfies the local, linearized dispersion relation of (5.7), namely

$$(5.8) \qquad D(\kappa; X) := \det\left[i\kappa_{\alpha} \mathcal{A}_0^{\alpha}(X) + \mathcal{G}(X) \right] = 0,$$

where

$$\mathcal{A}_0^{\alpha}(X) := \mathcal{A}(X, 0) \quad \text{and} \quad \mathcal{G}(X) := \nabla_{\mathbf{U}}\mathbf{g}(X, \mathbf{U})|_{\mathbf{U}=0}.$$

We suppose that there is a smooth function $\psi(X)$ such that

$$(5.9) \qquad \phi_1 + \phi_2 + \phi_3 = \frac{1}{2}\psi^2,$$

and $D\psi \neq 0$ in a neighborhood of $\psi = 0$. If necessary, we change the signs of all three phases, and make a constant phase shift. Equation (5.9) implies that the three waves come into resonance on the hypersurface $\psi = 0$, where

$$(5.10) \qquad \begin{aligned} \phi_1 + \phi_2 + \phi_3 &= 0, \\ \kappa_1 + \kappa_2 + \kappa_3 &= 0. \end{aligned}$$

We assume that (5.10) is the only resonance in a neighborhood of $\psi = 0$, i.e.

$$(5.11) \qquad D(2\kappa_j; X) \neq 0, \quad D(\kappa_p - \kappa_q; X) \neq 0 \quad \text{on} \quad \psi = 0.$$

To motivate the *ansatz* (5.12) below, we make an order of magnitude estimate of the strength of the nonlinear interaction. Let α be a dimensionless wave amplitude, λ a typical wavelength, and $L \gg \lambda$ a typical lengthscale for variations in the wavelength. Suppose two waves resonate near some hypersurface. The waves are in resonance for a distance of the order $d = (\lambda L)^{1/2}$ about the hypersurface. A typical quasi-linear term, $\mathbf{U} \cdot \nabla\mathbf{U}$, has magnitude α^2/λ. The amplitude of a wave generated by quadratically nonlinear interactions between the waves as they pass through resonance is of the order: (magnitude of nonlinearity)×(length of ray inside resonant region) or $\alpha^2 d/\lambda$. The wave generated will therefore have an amplitude of the same order of magnitude as the two original waves if $\alpha^2 d/\lambda = \alpha$, which imples that $\alpha = (\lambda/L)^{1/2}$. In the nondimensionalized equations (4.3), $\lambda = 1$ and $L = \epsilon^{-1}$, which implies that $\alpha = 1/d = \epsilon^{1/2}$.

We therefore look for an asymptotic solution of the form

$$(5.12) \qquad \mathbf{U}(\epsilon x; \epsilon) = \epsilon^{-1/2} \mathbf{U}_1 \left[\epsilon x, \epsilon^{1/2}\psi(\epsilon x), \epsilon^{-1}\phi_j(\epsilon x)\right]$$
$$+ \epsilon \mathbf{U}_2 \left[\epsilon x, \epsilon^{-1/2}\psi(\epsilon x), \epsilon^{-1}\phi_j(\epsilon x)\right] + 0(\epsilon^{3/2}).$$

Using (5.12) in (5.7), Taylor expanding, and equating coefficients of $\epsilon^{m/2}$ to zero implies that

$$(5.13) \qquad \sum_{j=1}^{3} \mathcal{H}_j \mathbf{U}_{1\theta_j} + \mathcal{G}\mathbf{U}_1 = 0,$$

$$(5.14) \qquad \sum_{j=1}^{3} \mathcal{H}_j \mathbf{U}_{2\theta_j} + \mathcal{G}\mathbf{U}_2 + I\mathbf{U}_{1\eta} +$$
$$\sum_{j=1}^{3} \kappa_{ja} D_{\mathbf{U}} \mathcal{A}^\alpha \cdot [\mathbf{U}_1]\mathbf{U}_{1\theta_j} + \frac{1}{2}\nabla_{\mathbf{U}}^2 \mathbf{g} \cdot [\mathbf{U}_1, \mathbf{U}_1] = 0,$$

where

$$(5.15) \qquad \mathcal{H}_j(X) = \phi_{jX^\alpha} \mathcal{A}_0^\alpha(X), \qquad I(X) = \psi_{X^\alpha} \mathcal{A}_0^\alpha(X).$$

A solution of (5.13) is

$$(5.16) \qquad \mathbf{U}_1 = \sum_{j=1}^{3} a_j(X, \eta) \exp(i\theta_j) \mathbf{R}_j(X) + c.c.,$$

where $\mathbf{R}_j$ is the null-vector of

$$(5.17) \qquad i\kappa_{ja} \mathcal{A}_0^\alpha(X) + \mathcal{G}(X).$$

In view of (5.9), we require that (5.14) has a solution for $\mathbf{U}_2$ which is a bounded function of θ_j on the surface

$$(5.18) \qquad \theta_1 + \theta_2 + \theta_3 = \frac{1}{2}\eta^2.$$

We do not require $\mathbf{U}_2$ to be bounded as $\eta \to \infty$. Therefore, our expansion is valid only near the resonance surface, when $\eta = 0(1)$ and $\psi = 0(\epsilon^{1/2})$. To derive the appropriate solubility condition, we consider the equation

$$(5.19) \quad \sum_{j=1}^{3} \mathcal{H}_j \mathbf{U}_{\theta_j} + \mathcal{G}\mathbf{U} = \sum_{j=1}^{3} f_j \exp(i\theta_j) + \sum_{p=1}^{3}\sum_{q=1}^{3} g_{pq} \exp\{-i(\theta_p + \theta_q)\}$$
$$+ \sum_{p=1}^{3}\sum_{q=1}^{3} h_{pq} \exp\{i(\theta_p - \theta_q)\} + c.c.$$

We average (5.19) with respect to θ_2 over the surface (5.18) keeping θ_1 and η fixed. The boundary terms vanish if U is bounded as $|\theta| \to \infty$. This gives

(5.20) $\qquad \mathcal{H}_1 < \mathbf{U} >_{\theta_1} + \mathcal{G} < \mathbf{U} >= \left[\mathbf{f}_1 + (\mathbf{g}_{23} + \mathbf{g}_{32}) \exp\{-i\eta^2/2\}\right] \exp i\theta_1 .$

In (5.20), $< \mathbf{U} > (\eta, \theta_1)$ is the average of $\mathbf{U}(\eta, \theta_1, \theta_2, \theta_3)$, defined by

$$< \mathbf{U} > (\eta, \theta) = \lim_{\Theta \to \infty} \frac{1}{2\Theta} \int_{-\Theta}^{+\Theta} \mathbf{U}(\eta, \theta, \xi, -\theta - \xi + \eta^2/2) d\xi .$$

Taking the inner product of (5.20) with $\exp(i\theta_1)\mathbf{L}_1$, where $\mathbf{L}_1$ is the left null vector of the matrix in (5.17) with $p = 1$, averaging the result with respect to θ_1, and integrating by parts on the left hand side, implies that

(5.21) $\qquad \mathbf{L}_1^* \cdot [\mathbf{f}_1 + (\mathbf{g}_{23} + \mathbf{g}_{32})e^{-i\eta^2/2}] = 0 .$

We use (5.16) in (5.14) and impose the solvability condition in (5.21), together with the solvability conditions obtained by cyclic permutations of $(1, 2, 3)$. This gives the following set of ODEs, in which the "slow" variables X occur as parameters,

(5.22)
$$a_{1\eta} = i\Lambda_1 e^{-i\eta^2/2} a_2^* a_3^* ,$$
$$a_{2\eta} = i\Lambda_2 e^{-i\eta^2/2} a_3^* a_1^* ,$$
$$a_{3\eta} = i\Lambda_3 e^{-i\eta^2/2} a_1^* a_2^* .$$

In (5.23),

(5.23) $\qquad \Lambda_j = (\mathbf{L}_j^* \cdot I \mathbf{R}_j)^{-1} \Gamma_j ,$

where Γ_j is the interaction coefficient for plane waves,

$$\Gamma_j = \kappa_{pa} \mathbf{L}_j^* \cdot \nabla_{\mathbf{U}} \mathcal{A}_0^\alpha \cdot [\mathbf{R}_q]\mathbf{R}_p + \kappa_{qa} \mathbf{L}_j^* \cdot \nabla_{\mathbf{U}} \mathcal{A}_0^\alpha \cdot [\mathbf{R}_p]\mathbf{R}_q + i\mathbf{L}_j^* \cdot \nabla_{\mathbf{U}}^2 \mathbf{g} \cdot [\mathbf{R}_p, \mathbf{R}_q] .$$

From (5.24),

$$\mathbf{L}_j^* \cdot I \mathbf{R}_j = \mathbf{L}_j^* \cdot \mathcal{A}_0^\alpha \mathbf{R}_j \partial_{X^\alpha} \psi = C_j^\alpha \partial_{X^\alpha} \psi ,$$

which is the derivative of ψ along the rays associated with ϕ_j. The interaction coefficient (5.23) is therefore infinite when the rays associated with ϕ_j are tangent to the resonance surface $\psi = 0$. In that case the resonance is sustained over longer distances than $d = (\lambda L)^{1/2}$, and the above expansion breaks down.

Equations (5.22) were derived be Ablowitz, Funk, and Newell (1973) for the slow passage of nonlinear oscillators through resonance, and by Grimshaw (1988) for dispersive waves. Grimshaw (1987) studies these equations for the "explosive" case, in which the Λ_j's have the same sign. In that problem the tendency of the interaction to cause blow up is opposed by the localization of the resonance.

If the nonuniformity is weaker than in (5.7), then it causes a small amount of detuning during a resonant interaction. Suppose that

(5.24) $\qquad \mathcal{A}^\alpha(\epsilon^2 x, \mathbf{U})\mathbf{U}_{x^\alpha} + \mathbf{g}(\epsilon^2 x, \mathbf{U}) = 0 ,$

where $g(Y,0) = 0$. We introduce slow variables $X = \epsilon x$, and expand $\mathbf{U}$ as

$$\mathbf{U} = \epsilon \mathbf{U}_1(X,\theta_j) + \epsilon^2 \mathbf{U}_2(X,\theta_j) + 0(\epsilon^3), \quad \text{as} \quad \epsilon \to 0 \quad \text{with} \quad X = 0(1).$$

Here, the phase variables are $\theta_j = \kappa_j \cdot x \ (j = 1,2,3)$ where

$$\kappa_1 + \kappa_2 + \kappa_3 = 0,$$

and

$$\mathbf{U}_1 = \sum_{j=1}^{3} a_j(X)\mathbf{R}_j \exp(i\theta_j) + c.c.$$

Taylor expanding the coefficients $A^\alpha(Y,\mathbf{U})$ and $g(Y,\mathbf{U})$ gives

$$A^\alpha(\epsilon^2 x, \mathbf{U}) = A_0^\alpha + \epsilon \nabla_Y A_0^\alpha \cdot [\mathbf{U}_1] + \epsilon \nabla_{\mathbf{U}} A_0^\alpha \cdot [X] + 0(\epsilon^2),$$

$$g(\epsilon^2 x, \mathbf{U}) = \epsilon g \mathbf{U}_1 + \epsilon^2 g \mathbf{U}_2 + \epsilon^2 \nabla_Y^2 \mathbf{U} g_0 \cdot [X, \mathbf{U}_1] + \frac{1}{2}\epsilon^2 \nabla_{\mathbf{U}}^2 g_0 \cdot [\mathbf{U}_1, \mathbf{U}_1],$$

where the subscript zero stands for evaluation at $Y = 0$ and $\mathbf{U} = 0$. Using these equations in (5.24) gives the following equations for the amplitudes,

(5.25) $$\qquad C_j^\alpha \partial_{X^\alpha} a_j + i\delta_{ja} X^\alpha a_j + i\Gamma_j a_p^* a_q^* = 0.$$

Here, the group velocities C_j and the interaction coefficients Γ_j are the same as in (3.25). The effect of the weak inhomogeneity is described by

$$\delta_{ja} = \mathbf{L}_j^* \cdot [-\kappa_{j\beta} \partial_{Y^\alpha} A_0^\beta + i\partial_{Y^\alpha} \nabla_{\mathbf{U}} g_0]\mathbf{R}_j.$$

Equation (5.25) can be transformed into the TWRI equations for a homogeneous medium when there are real matrices P_j such that

$$P_1 + P_2 + P_3 = 0, \quad P_j C_j = \delta_j.$$

The transformed amplitudes

$$\bar{a}_j = \exp\left[\frac{1}{2} iX \cdot P_j X\right] a_j(X)$$

satisfy (3.25). For example, Reiman (1979) uses this transformation to reduce the TWRI for a weakly inhomogeneous stationary medium in one space dimension, to the homogeneous TWRI equations.

REFERENCES

- M. J. Ablowitz, B. A. Funk and A. C. Newell, *Semi-resonant interactions and frequency dividers.* Stud. Appl. Math. **52**, 51 (1973).
- M. J. Ablowitz and H. Segur, *Solitons and the Inverse Scattering Transform,* SIAM, Philadelphia (1981).
- D. J. Benney, *A general theory for interactions between long and short waves,* Stud. Appl. Math. **56**, 81 (1977).
- D. J. Benney and P. G. Saffman. *Nonlinear interactions of random waves in a dispersive medium,* Proc. Roy. Soc. London **A 289**, 301 (1966).
- D. J. Benney and A. C. Newell, *Random wave closures,* Stud. Appl. Math. **48**, 29 (1969).
- A. D. D. Craik, *Wave Interactions and Fluid Flows,* Cambridge University Press, Cambridge (1985).
- R. K. Dodd, J. C. Eilbeck, J. D. Gibbon and H. C. Morris, *Solitons and Nonlinear Wave Equations,* Accademic Press, London (1982).
- R. Grimshaw, *Wave action and wave-mean flow interaction, with application to stratified shear flows,* Ann. Rev. Fluid Mech. **16**, 11 (1984).
- R. Grimshaw, *Triad resonance for weakly coupled slowly varying oscillators,* Stud. Appl. Math. **77**, 1 (1987).
- R. Grimshaw, *Resonant wave interactions in a stratified shear flow,* J. Fluid. Mech. **190**, 357 (1988).
- H. Hasimoto, Kagako **40**, 401 [In Japanese] (1970).
- J. K. Hunter and J. Scheurle, *Existence of perturbed solitary wave solutions to a model equation for water waves,* Physica D **32**, 253 (1988).
- K. Hasselmann, *Feynman diagrams and interaction rules of wave-wave scattering processes,* Rev. Geophysics. **4**, 1 (1966).
- T. Kakutani and H. Ono, *Weak nonlinear hydromagnetic waves in a cold collision-free plasma,* J. Phys. Soc. Japan **26**, 1305 (1969).
- D. J. Kaup, *The inverse scattering solution for the full three dimensional three-wave resonant interaction,* Physica D **1**, 45 (1980).
- D. J. Kaup, A. Reiman and A. Bers, *Space-time evolution of nonlinear three wave interactions, I: Interactions in a homogeneous medium,* Rev. Mod. Phys. **51**, 275 (1979).
- E. Knobloch and J. De Luca, *Amplitude Equations for travelling Wave Convection,* Nonlinearity **3**, 975 (1990).
- M. J. Lighthill, *Waves in Fluids,* Cambridge University Press, Cambridge (1978).
- C. H. McComas and F. P. Bretherton, *Resonant interaction of oceanic internal waves,* J. Geophysical Research **82** (9), 1397 (1977).
- O. M. Phillips, *The interaction trapping of internal gravity waves,* J. Fluid Mech. **34**, 407 (1968).
- A. Reiman, *Space-time evolution of nonlinear three wave interactions, II: Interactions in an inhomogenous medium,* Rev. Mod. Phy. **51**, 311 (1979).
- P. Ripa, *On the theory of nonlinear wave-wave interactions among geophysical waves,* J. Fluid Mech. **103**, 87 (1981).

- H. Segur, *Toward a new kinetic theory for resonant triads*, Contemporary Mathematics. **28**, 281 (1984).
- W. F. Simmons, *A variational method for weak resonant wave interactions*, Proc. Roy. Soc. A **309**, 551 (1969).
- J. Weiland and H. Wilhelmsson, *Coherent Nonlinear Interaction of Waves in Plasmas*, Pergamon, Oxford (1977).
- J. A. Zufiria, *Symmetry breaking in periodic and solitary waves on water of finite depth*, J. Fluid Mech. **184**, 183 (1987).

INTERACTION OF
HYPERBOLIC WAVES

Introduction

In Chapter 10, we derived asymptotic equations describing the resonant interaction
of weakly nonlinear dispersive waves. In this chapter, we analyse the resonant
interaction of hyperbolic waves. Hyperbolic waves are nondispersive. As a result
they resonate much more easily than dispersive waves. The nonlinear distortion
of the wave forms have to be determined simultaneously with the evolution of
modulations in the wave amplitudes. In contrast, the wave forms of dispersive
waves remain sinusoidal.

In Section 1, we derive the tri-resonance condition for scale invariant waves, of
which hyperbolic waves the main an example. In Section 2, we derive the resonant interaction equation for hyperbolic waves in one space dimension. Section 3
and 4 describe the extension to oblique plane wave interactions in several space
dimensions. In Section 5, we apply the general theory to gas dynamics. The resulting equations describe the resonant reflection of sound waves off periodic entropy
perturbation. Finally, in Section 6, we study the interaction of longitudinal and
transverse waves in nonlinear elasticity.

The asymptotic equations for resonantly interacting hyperbolic waves were
derived by Majda and Rosales (1984), for one space dimension, and by Hunter,
Majda and Rosales (1986), for several space dimensions. They were obtained independently in the Soviet Union by Maslov (1983$_a$), (1983$_b$), (1986). The interaction
of hyperbolic surface waves leads to equations of the same form as the ones derived here for hyperbolic waves in an unbounded medium. We do not discuss such
problems in this book (see Artola and Majda (1989) for the interaction of waves
on a supersonic vertex sheet). Passage through resonance of interacting hyperbolic
waves is analysed by Almgren (1990) and by Joly, Metevier and Rauch (1990).

1 The tri-resonance condition for scale invariant wave motions

The dispersion relation of a hyperbolic PDE without lower order terms,

(1.1) $$A^0 \mathbf{U}_t + A^j \mathbf{U}_{x^j} = 0,$$

is $D(\omega, \vec{k}) = 0$, where

$$D(\omega, \vec{k}) = \det(k_j A^j - \omega A^0).$$

We denote a typical root for ω by

(1.2) $$\omega = W(\vec{k}).$$

Since $D(\alpha\omega, \alpha\vec{k}) = \alpha^N D(\omega, \vec{k})$, W is a homogeneous function of degree one in $\vec{k}$ meaning that

(1.3) $$W(\alpha\vec{k}) = \alpha W(\vec{k}) \quad \alpha > 0.$$

In this section we describe the tri-resonance condition for quadratically nonlinear interaction of waves whose dispersion relations have the property in (1.3).

Before doing this, we comment further on how (1.3) arises. It is a consequence of the invariance of (1.1) under the scaling transformations

$$\bar{x}^j = \alpha x^j, \quad \bar{t} = \alpha t.$$

The dispersion relation of any wave motion with this scale invariance property satisfies (1.3). Waves modelled by hyperbolic PDEs without source terms are the most common type of scale invariant waves, but they are not the only ones. Surface waves on a half-space, that are governed by equations like (1.1) inside the half-space, are scale invariant because the geometry does not define a length scale. Rayleigh waves in elasticity are one example. There are also some exceptional waves, like Kelvin waves in a rotating fluid, whose dispersion relations satisfy (1.3) even though the equation of motion are not scale invariant.

The dimensioned parameters characteristic of a scale invariant wave motion can include velocities, but no lengths, times, or accelerations. Thus, gravity water waves on deep water are characterised by the gravitational acceleration g. The dispersion relation is $\omega^2 = gk$, which does not satisfy (1.3).

The dispersion relation of hyperbolic equations with lower order terms need not satisfy (1.3). For example, the *Klein-Gordon equation*,

$$\begin{pmatrix} u \\ v \end{pmatrix}_t + \begin{pmatrix} c_0 & 0 \\ 0 & -c_0 \end{pmatrix} \begin{pmatrix} u \\ v \end{pmatrix}_x + \begin{pmatrix} 0 & \omega_0 \\ -\omega_0 & 0 \end{pmatrix} \begin{pmatrix} u \\ v \end{pmatrix} = 0,$$

has the dispersion relation

$$\omega^2 = c_0^2 k^2 + \omega_0^2.$$

Waves with frequencies of the order ω_0 are strongly dispersive. When $\omega \gg \omega_0$, the dispersion relation is approximately $\omega^2 \simeq c_0^2 k^2$. The waves are asymptotically scale invariant as $\omega \to \infty$.

More generally, waves whose frequency and wavelength are much greater than or much smaller than any characteristic frequencies and length scales of the wave motion can satisfy (1.3) asymptotically even if the full problem is not scale invariant. Examples include shallow water waves, whose wavelength is much greater than the depth of the water, and ion acoustic waves in a plasma with frequency much less than the ion plasma frequency.

Equation (1.3) states that the phase speed ω/k is independent of frequency for waves propagating in a given direction. This implies that the wave motion is nondispersive. For anisotropic wave motions, it is not true that the phase velocity $k^{-2}\vec{k}W(\vec{k})$ and the group velocity $\nabla_{\vec{k}}W$ are equal.

There are nondispersive wave motions which are not scale invariant. Convected waves are one example. They have dispersion relations of the form

$$\omega = \vec{u} \cdot \vec{k} + \omega_0 .$$

Such waves are completly nondispersive since the dispersion tensor $\nabla_{\vec{k}}^2\omega$ is identically zero. They are not scale invariant if $\omega_0 \neq 0$.

Now we look at the consequences of (1.3) for resonant wave interactions. First, we consider a single wave with frequency ω and wave number $\vec{k}$, which is initially sinusoidal. Equation (1.3) implies that second harmonic resonance occurs. We assume throughout this section that the appropriate interaction coefficients are nonzero. The resonant self-interaction of the wave therefore produces a harmonic with frequency 2ω and wave number $2\vec{k}$. This second harmonic resonates with the fundamental to produce third harmonics, and so on. This process is dearly described by Rudenleo and Soluyan (1977).

To make a rough estimate of the growth of the harmonics, suppose that the dimensionless amplitude of the fundamental is of the order $\alpha \ll 1$. The characteristic time scale of the nonlinearity is ω^{-1}, since for a scale invariant problem, this time scale must come from the wave itself.

Quadratically nonlinear interactions therefore lead to a growth rate for the second harmonic of the order $\alpha^2\omega$. After time t, the amplitude of the second harmonic is of the order $\alpha^2\omega t$. Similarly, the amplitude of the nth harmonic is of the order

$$\alpha^n (\omega t)^{n-1} .$$

For $\omega t \ll \alpha^{-1}$, the amplitude of the higher harmonics is much smaller than that of the fundamental. Their amplitudes become comparable after times of the order $T_N = \alpha^{-1}\omega^{-1}$.

The generation of higher harmonics implies a progressive distortion of the waveform. The waveform is approximately sinusoidal only for $\omega t \ll \alpha^{-1}$.

Typically, the waveform steepens and the slope becomes infinite somewhere at a time t_* of the order $\alpha^{-1}\omega^{-1}$. If the wave is modelled by a hyperbolic system of conservation laws, the solution may be continued beyond t_* by introducing a shock wave. Otherwise, the basic assumptions of the model must be re-examined. For example, in the case of the shallow water wave equations, the small slope assumption used to derive them breaks down at t_*, and additional dispersive effects must then be included.

As an example of the above rough estimates, we consider sound waves. An appropriate measure of the amplitude of a sound wave is the acoustic Mach number, $M_a = c_0^{-1}u_{\max}$, where c_0 is the undisturbed sound speed and $u_{\max}$ is the maximum value of the velocity perturbations carried by the wave. The cumulative effect of nonlinearity on a sound wave with wave number k is important over propagation distances $L_N = O(M_a k)^{-1}$. More precisely, we can take L_N to be the shock formation distance for the wave. If the maximum initial slope of the velocity profile is kM_a, then $L_N = (M_akG)^{-1}$, where G is an order one parameter called the parameter of nonlinearity of the fluid (see Section (11.5)). For example, suppose that the sound wave is a 20 kHz ultrasonic wave in water (at standard conditions) with strength $M_a = 10^{-4}$. This corresponds to a wavelength of about 8 cm and maximum overpressures of two atmospheres. Then the discontinuity distance is $L_N \simeq 35$ m.

A simple problem that illustrates the nonlinear production of harmonics is the inviscid Burgers equation with sinusoidal initial data,

(1.4)
$$u_t + c_0 u u_x = 0,$$
$$u(x,0) = \alpha \sin(kx).$$

The Fubini solution of (1.4) is

(1.5)
$$u(x,t) = \sum_{n=1}^{\infty} \frac{2(-1)^{n+1}}{nkt} J_n(\alpha n \omega t) \sin(nkx)$$

where $\omega = c_0 k$ and J_n is the nth order Bessel function. The series in (1.5) converges to the solution of (1.4) for $t < t_* = (\alpha \omega)^{-1}$, at which time shocks form. For $t \ll (n\alpha\omega)^{-1}$, the amplitude of the nth harmonic is approximately $K_n \alpha^n (\omega t)^{n-1}$, where

$$K_n = \frac{1}{n!} \left[-\frac{n}{2} \right]^{n-1},$$

in agreement with the rough estimate given above.

Summarising, the first consequence of (1.3) is waveform distortion and the possible appearance of shocks. This behaviour is different from that of weakly nonlinear dispersive waves. In the absence of second harmonic resonance, the waveform of dispersive waves is sinusoidal; nonlinearity causes the wave amplitude and the phase to vary slowly.

Next, consider the interaction of two periodic waves, with frequencies $\{\omega_1, \omega_2\}$ and wave numbers $\{\vec{k}_1, \vec{k}_2\}$. The self-interactions couple the fundamental and the higher harmonics of the waves. A pair of these harmonics resonate if

$$D(m_1\omega_1 + m_2\omega_2, m_1\vec{k}_1 + m_2\vec{k}_2) = 0$$

for some integers m_1, m_2. The wave produced by this resonance has frequency $m_1\omega_1 + m_2\omega_2$. Thus, three periodic waves resonate if their frequencies and wave numbers are rationally dependent, meaning that

(1.6)
$$m_1\omega_1 + m_2\omega_2 + m_3\omega_3 = 0,$$
$$m_1\vec{k}_1 + m_2\vec{k}_2 + m_3\vec{k}_3 = 0,$$
$$D(\omega_j, \vec{k}_j) = 0,$$

for some nonzero integers m_1, m_2, and m_3.

More generally, consider the interaction of almost periodic waves, for which the wave amplitudes $a_j[\vec{x}, t, \epsilon^{-1}(\vec{k}_j \cdot \vec{x} - \omega_j t)]$ have the expansion

$$(1.7) \qquad a_j(\vec{x}, t, \theta) = \sum_{n=-\infty}^{\infty} a_{nj}(\vec{x}, t) \exp(i\nu_{nj}\theta).$$

The condition for quadratically nonlinear resonant interactions to occur between such waves is that

$$\nu_{n_1 1}\omega_1 + \nu_{n_2 2}\omega_2 + \nu_{n_3 3}\omega_3 = 0,$$
$$\nu_{n_1 1}\vec{k}_1 + \nu_{n_2 2}\vec{k}_2 + \nu_{n_3 3}\vec{k}_3 = 0,$$

for some integers n_1, n_2, n_3. In general, without specific information about the spectrum of the waveforms, resonance is possible whenever

$$\mu_1\omega_1 + \mu_2\omega_2 + \mu_3\omega_3 = 0,$$
$$(1.8) \qquad \mu_1\vec{k}_1 + \mu_2\vec{k}_2 + \mu_3\vec{k}_3 = 0,$$
$$D(\omega_j, \vec{k}_j) = 0,$$

for some nonzero real numbers μ_1, μ_2, and μ_3. Geometrically, (1.8) states that the frequency-wave number vectors of the three waves are coplanar. This interpretation is explored in greater detail in Section 11.3. Whether or not resonant interactions in fact occur when (1.8) is satisfied depends on the spectral content of the waveforms and on the nonvanishing of the appropriate interaction coefficients.

Clearly, (1.8) is much less restrictive than the corresponding condition for dispersive waves, which has $\mu_j = \pm 1$. One consequence is that cubically nonlinear (four-wave) interactions are of lesser importance for scale invariant waves, because three wave interactions nearly always occur. Shocks typically cause a wave to decay long before four-wave interactions can have any significant effect.

The generation of higher harmonics means that Fourier decomposing the waveform is not the simplest approach to adopt. Moreover, once shocks appear, the Fourier representation gives incorrect answers (cf. the Fubini solution (1.5)); however, the resonance condition (1.6) remains valid for periodic solutions which contain shocks. Instead, we shall obtain asymptotic equations for the wave amplitudes $a_j(\vec{x}, t, \theta)$ in (1.7) directly. The waveform is described by the way in which a_j depends on the phase variable

$$\theta = \epsilon^{-1}[\vec{k}_j \cdot \vec{x} - \omega_j t].$$

The θ-dependence of $a_j(\vec{x}, t, \theta)$ changes as the "slow" space and time variables $\vec{x}$ and t change, and this describes the phenomenon of waveform distortion.

The analysis of single hyperbolic waves does not depend critically on the waveform of the waves. Oscillatory waves, pulses, shock-type waves etc., are all described asymptotically by the inviscid Burgers equation. However, for interaction problems, the type of the waveform is important. Interactions between localized waves—like pulses or shocks—are weak because if the waves have different group velocities the interactions only take place along a small part of each ray. Thus, to leading order in the wave amplitudes, the waves decouple, and single wave theory suffices. We shall therefore consider oscillatory wave interactions.

We end this section by applying the tri-resonance condition to hyperbolic PDEs in one space dimension. Suppose that the linearized equations are

$$(1.9) \qquad \mathbf{U}_t + \mathcal{A}\mathbf{U}_x = 0 \,,$$

where $\mathbf{U}(x,t) \in \mathbb{R}^N$, and $\mathcal{A}$ is an $N \times N$ matrix with eigenvalues

$$\lambda_1 < \lambda_2 < \ldots < \lambda_N \,.$$

The linearized dispersion relation of the jth wave mode is

$$\omega = \lambda_j k \,.$$

Consider three waves with frequencies $\{\omega_j, \omega_p, \omega_q\}$ and wave numbers $\{k_j, k_p, k_q\}$, then

$$(1.10) \qquad \begin{aligned} \nu_{pq}\omega_j + \nu_{qj}\omega_p + \nu_{jp}\omega_q &= 0 \,, \\ \nu_{pq}k_j + \nu_{qj}k_p + \nu_{jp}k_q &= 0 \,, \\ \omega_j = \lambda_j k_j \,, \quad \omega_p = \lambda_p k_p \,, \quad \omega_q &= \lambda_q k_q \,, \end{aligned}$$

where

$$\nu_{pq} = k_p k_q (\lambda_p - \lambda_q) \,.$$

Equation (1.10) may also be written in the form

$$(1.11) \qquad \begin{aligned} \omega_j &= \mu_{jqp}\omega_p + \mu_{jpq}\omega_q \,, \\ k_j &= \mu_{jqp}k_p + \mu_{jpq}k_q \,, \end{aligned}$$

where

$$\mu_{jpq} = -\frac{\nu_{jp}}{\nu_{qj}} \,.$$

Thus, any pair of characteristic fields resonates with all other characteristic fields. It follows from (1.10) and (1.6) that the condition for two periodic waves on the pth and q^{th} characteristics to produce a wave on the jth characteristic is

$$(1.12) \qquad \frac{k_p(\lambda_j - \lambda_p)}{k_q(\lambda_j - \lambda_q)} \quad \text{is rational.}$$

For example, in one-dimensional, nonisentropic gas dynamics, the wave velocities in a stationary gas with sound speed c_0 are: $\lambda_1 = -c_0$ (left moving sound wave); $\lambda_2 = 0$ (entropy wave); $\lambda_3 = c_0$ (right moving sound wave). The condition for a periodic sound wave with wave number k_3 to reflect resonantly off a periodic entropy wave with wave number k_2 is that

$$(1.13) \qquad \frac{k_2}{2k_3} \quad \text{is rational.}$$

The strongest interaction is when the fundamental harmonics of both waves resonate directly. This occurs when $k_2 = 2k_3$.

A second example is one dimensional elasticity. In that case, the wave velocities
are

$$\lambda_1 = -c_p, \quad \lambda_2 = -c_s, \quad \lambda_3 = c_s, \quad \lambda_4 = c_p,$$

where c_p is the sound speed and c_s is the shear speed. From (1.12), two shear
waves with wave numbers k_2, k_3 produce a sound wave with velocity c_p if

(1.14)
$$\frac{k_2(c_p + c_s)}{k_3(c_p - c_s)} \quad \text{is rational.}$$

The strongest interaction occurs when the quantity in (1.13) is one. The
resulting wave pattern is spatially periodic only if $(c_p + c_s)/(c_p - c_s)$ is rational;
otherwise, k_2 and k_3 are incommensurable. Similarly, the shear waves produce a
sound wave with velocity $-c_p$ if

(1.15)
$$\frac{k_2(c_p - c_s)}{k_3(c_p + c_s)} \quad \text{is rational.}$$

Thus, dividing (1.14) by (1.15), sound waves propagating in both directions
are only generated if the wave speeds satisfy

$$\left(\frac{c_p - c_s}{c_p + c_s} \right)^2 \quad \text{is rational.}$$

These delicate rationality conditions may suggest that there are small divisor problems in the asymptotic expansion. However, this is not the case. Small divisor
problems arise when trying to construct expansions that are uniformly valid as
$t \to +\infty$. The expansion we shall construct is valid only for times of the order
wave-period/amplitude.

What is true is that if $k_2/2k_3$ in (1.13) is irrational, say, the range of amplitudes
for which the asymptotic theory gives a good approximation will depend on how
well $k_2/2k_3$ is approximated by rational numbers (cf. the discussion of detuning
in Section (10.2)).

2 Resonant interactions of hyperbolic waves in one space dimension

The purpose of this section is to derive asymptotic equations that describe the
resonant interaction of small amplitude, high-frequency hyperbolic waves. We
suppose that the wave motion is modelled by a system of conservation laws,

(2.1)
$$\mathbf{U}_t + \mathbf{f}(\mathbf{U})_x + \mathbf{g}(\mathbf{U}) = 0.$$

In (2.1), $\mathbf{U}(x,t) \in \mathbb{R}^N$ is the vector of conserved quantities, $\mathbf{f} \colon \mathbb{R}^N \longrightarrow \mathbb{R}^N$ is the
flux vector, and $\mathbf{g} \colon \mathbb{R}^N \longrightarrow \mathbb{R}^N$ is a source term. We assume that

$$\mathbf{g}(\mathbf{U}_0) = 0,$$

so that $\mathbf{U} = \mathbf{U}_0$ is a constant solution of (2.1), and that (2.1) is strictly hyperbolic at $\mathbf{U}_0$. We denote the eigenvalues of $\mathcal{A} = \nabla \mathbf{f}(\mathbf{U}_0)$ by

$$\lambda_1 < \lambda_2 < \ldots < \lambda_N .$$

The associated left and right eigenvectors denoted by $\mathbf{L}_j$ and $\mathbf{R}_j$, satisfy

$$\mathbf{L}_j \cdot (\mathcal{A} - \lambda_j I) = 0, \quad (\mathcal{A} - \lambda_j I)\mathbf{R}_j = 0 .$$

We normalize $\mathbf{L}_j$ so that

$$\mathbf{L}_p \cdot \mathbf{R}_q = \delta_{pq} .$$

We use the method of multiple scales to derive the asymptotic solution.
    We look for a solution to (2.1) of the form

(2.2) $$\mathbf{U}(x,t;\epsilon) = \mathbf{U}_0 + \epsilon \mathbf{u}(x,t,\epsilon^{-1}x,\epsilon^{-1}t;\epsilon) .$$

$$\epsilon \to 0 \quad \text{with} \quad (x,t) = O(1) .$$

This form corresponds to a high-frequency, "geometrical-optics" expansion. In rescaled variables, $\bar{x} = \epsilon^{-1}x, \bar{t} = \epsilon^{-1}t$, (2.1) becomes

$$\mathbf{U}_{\bar{t}} + \mathbf{f}(\mathbf{U})_{\bar{x}} + \epsilon \mathbf{g}(\mathbf{U}) = 0 .$$

The equivalent expansion to (2.2) is

$$\mathbf{U}(\bar{x},\bar{t};\epsilon) = \mathbf{U}_0 + \epsilon \mathbf{u}(\epsilon\bar{x}, \epsilon\bar{t}, \bar{x}, \bar{t}; \epsilon), \quad \epsilon \to 0+ \quad \text{with} \quad (\bar{x}, \bar{t}) = O(\epsilon^{-1}) .$$

This corresponds to a long-time, far-field expansion.
    Substituting (2.2) into (2.1) implies that $\mathbf{u}(x,t,\xi,\tau;\epsilon)$ satisfies

(2.3) $$\mathbf{u}_\tau + \mathbf{f}(\mathbf{U}_0 + \epsilon \mathbf{u})_\xi + \epsilon[\mathbf{u}_t + \mathbf{f}(\mathbf{U}_0 + \epsilon \mathbf{u})_x + \mathbf{g}(\mathbf{U}_0 + \epsilon \mathbf{u})] = 0 ,$$

when $\xi = \epsilon^{-1}x$, and $\tau = \epsilon^{-1}t$. We require $\mathbf{u}(x,t,\xi,\tau)$ to satisfy (2.3) for all values of (x,t,ξ,τ). Thus, when we solve (2.3) we treat the *fast* variables (ξ,τ) as independent of the *slow* variables (x,t), and evaluate (ξ,τ) at $(\epsilon^{-1}x, \epsilon^{-1}t)$ in the final solution. This has the apparent disadvantage of replacing a problem in two independent variables, (2.1), by a problem in four independent variables, (2.3). The advantage is that (2.3), unlike (2.1), can be solved by a simple power series expansion in ϵ. A peculiarity of the method of multiple scales is that although we find $\mathbf{u}$ for all values of (x,t,ξ,τ), it is only the values on the subdomain $(\xi,\tau) = (\epsilon^{-1}x, \epsilon^{-1}t)$ which affect the final answer.
    To solve (2.3) we expand $\mathbf{u}$ in a power series,

(2.4) $$\mathbf{u} = \mathbf{U}_1(x,t,\xi,\tau) + \epsilon \mathbf{U}_2(x,t,\xi,\tau) + O(\epsilon^2) \quad \text{as} \quad \epsilon \to 0+ ,$$

and we Taylor expand the flux and source terms,

(2.5)
$$f(\mathbf{U}_0 + \epsilon \mathbf{u}) = \mathbf{f}(\mathbf{U}_0) + \epsilon \mathcal{A}\mathbf{u} + \frac{1}{2}\epsilon^2 \mathcal{B} \cdot \mathbf{u}\mathbf{u} + O(\epsilon^3) ,$$

$$\mathbf{g}(\mathbf{U}_0 + \epsilon \mathbf{u}) = \epsilon \mathcal{G}\mathbf{u} + O(\epsilon^2) ,$$

In (2.5), $\mathcal{B} = \nabla^2 \mathbf{f}(\mathbf{U}_0)$ and $\mathcal{G} = \nabla \mathbf{g}(\mathbf{U}_0)$. Using (2.4) and (2.5) in (2.3) and equating coefficients of ϵ and ϵ^2 to zero in the result implies that

$$(2.6) \qquad\qquad \mathbf{U}_{1\tau} + \mathcal{A}\mathbf{U}_{1\xi} = 0,$$

$$(2.7) \qquad\qquad \mathbf{U}_{2\tau} + \mathcal{A}\mathbf{U}_{2\xi} + \mathbf{F}(x,t,\xi,\tau) = 0,$$

where

$$(2.8) \qquad\qquad \mathbf{F} = \mathbf{U}_{1t} + \mathcal{A}\mathbf{U}_{1x} + \frac{1}{2}\partial_\xi \mathcal{B}\cdot\mathbf{U}_1\mathbf{U}_1 + \mathcal{G}\mathbf{U}_1.$$

A solution of (2.6) is

$$(2.9) \qquad\qquad \mathbf{U}_1 = \sum_{j=1}^{M} a_j(x,t,k_j\xi - \omega_j\tau)\mathbf{R}_j,$$

where the frequencies ω_j and the wave number k_j satisfy

$$(2.10) \qquad\qquad \omega_j = \lambda_j k_j, \quad j = 1,\dots,M.$$

We suppose that the wave amplitudes $a_j(x,t,\theta)$ and their derivatives with respect to x,t, and θ are periodic or almost periodic functions of θ. For simplicity, we also assume that the wave amplitudes have zero mean with respect to θ, meaning that

$$(2.11) \qquad\qquad \lim_{T\to\infty} T^{-1}\int_0^T a_j(x,t,\theta)d\theta = 0.$$

It is straightforward to allow for nonzero means and the results are not essentially different. The solution (2.9) is a sum of rapidly oscillating waves. There is one wave for each characteristic field of (2.1). In future, we will not show explicitly any dependence on (x,t) unless it is necessary to do so.

　　To obtain equations for the wave amplitudes, we eliminate secular terms from $\mathbf{U}_2$ in the standard way. We shall require $\mathbf{U}_2$ to be sublinear in τ, meaning that

$$(2.12) \qquad\qquad \lim_{\tau\to+\infty} \tau^{-1}\mathbf{U}_2(\xi,\tau) = 0.$$

This condition ensures that $\epsilon^2\mathbf{U}_2 \ll \epsilon\mathbf{U}_1$ when $\tau = O(\epsilon^{-1})$. To see what (2.12) implies about the wave amplitudes, we compute $\mathbf{U}_2$ explicitly by using the method of characteristics. We expand $\mathbf{U}_2$ in the basis of eigenvectors of $\mathcal{A}$,

$$(2.13) \qquad\qquad \mathbf{U}_2 = \sum_{j=1}^{M} b_j(k_j\xi - \omega_j\tau,\tau)\mathbf{R}_j,$$

$$b_j = \mathbf{L}_j \cdot \mathbf{U}_2.$$

Taking the scalar product of (2.7) with $\mathbf{L}_j$, and using (2.13) implies that b_j satisfies

$$(2.14) \qquad b_{j\tau}(\theta,\tau) + \mathbf{L}_j \cdot \mathbf{F}(k_j^{-1}\theta + \lambda_j\tau,\tau) = 0.$$

Using (2.8) in (2.14) and integrating the result, we find that

$$(2.15)\; b_j(\theta,\tau) = -\tau[a_{jt}(\theta) + \lambda_j a_{jx}(\theta) + k_j \mathbf{L}_j \cdot \mathcal{B} \cdot \mathbf{R}_j \mathbf{R}_j a_j(\theta) a_{j\theta}(\theta) + \mathbf{L}_j \cdot \mathcal{G} \mathbf{R}_j a_j(\theta)]$$

$$-\overset{(j)}{\underset{p}{\sum}}\,\overset{(j)}{\underset{q}{\sum}} k_q \mathbf{L}_j \cdot \mathcal{B} \cdot \mathbf{R}_p \mathbf{R}_q \int_0^\tau a_p[k_j^{-1}(k_p\theta + \nu_{jp}\sigma)]a_q'[k_j^{-1}(k_q\theta + \nu_{jq}\sigma)]d\sigma$$

$$-\overset{(j)}{\underset{p}{\sum}} \mathbf{L}_j \cdot \mathcal{G} \mathbf{R}_p \int_0^\tau a_p[k_j^{-1}(k_p\theta + \nu_{jp}\sigma)]d\sigma + \beta_j(\theta).$$

Here, $\beta_j(\theta)$ is an arbitrary function of integration,

$$a_q'(x,t,\theta) = \partial_\theta a_q(x,t,\theta),$$

ν_{jp} is defined below (1.10), and

$$\overset{(j)}{\underset{p}{\sum}}$$ stands for the sum over p with $1 \le p \le M$ and $p \ne j$,

The sublinearity condition (2.12) implies that

$$(2.16) \qquad \lim_{\tau\to\infty} \tau^{-1} b_j(\theta,\tau) = 0.$$

Using (2.15) in (2.16) leads to a set of integro-differential equations for $\{a_1,\cdots,a_M\}$:

$$(2.17) \qquad a_{jt}(\theta) + \lambda_j a_{jx}(\theta) + M_j a_j(\theta) a_{j\theta}(\theta) + G_j a_j(\theta)$$

$$+ \overset{(j)}{\underset{p\ne q}{\sum}} \tilde{\Gamma}_{jpq} \lim_{T\to\infty} \frac{1}{T}\int_0^T a_p(\mu_{pqj}\theta + \mu_{pjq}\sigma)a_{q\sigma}(\sigma)d\sigma = 0.$$

In (2.17),

$$\overset{(j)}{\underset{p\ne q}{\sum}}$$ stands for the sum over all $1 \le p,\ q \le M$ with
$p \ne q,\ p \ne j$ and $q \ne j$.
The coefficients are

$$(2.18) \qquad \begin{aligned} M_j &= k_j \mathbf{L}_j \cdot \mathcal{B} \cdot \mathbf{R}_j \mathbf{R}_j, \\ \tilde{\Gamma}_{jpq} &= k_q \mathbf{L}_j \cdot \mathcal{B} \cdot \mathbf{R}_p \mathbf{R}_q, \\ G_j &= \mathbf{L}_j \cdot \mathcal{G} \mathbf{R}_j, \\ \mu_{pqj} &= \frac{k_p(\lambda_p - \lambda_q)}{k_j(\lambda_j - \lambda_q)}. \end{aligned}$$

Using integration by parts to simplify the interaction terms, (2.17) can be written

$$a_{jt}(\theta) + \lambda_j a_{jx}(\theta) + M_j a_j(\theta) a_{j\theta}(\theta) + G_j a_j(\theta)$$

$$+ \sum_{p<q}^{(j)} \overline{\Gamma}_{jpq} \lim_{T\to\infty} \frac{1}{T} \int_0^T a_p[\mu_{pqj}\theta + \mu_{pjq}\sigma]a_{q\sigma}[\sigma]d\sigma = 0,$$

where

$$\sum_{p<q}^{(j)}$$ stands for the sum over all $1 \le p < q \le M$ with $p \ne j$ and $q \ne j$, and

$$\overline{\Gamma}_{jpq} = \mu_{qpj}k_j\mathbf{L}_j \cdot \mathcal{B} \cdot \mathbf{R}_p\mathbf{R}_q.$$

The conservation form of (2.17) is

$$(2.19)\, \partial_t a_j(\theta) + \partial_x[\lambda_j a_j(\theta)] + \partial_\theta \left\{ \frac{1}{2} M_j a_j^2(\theta) \right.$$

$$\left. + \sum_{p<q}^{(j)} \Gamma_{jpq} \lim_{T\to\infty} \left[\frac{1}{T} \int_0^T a_p(\mu_{pqj}\theta + \mu_{pjq}\sigma)a_q(\sigma)d\sigma \right] \right\} + G_j a_j(\theta) = 0.$$

In (2.19), Γ_{jpq} is the symmetric interaction coefficient,

$$\Gamma_{jpq} = k_j\mathbf{L}_j \cdot \mathcal{B} \cdot \mathbf{R}_p\mathbf{R}_q.$$

Equations (2.19) remain valid in the weak sense after shocks form (Chelesky and Rosales, 1986). If a shock in the jth wave has position $\theta = h(x,t)$, the jump condition is that

$$h_t + \lambda_j h_x = \frac{1}{2} M_j(a_{j+} + a_{j-}),$$

where

$$a_{j+} = \lim_{\theta\downarrow h} a_j(\theta), \qquad a_{j-} = \lim_{\theta\uparrow h} a_j(\theta).$$

The shock is admissible if

$$M_j\{a_{j-} - a_{j+}\} \ge 0.$$

Taking the mean of (2.19) with respect to θ and using the boundedness of the a_j's shows that

$$\overline{a}_{jt} + \lambda_j \overline{a}_{jx} + G_j \overline{a}_j = 0,$$

where

$$\overline{a} = \lim_{T\to\infty} \frac{1}{T} \int_0^T a_j(\theta)d\theta.$$

Thus, equations (2.19) are consistent with the zero-mean assumption in (2.11).

Equation (2.17) is a system of inviscid Burgers equations coupled together by integral averages. The coefficient M_j is an interaction coefficient for quadratically nonlinear self-interactions of the jth wave. It follows from (2.18)$_1$ that

$$M_j = \nabla_{\mathbf{U}}\lambda_j(\mathbf{U}_0) \cdot \mathbf{R}_j(\mathbf{U}_0),$$

where $\lambda_j(\mathbf{U})$ is the jth eigenvalue of $\nabla \mathbf{f}(\mathbf{U})$, and $\mathbf{R}_j(\mathbf{U})$ is the associated eigenvector. The jth characteristic is said to be *genuinely nonlinear* (Lax, 1973) if

$$\nabla \lambda_j(\mathbf{U}) \cdot \mathbf{R}_j(\mathbf{U}) \neq 0 \qquad \text{for all} \qquad \mathbf{U} \in \mathbb{R}^N,$$

and *linearly degenerate* if

$$\nabla \lambda_j(\mathbf{U}) \cdot \mathbf{R}_j(\mathbf{U}) = 0 \qquad \text{for all} \qquad \mathbf{U} \in \mathbb{R}^N.$$

Thus, $M_j \neq 0$ if the jth characteristic is genuinely nonlinear, and $M_j = 0$ if the jth characteristic is linearly degenerate.

The coefficient Γ_{jpq} describes the strength of quadratically nonlinear forcing on the jth wave by interactions between waves on the pth and q^{th} characteristics. If the jth characteristic field of (1.1) has a Riemann invariant then $\Gamma_{jpq} = 0$ for all pairs (p, q) distinct from j. The equation for a_j then decouples from the equations for the other wave amplitudes. An example of this occurs in gas dynamics where entropy is a Riemann invariant (see Section (11.5)).

If (2.1) is a 2×2 system, then no correlations appear in (2.17). The waves on each characteristic satisfy decoupled inviscid Burgers equations. The 2×2 case (e.g. isentropic gas dynamics) case is therefore essentially simpler than the $N \times N$ case with $N \geq 3$ (e.g. nonisentropic gas dynamics).

When deriving the resonant interaction equations for specific systems, the algebra is often simplified by taking the equations in nonconservative form. Suppose the equations are

$$(2.20) \qquad A^0(\mathbf{U})\mathbf{U}_t + A^1(\mathbf{U})\mathbf{U}_x + \mathbf{g}(\mathbf{U}) = 0, \qquad \mathbf{g}(\mathbf{U}_0) = 0.$$

We assume that t is a noncharacteristic direction of (2.20). The characteristic velocities λ_j and the associated eigenvectors $\mathbf{R}_j$ and $\mathbf{L}_j$ satisfy,

$$[A^1(\mathbf{U}_0) - \lambda_j A^0(\mathbf{U}_0)]\mathbf{R}_j = 0,$$
$$\mathbf{L}_j \cdot [A^1(\mathbf{U}_0) - \lambda_j A^0(\mathbf{U}_0)] = 0,$$
$$\mathbf{L}_j \cdot A^0(\mathbf{U}_0)\mathbf{R}_j = 1.$$

Using the expansion (2.4) and (2.9) in (2.20) leads to (2.17) with

$$M_j = k_j \mathbf{L}_j \cdot \nabla A^1(\mathbf{U}_0) \cdot \mathbf{R}_j \mathbf{R}_j - \omega_j \mathbf{L}_j \cdot \nabla A^0(\mathbf{U}_0) \cdot \mathbf{R}_j \mathbf{R}_j,$$
$$\tilde{\Gamma}_{jpq} = k_q \mathbf{L}_j \cdot \nabla A^1(\mathbf{U}_0) \cdot \mathbf{R}_p \mathbf{R}_q - \omega_q \mathbf{L}_j \cdot \nabla A^0(\mathbf{U}_0) \cdot \mathbf{R}_p \mathbf{R}_q.$$

For weak solutions, some care is required in using nonconservative equations. A nonlinear change of variables that preserves smooth solutions need not preserve weak solutions (Whitham, 1974). Here, however, this is not a difficulty. The same asymptotic equations are obtained from the conservative equations (2.1) and any trasformed equations (2.20). The reason for this is that a smooth nonlinear change of variables $\overline{\mathbf{U}} = \mathbf{h}(\mathbf{U})$ yields only a linear change of variables at first order in the wave amplitude. If $\mathbf{U} = \mathbf{U}_0 + \epsilon \mathbf{U}_1 + O(\epsilon^2)$ and $\overline{\mathbf{U}} = \overline{\mathbf{U}}_0 + \epsilon \overline{\mathbf{U}}_1 + (\epsilon^2)$, then $\overline{\mathbf{U}}_0 = \mathbf{h}(\mathbf{U}_0)$ and $\overline{\mathbf{U}}_1 = \nabla \mathbf{h}(\mathbf{U}_0) \cdot \mathbf{U}_1$. Linear changes variables preserve weak solutions. In fact, if $\mathbf{U}_1 = \Sigma a_j \mathbf{R}_j$, then $\overline{\mathbf{U}} = \Sigma a_j \overline{\mathbf{R}}_j$ where $\overline{\mathbf{R}}_j = \nabla \mathbf{h}(\mathbf{U}_0) \cdot \mathbf{R}_j$, so that the wave amplitudes are identical.

Additional effects, such as weak dissipation or dispersion, and slow transverse diffraction, can be incorporated into the above expansion. We shall give some examples in Sections (11.5) and (11.6).

3   The tri-resonance condition for hyperbolic waves

In this section we describe a geometrical interpretation of the tri-resonance condition (1.10) for hyperbolic waves. The dispersion relation of (1.1) is

$$(3.1) \qquad\qquad D(\omega, \vec{k}) = 0 \,,$$

where D is defined below (1.1). The surface in $\mathbb{R}^{m+1}$ with equation (4.1) is called the *characteristic variety* of (1.1). Since D is a homogeneous function of $(\omega, \vec{k})$, it suffices to consider $\vec{\xi} = \omega^{-1}\vec{k}$, which is called the *slowness vector* of the wave. This name comes from the fact that $\vec{\xi}$ is parallel to the phase velocity and has magnitude one over the phase speed. The slowness vector lies on a surface in real m-dimensional projective space with equation

$$(3.2) \qquad\qquad D(1, \vec{\xi}) = 0 \,.$$

This surface is called the *normal* or *slowness surface* of (1.1). The tri-resonance condition (1.10) implies that the slowness vectors of three resonantly interacting waves lie on a straight line. This gives the following result.

PROPOSITION (3.1). *The slowness vectors of waves produced by quadratically non-linear interactions between two waves with slowness vectors $\vec{\xi}_1$ and $\vec{\xi}_2$, lie on the intersections of the normal surface (4.2) with the straight line through $\{\vec{\xi}_1, \vec{\xi}_2\}$.*

To implement this result algebraically, we find the roots of the polynomial

$$P(\mu) = D(1, \mu\vec{\xi}_1, +(1-\mu)\vec{\xi}_2) \,.$$

If $P(\mu_*) = 0$, where $\mu_* \neq 0, 1$, then interactions between $\{\vec{\xi}_1, \vec{\xi}_2\}$ can produce a new wave with slowness vector $\vec{\xi} = \mu_*\vec{\xi}_1 + (1-\mu_*)\vec{\xi}_2$. For example, if the original equations are 2×2 first order system, then P is quadratic, and no new waves are produced.

Proposition (3.1) allows one to determine the directions of any new waves produced by the interaction. Taking account of the spectral content of the wave forms leads to a stronger resonance condition. For simplicity, suppose that the waves are periodic. The tri-resonance condition (1.6) implies that

$$\vec{\xi}_3 = \mu_1\vec{\xi}_1 + \mu_2\vec{\xi}_2 \,,$$

where
$$\mu_1 = \frac{m_1\omega_1}{m_1\omega_1 + m_2\omega_2}, \qquad \mu_2 = 1 - \mu_1 = \frac{m_2\omega_2}{m_1\omega_1 + m_2\omega_2} \,.$$

Here, m_1 and m_2 are arbitrary integers. It follows that a new wave corresponding to the root $P(\mu_1) = 0$ is only generated if

$$(3.3) \qquad\qquad \frac{\mu_1\omega_1}{\mu_2\omega_2} \quad \text{is rational} \,.$$

Next we illustrate this result with two examples. The first example is gas dynamics. It suffices to consider the two dimensional equations, since the wave numbers of any three resonant waves are coplanar. For waves propagating through a gas with sound speed c_0 and flow velocity $\vec{u} = (u, v)$, the dispersion relation is

$$(3.4) \qquad (\omega - \vec{u} \cdot \vec{k})^2 = c_0^2 |\vec{k}|^2 \, ,$$

$$(3.5) \qquad \omega - \vec{u} \cdot \vec{k} = 0 \, .$$

Equation (3.4) is the dispersion relation of sound waves. Equation (3.5) is the dispersion relation of vorticity waves or entropy waves, which are advected by the background flow. For definiteness, we shall only consider entropy waves. This is permissible because a detailed analysis of the gas dynamics equations shows that no vorticity is generated by the wave interactions at leading order in the wave amplitude (cf. Section (11.5)).

We denote the components of the slowness vector by $\xi = (\varsigma, \eta)$. The equation of the normal surface is

$$(3.6) \qquad (c_0^2 - u^2)\varsigma^2 + (c_0^2 - v^2)\eta^2 + 2u\varsigma + 2v\eta - 1 = 0 \, .$$

$$(3.7) \qquad u\varsigma + v\eta = 1 \, .$$

Equation (3.6) is an ellipse for a subsonic background flow. Equation (3.7) is a line. If $u = 0$, then $t = $ constant are characteristic surfaces and (3.7) is the line at infinity in $P\mathbb{R}^2$. Any resonant triad consists of two sound waves $\{P_1, P_2\}$ on two ellipse (3.6) and an entropy wave $\{P_0\}$ on the line (3.7). From Proposition (3.1), each entropy-sound wave pair $\{P_0, P_1\}$ resonates with one reflected sound wave, $\{P_2\}$. There are no resonant triads that involve three oblique sound waves.

Making a Galilean transformation, we can assume that $u = 0$ without loss of generality. We also rotate coordinates so that the entropy wavefronts are $x = $ constant. The frequency and wave number vector of the entropy wave are then

$$(3.8) \qquad \omega_0 = 0 \, , \qquad \vec{k}_0 = k_0 \begin{pmatrix} 1 \\ 0 \end{pmatrix} \, .$$

The frequency and wave number vector of a sound wave incident at angle $\pi/2 - \alpha$ on the entropy wave are

$$(3.9) \qquad \omega_1 = c_0 k_1 \, , \qquad \vec{k}_1 = k_1 \begin{pmatrix} \cos \alpha \\ \sin \alpha \end{pmatrix} \, .$$

Here, the dispersion relation (3.6), (3.7) is simple enough that we can solve the tri-resonance condition by inspection. The reflected sound wave is at an angle $\pi - \alpha$ to the entropy wave,

$$(3.10) \qquad \omega_2 = c_0 k_2 \, , \qquad \vec{k}_2 = k_2 \begin{pmatrix} -\cos \alpha \\ \sin \alpha \end{pmatrix} \, .$$

Thus, in this case, the tri-resonance condition simply implies that the linearized rays satisfy: angle of incidence = angle of reflection.

The frequencies and wave vectors in (3.8-9) are related by

$$\omega_2 = \frac{k_2}{k_1}\omega_1 - \frac{2k_2 \cos \alpha}{k_0}\omega_0,$$

$$\vec{k}_2 = \frac{k_2}{k_1}\vec{k}_1 - \frac{2k_2 \cos \alpha}{k_0}\vec{k}_0.$$

If the entropy wave and the incident sound wave are periodic, then the condition (1.6) for resonance implies that

(3.11) $$\frac{2k_1 \cos \alpha}{k_0} = \frac{m}{n} \text{ is rational.}$$

In that case, taking m and n relatively prime, the frequency of the reflected sound wave is $\omega_2 = n\omega_1$. For normal incidence ($\alpha = 0$), (3.11) reduces to the one dimensional condition (1.13). The interaction coefficient for the interaction: sound + entropy $\to$ sound, is nonzero unless $\alpha = \pi/4$ (cf. Section (11.5)). Therefore, provided that all the harmonics of the incident sound and entropy wave amplitudes are nonzero, a reflected sound wave is generated for any $\alpha \neq \pi/4$ that satisfies (3.11). Thus, if the wave numbers k_0 and k_1 are fixed and the angle of incidence is continuously varied, a reflected sound wave is generated for a countable, dense set of angles α.

Our second example is elasticity. The dispersion relation for waves in an isotropic elastic medium is

$$\omega^2 = c_s^2 k^2,$$

$$\omega^2 = c_p^2 k^2,$$

where c_s is the shear or s-wave speed, and c_p is the longitudinal or p-wave speed ($c_p > c_s > 0$). In two space dimensions, the slowness surface consists of two concentric circles

$$\xi^2 = c_s^{-2}, \qquad \xi^2 = c_p^{-2}.$$

The surface is shown in Figure (11.1).

For the sake of definiteness, let us consider the interaction of two s-waves. From Figure (11.1), s-waves on the outer circle produce two p-waves unless their directions are too close together. Suppose that the frequency, wavenumber, and slowness vectors of the s-waves are

$$\omega_j = c_s k_j, \quad \vec{k}_j = k_j \begin{pmatrix} \cos \alpha \\ \sin \alpha \end{pmatrix}, \quad \vec{\xi}_j = \frac{1}{c_s}\begin{pmatrix} \cos \alpha \\ \sin \alpha \end{pmatrix}, \quad j = 1, 2.$$

The slowness vectors of the p-waves produced by the interaction satisfy

$$\vec{\xi} = \mu\vec{\xi}_1 + (1 - \mu)\vec{\xi}_2, \quad \xi^2 = c_p^{-2}.$$

The equation for μ is therefore

(3.12)
$$\mu^2 - \mu + \frac{1}{4}\beta = 0,$$

where

(3.13)
$$\beta = \frac{c_p^2 - c_s^2}{c_p^2 \sin^2 \left[\frac{1}{2}(\alpha_1 - \alpha_2)\right]}.$$

The solutions of (3.12) are

(3.14)
$$\mu = \frac{1}{2}\left[1 \pm (1 - \beta)^{1/2}\right].$$

There is no resonant interaction if $\beta > 1$. From (3.13), this occurs when the angle $|\alpha_1 - \alpha_2|$ between the s-waves is less than a critical angle α_*, where

$$\alpha_* = 2\sin^{-1}\{c_p^{-1}\left[c_p^2 - c_s^2\right]^{1/2}\}.$$

Using (3.3) and (3.14), the condition for periodic waves to resonate is

$$[1 \pm (1 - \beta)^{1/2}]^2 \frac{\omega_1}{\beta\omega_2} \quad \text{is rational.}$$

The p-waves corresponding to both choices of sign in (3.14) are only produced if the ratio of these quantities is rational, i.e. if

$$\left\{\frac{1 - (1 - \beta)^{1/2}}{1 + (1 - \beta)^{1/2}}\right\}^2 \quad \text{is rational.}$$

For one dimensional waves, when $\alpha_2 - \alpha_1 = \pi$, this condition reduces to (1.15).

If we start with a pair of waves, then the number of waves propagating in different directions that are produced by resonant interactions is less than or equal to the order of the hyperbolic system. However, it turns out that three waves can generate a countably infinite set of new waves through successive resonant interactions. We illustrate this for the characteristic variety of isotropic elasticity using an idea of Joly and Rauch (1991). We start with an equilateral triangle of three points on the inner circle ("p-waves"). Using complex notation, we denote these points by $1, \Omega$ and Ω^2, where $\Omega = (-1/ + i\sqrt{3})/2$ is a cube root of unity (see Figure (11.1)), and we take $c_p = 1$, $c_s - c < 1$ without loss of generality. Pairwise interactions between

$$\{1, \Omega\}, \{\Omega, \Omega^2\}, \quad \text{and} \quad \{\Omega^2, 1\}$$

generate the points $z, \Omega z, \Omega^2 z$ on the outer circle ("s-waves"), where

(3.15)
$$z = \mu + (1 - \mu)\Omega, \qquad |z|^2 = c^{-2}.$$

Pairwise interactions between $\{z, \Omega^2\}, \{\Omega z, 1\}$ and $\{\Omega^2 z, \Omega\}$ generate the points $\omega, \Omega\omega$, and $\Omega^2\omega$ on the inner circle, where

(3.16) $$\omega = \nu z + (1 - \nu)\Omega^2 \,, |w|^2 = 1 \,.$$

If μ and ν are rational, and the angle between 1 and ω is an irrational multiple of 2π, then successive applications of this procedure constructs a dense set of slowness vectors on both inner and outer circles, all of which are generated by pairwise interactions from the three original waves.

    To show that this can be done, we eliminate z from (3.15), which impies that $3\mu^2 - 3\mu + 1 - c^{-2} = 0$. Then (3.15) and (3.16) imply that

$$\nu = \frac{3c^2}{c^2 + 1} \,.$$

The angle ϕ between 1 and ω is

$$\phi = \tan^{-1}\left[\frac{\sqrt{3}\mu(2 - \mu)}{\mu^2 + 2\mu - 2}\right] \,.$$

If we choose $c^{-2} = 13/4$, then $\nu \in \mathbb{Q}$, and $\mu = 3/2 \in \mathbb{Q}$. The corresponding angle is $\phi = \tan^{-1} 5\sqrt{3}/11$, and this is an irrational multiple of π (Joly and Rauch, 1991).

    Thus, provided the appropriate interaction coefficients are nonzero (which might not be true elasticity), and the ratios of the frequencies of the original waves are rational, the three original waves generate new waves propagating in a dense set of directions at both wave-speeds.

    In the next section we construct an asymptotic solution assuming that only finitely many waves are produced by quadratically nonlinear resonant interactions.

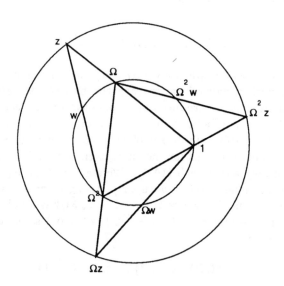

FIGURE (11.1)

Elasticity. Slowness surface in two dimensions.

4 Resonant interaction of hyperbolic waves in several space dimensions

Suppose that $\mathbf{U}: \mathbb{R}^{m+1} \longrightarrow \mathbb{R}^N$ satisfies a system of conservation laws in $m+1$ space-time dimensions,

$$(4.1) \qquad \partial_\alpha \mathbf{f}^\alpha(\mathbf{U}) + \mathbf{g}(\mathbf{U}) = 0,$$

where $\mathbf{f}^\alpha, \mathbf{g}: \mathbb{R}^N \longrightarrow \mathbb{R}^N, \partial_\alpha$ is the derivative with respect to x^α, and summation of α from 0 to m is implied. Here $x = (x^0, \cdots, x^m) \in \mathbb{R}^{m+1}$.

We look for solutions of (4.1) with an asymptotic expansion

$$(4.2) \qquad \mathbf{U} = \mathbf{U}_0 + \epsilon \mathbf{U}_1(x, \epsilon^{-1}x) + \epsilon^2 \mathbf{U}_2(x, \epsilon^{-1}x) + O(\epsilon^3).$$

Here, $\mathbf{U}_0$ is a constant solution of (4.1), so that $\mathbf{g}(\mathbf{U}_0) = 0$. Using (4.2) in (4.1), expanding, and equation coefficients of ϵ^n to zero shows that $\mathbf{U}_1(x, \xi)$ and $\mathbf{U}_2(x, \xi)$ satisfy

$$(4.3) \qquad \mathcal{A}_0^\alpha \partial_{\xi^\alpha} \mathbf{U}_1 = 0,$$
$$(4.4) \qquad \mathcal{A}^\alpha \partial_{\xi^\alpha} \mathbf{U}_2 + \mathbf{F}_1 = 0,$$

where

$$(4.5) \qquad \mathbf{F}_1 = \mathcal{A}_0^\alpha \partial_{x^\alpha} \mathbf{U}_1 + \frac{1}{2} \nabla^2 \mathbf{f}^\alpha \cdot \partial_{\xi^\alpha}(\mathbf{U}_1 \mathbf{U}_1) + \mathcal{G} \mathbf{U}_1.$$

Here,

$$\mathcal{G} = \nabla \mathbf{g}(\mathbf{U}_0), \quad \mathcal{A}_0^\alpha = \nabla \mathbf{f}^\alpha(\mathbf{U}_0) \text{ and } \nabla^2 \mathbf{f}_0 = \nabla^2 \mathbf{f}(\mathbf{U}_0).$$

A solution of (4.3) is

$$(4.6) \qquad \mathbf{U}_1 = \sum_{j=1}^J a_j(\kappa_j \cdot \xi, x) \mathbf{R}_j.$$

In (4.6), the wave vectors $\kappa_j = (\kappa_{j0}, \ldots, \kappa_{jm})$ satisfy the linearized dispersion relation

$$(4.7) \qquad \Delta(\kappa_j) = 0, \qquad \Delta(\kappa) = \det[\kappa_\alpha \mathcal{A}_0^\alpha].$$

Also, $\mathbf{R}_j$ and $\mathbf{L}_j$ are associated right and left null vectors,

$$\mathbf{R}_j = \mathbf{R}(\kappa_j), \quad \mathbf{L}_j = \mathbf{L}(\kappa_j),$$
$$\kappa_\alpha \mathcal{A}_0^\alpha \mathbf{R}(\kappa) = 0, \quad \mathbf{L}(\kappa) \cdot (\kappa_\alpha \mathcal{A}_0^\alpha) = 0.$$

We make the following assumptions concerning the wave vectors in (4.6):

(a) Zero is a simple eigenvalue of $\kappa_{j\alpha} A_0^\alpha$ for each j;

(b) The wave numbers $\{\kappa_1, \ldots, \kappa_J\}$ are a nondegenerate set, meaning that $\kappa_j \neq 0$ and $\kappa_p \neq \mu \kappa_q$ for any scalar μ and $1 \leq p \neq q \leq J$;

(c) The wave numbers $\{\kappa_1, \ldots, \kappa_J\}$ are a closed set, meaning that if

$$\Delta(\mu_p \kappa_q + \mu_q \kappa_q) = 0$$

for any $\mu_p, \mu_q \in \mathbb{R}$, then there is a $1 \leq j \leq J$ and a scalar $\mu_j \in \mathbb{R}$ such that $\mu_p \kappa_p + \mu_q \kappa_q = \mu_j \kappa_j$.

The last assumption is the most crucial one; it implies that any waves which can be generated by pairwise resonant interactions are included in the leading order solution (4.6).

We assume that the wave amplitudes $a_j(\theta, x)$ and their derivatives with respect to θ and x are smooth almost periodic functions of θ. For simplicity, we also suppose that they have zero mean with respect to θ.

$$(4.8) \qquad\qquad < a_j > = \lim_{T \to \infty} \int_0^T a_j(\theta, x) d\theta = 0.$$

It is straightforward to allow nonzero means. We assume that we can exchange the order of differentiating by x and taking means with respect to θ.

The equations for the wave amplitudes $a_j(\theta, x)$ follow by eliminating secular terms from U_2, just as in Section (11.2). The equation for U_2 is of the form

$$(4.9) \qquad\qquad A_0^\alpha \partial_{\xi^\alpha} U = F(\xi).$$

To state the solvability conditions for (4.9), we introduce some notation. We denote the mean of F with respect to ξ by $< F >$,

$$< F > = \lim_{T \to \infty} T^{-(m+1)} \int_0^T \ldots \int_0^T F(\xi) d\xi^0 \ldots d\xi^m.$$

For $\kappa \neq 0$, we denote the mean of F with respect to ξ over the hyperplane $\kappa \cdot \xi = \theta$ by $< F >^\kappa (\theta)$. That is, using an invertible linear change of coordinates $\overline{\xi} = L\xi$ with $\overline{\xi}^0 = \kappa \cdot \xi$ and $\overline{F}(\overline{\xi}) = F(\xi)$, we define

$$< F >^\kappa (\overline{\xi}^0) = \lim_{T \to \infty} T^{-m} \int_0^T \ldots \int_0^T \overline{F}(\overline{\xi}) d\overline{\xi}^1 \ldots d\overline{\xi}^m.$$

These means exist if, for example, F is an almost periodic function of ξ.

PROPOSITION (4.1). *Suppose that F is an almost periodic function of ξ. Necessary conditions for (4.9) to have a solution for U which is sublinear in ξ are that*

$$(4.10) \qquad \begin{array}{ll} (a) & < F > = 0; \\ (b) & L(\kappa) \cdot < F >^\kappa (\theta) = 0 \ \text{for all} \ \theta \in \mathbb{R} \ \text{and} \ \kappa \neq 0 \ \text{such that} \Delta(\kappa) = 0. \end{array}$$

PROOF. Condition (a) follows by averaging (4.9) with respect to ξ. The average of the left hand side is zero if $\mathbf{U}$ is sublinear. Condition (b) follows similarly by averaging (4.9) with respect to ξ over $\kappa \cdot \xi = \theta$, and taking the scalar product of the result with $\mathbf{L}(\kappa)$.   □

The conditions in Proposition (4.1) are also sufficient for (4.9) to have a sublinear solution, but we shall not prove that fact here.

Next, we compute these solvability conditions for (4.4–5). From (4.8), $\mathbf{U}_1$ has zero mean with respect to ξ. In that case, (4.10)(a) is automatically satisfied.

To apply (4.10)(b), we observe that the forcing term $\mathbf{F}_1$ in (4.4–5) is quadratic in $\mathbf{U}_1$. From (4.6) and (4.8), $\mathbf{F}_1$ is a sum of terms of the form

$$G(\kappa_j \cdot \xi), \quad H(\kappa_p \cdot \xi, \kappa_q \cdot \xi),$$

where $G(\theta_j), H(\theta_p, \theta_q)$ have zero mean with respect to θ_j, θ_p, and θ_q. It follows from the definition of $< \cdot >^\kappa$ that

$$< G(\kappa_j \cdot \xi) >^\kappa (\theta) = G(\kappa_j \cdot \xi)(\mu^{-1}\theta)$$

if $\kappa = \mu\kappa_j$ for some $\mu \in \mathbb{R} \setminus \{0\}$, and is zero otherwise. Similarly,

$$< H(\kappa_p \cdot \xi, \kappa_q \cdot \xi) >^\kappa (\theta) = \lim_{T \to \infty} \frac{1}{T} \int_0^T H(\nu\xi, \nu^{-1}\theta - \mu\xi)d\xi$$

if $\kappa = \mu\kappa_p + \nu\kappa_q$ for some $\mu, \nu \in \mathbb{R} \setminus \{0\}$ and is zero otherwise.

There are now two cases to consider. First, suppose that $\kappa \neq \mu\kappa_p + \nu\kappa_q$ for any $\mu, \nu \in \mathbb{R}$ and $1 \leq p, q, \leq J$. Then, it follows from above that $< \mathbf{F}_1 >^\kappa = 0$. Second, suppose that $\kappa = \mu\kappa_p + \nu\kappa_q$ for some $\mu, \nu \in \mathbb{R}$ and some j. Therefore, since $< \mathbf{F} >^\kappa (\mu\theta) = < \mathbf{F} >^{\mu\kappa} (\theta)$ for any $\mu \in \mathbb{R} \setminus \{0\}$, it suffices to impose the condition $(4.10)_1$ only for $\{\kappa_1, \ldots, \kappa_J\}$. The solvability conditions for (4.4–5) are then

(4.11) $< \mathbf{F}_1 >^j = 0, \quad j = 1, \ldots, J,$

where $< \cdot >^j = < \cdot >^{\kappa_j}$. We use (4.5) and (4.6) in (4.11) and simplify the result. This gives a set of equations for the wave amplitudes,

(4.12) $C_j^\alpha \partial_{x^\alpha} a_j(\theta) + M_j a_j(\theta) a_{j\theta}(\theta) + G_j a(\theta)$

$$+ \sum_{p \neq q}^{(j)} \tilde{\Gamma}_{jpq} \lim_{T \to \infty} \frac{1}{T} \int_0^T a_p(\mu_{pqj}\theta + \mu_{pqj}\sigma)a_{q\sigma}(\sigma)d\sigma = 0.$$

In (4.12), $\mathbf{C}_j = (C_j^0 \ldots, C_j^m)$ is the group velocity vector of the jth wave,

$$C_j^\alpha = \mathbf{L}_j \cdot A_0^\alpha \mathbf{R}_j.$$

The interaction coefficients are

$$\tilde{\Gamma}_{jpq} = \mathbf{L}_j \cdot \kappa_{q\alpha} \nabla^2 \mathbf{f}_0^\alpha \cdot \mathbf{R}_p \mathbf{R}_q,$$

$$M_j = \tilde{\Gamma}_{jjj}.$$

The summation is defined by

$$\overset{(j)}{\underset{p \neq q}{\sum}} \text{ is the sum over all } p, q \text{ with } p \neq q, p \neq j, \text{ and } q \neq j,$$

such that $\kappa_j = \mu_{jqp}\kappa_p + \mu_{jpq}\kappa_q$ for some $\mu_{jpq}, \mu_{jqp} \in \mathbb{R} \setminus \{0\}$.

The equations (4.12) are identical in form to the equations for one space dimension (2.17). Equation (4.12) can be put in conservation form. The interaction terms are the same as in (2.19). The symmetric interaction coefficients are given by

$$\Gamma_{jpq} = \mu_{jpq}\tilde{\Gamma}_{jpq} + \mu_{jqp}\tilde{\Gamma}_{jqp} = \mathbf{L}_j \cdot \kappa_{j\alpha}\nabla^2 \mathbf{f}_0^\alpha \cdot \mathbf{R}_p\mathbf{R}_q \,.$$

5 Gas dynamics

One space dimension

The equations of motion for the one-dimensional flow of a compressible gas are

(5.1)
$$\rho_t + (\rho v)_x = 0 \,,$$
$$(\rho v)_t + (\rho v^2 + P - \bar{\mu}v_x)_x = 0 \,,$$
$$\left(\rho e + \frac{1}{2}\rho v^2\right)_t + \left(\rho e v + \frac{1}{2}\rho v^3 + P v - \bar{\mu}v v_x - \kappa T_x\right)_x = 0 \,.$$

Here, ρ is the mass density, v is the flow velocity, P is the gas pressure, e is the internal energy per unit mass, and T is the temperature. The coefficient κ is the thermal conductivity of the gas, and

$$\bar{\mu} = \frac{4}{3}\mu + \mu_b \,,$$

where μ is the shear viscosity and μ_b is the bulk viscosity. The equations in (5.1) express conservation of mass, momentum and energy, respectively. We assume that body forces and energy sources may be neglected.

Equation (5.1) is supplemented by an equation of state for the internal energy, $e = e(\rho, s)$, and constitutive relations for the viscosity and thermal conductivity,

$$\bar{\mu} = \bar{\mu}(\rho, s) \,, \qquad \kappa = \kappa(\rho, s) \,.$$

Here, s is the specific entropy. The thermodynamic relation

$$de = T ds - P d(\rho^{-1})$$

implies that $P = \rho^2 e_\rho$, $T = e_s$. The sound speed $c(\rho, s)$ is defined by

$$c = (P_\rho)^{1/2} \,.$$

The specific heats at constant volume and pressure, c_V and c_P, are

$$c_V = T \left. \frac{\partial s}{\partial T} \right|_\rho , \qquad c_P = T \left. \frac{\partial s}{\partial T} \right|_P .$$

(Here, the subscripts V and P do not stand for partial derivatives. We shall continue to use subscripts for derivatives with respect to ρ keeping s fixed, and derivatives with respect to s keeping ρ fixed.)

Several dimensionless parameters can be obtained from the equation of state (Menikoff and Plohr, 1989). It is convenient to write $e(\rho, s) = E(V, s)$ where $V = \rho^{-1}$ is the specific volume. The adiabatic exponent γ, the Gruneisen coefficient G, and a dimensionless specific heat g are defined by

(5.2)
$$\gamma = \left. \frac{V}{P} \frac{\partial^2 E}{\partial V^2} \right|_s , \qquad G = - \frac{V}{T} \frac{\partial^2 E}{\partial s \partial V} , \qquad g = \left. \frac{PV}{T^2} \frac{\partial^2 E}{\partial s^2} \right|_V .$$

We shall also use two dimensionless third derivatives,

(5.3)
$$\mathcal{G} = -\frac{1}{2} V \frac{\partial^3 E / \partial V^3 |_s}{\partial^2 E / \partial V^2 |_s} , \qquad \mathcal{H} = \frac{V^2}{2\gamma T} \frac{\partial^3 E}{\partial s \partial^2 V} .$$

For an ideal gas, $\gamma = c_P / c_V$ is the ratio of specific heats (which, in general, is a function of temperature) and

$$G = g = \gamma - 1 ,$$

$$\mathcal{G} = \frac{1}{2} \left\{ \gamma + 1 + (\gamma - 1) \frac{T}{\gamma} \frac{d\gamma}{dT} \right\}$$

$$\mathcal{H} = \frac{\gamma - 1}{2} \left\{ 1 + \frac{T}{\gamma} \frac{d\gamma}{dT} \right\} .$$

A convenient choice of the vector of dependent variables is $\mathbf{U} = (\rho, v, s)^t$. Equation (5.1) may be written in the non-conservative form

(5.4)
$$\begin{pmatrix} \rho \\ v \\ s \end{pmatrix}_t + \begin{pmatrix} v & \rho & 0 \\ \rho^{-1}c^2 & v & \rho^{-1}P_s \\ 0 & 0 & v \end{pmatrix} \begin{pmatrix} \rho \\ v \\ s \end{pmatrix}_x = \begin{pmatrix} 0 \\ \rho^{-1}\partial_x(\bar{\mu}v_x) \\ \rho^{-1}T^{-1}[\bar{\mu}v_x^2 + \partial_x(\kappa T_x)] \end{pmatrix} .$$

We shall analyze the interaction of waves propagating through a homogeneous gas at rest with density ρ_0, sound speed c_0, and entropy s_0.

First, let us consider the linearized equations. These are just the usual equations of acoustics. We let

(5.5)
$$\begin{pmatrix} \rho \\ v \\ s \end{pmatrix} = \begin{pmatrix} \rho_0 \\ 0 \\ s_0 \end{pmatrix} + \begin{pmatrix} \rho' \\ v' \\ s' \end{pmatrix} .$$

We use (5.5) in (5.4) and neglect quantities that are second order in the primed variables. We also neglect thermo-viscous effects ($\bar{\mu} = \kappa = 0$). This gives

(5.6)
$$\begin{pmatrix} \rho' \\ v' \\ s' \end{pmatrix}_t + \begin{pmatrix} 0 & \rho_0 & 0 \\ \rho_0^{-1}c_0^2 & 0 & \rho^{-1}P_{s0} \\ 0 & 0 & 0 \end{pmatrix} \begin{pmatrix} \rho' \\ v' \\ s' \end{pmatrix}_x = 0.$$

In (5.6), a subscript zero on a thermodynamic variables means that it is evaluated at $\rho = \rho_0$ and $s = s_0$. Equation (5.6) is strictly hyperbolic, with characteristic velocities

$$\lambda_1 = -c_0, \qquad \lambda_2 = 0, \qquad \lambda_3 = c_0.$$

Corresponding right eigenvectors are

$$\mathbf{R}_1 = \begin{pmatrix} \rho_0 \\ -c_0 \\ 0 \end{pmatrix}, \qquad \mathbf{R}_2 = \begin{pmatrix} P_{s0} \\ 0 \\ -c_0^2 \end{pmatrix}, \qquad \mathbf{R}_3 = \begin{pmatrix} \rho_0 \\ c_0 \\ 0 \end{pmatrix}.$$

Thus, the 1-wave is a sound wave propagating to the left. It carries density and velocity perturbations at constant entropy. The 3-wave is a sound wave propagating to the right. The 2-wave consists of stationary entropy perturbations at constant pressure.

Before stating the weakly nonlinear equations that describe the interaction between such waves, we nondimensionalize (5.4). Suppose that the dimensionless wave amplitude (measured, for example, by the ratio of density perturbations to ρ_0) is of the order $\epsilon \ll 1$. Also suppose that the wavelength is of the order λ. We define dimensionless variables (superscripted with a $*$) by

(5.7)
$$\rho = \rho_0 \rho^*, \qquad v = c_0 v^*, \qquad s - s_0 = c_{P0} s^*,$$
$$x = \epsilon^{-1}\lambda x^*, \qquad t = \epsilon^{-1}c_0^{-1}\lambda t^*,$$
$$P = \rho_0 c_0^2 P^*, \qquad c = c_0 c^*, \qquad T = c_0^2 c_{P0}^{-1} T^*.$$

We have nondimensionalized lengths by $\epsilon^{-1}\lambda$, which is the length scale for nonlinear effects to become important—we could equally well have used the wavelength λ. The choice of $\epsilon^{-1}\lambda$ is consistent with our notation in Section (11.2).

We assume that the acoustic Reynolds number

$$Re = \frac{\rho_0 c_0 \lambda}{\bar{\mu}}$$

is of the order ϵ^{-1} as $\epsilon \to 0$, and that the Prandtl number

$$Pr = \frac{c_{P0}\bar{\mu}}{\kappa}$$

is of the order one. This choice of scaling allows a balance between weakly nonlinear effects and weak dissipation. We therefore let

(5.8)
$$\bar{\mu} = \epsilon \rho_0 c_0 \lambda \hat{\mu}, \qquad \kappa = \epsilon \rho_0 c_0 \lambda c_{P0} \hat{\kappa}.$$

where $\hat{\mu}$ and $\hat{\kappa}$ are dimensionless order one parameters.

Using (5.7) and (5.8) in (5.4), and dropping *'s on the dimensionless variables, leads to nondimensionalized equations of the same form as (5.4),

$$
\begin{pmatrix} \rho \\ v \\ s \end{pmatrix}_t + \begin{pmatrix} v & \rho & 0 \\ \rho^{-1}c^2 & v & \rho^{-1}P_s \\ 0 & 0 & v \end{pmatrix} \begin{pmatrix} \rho \\ v \\ s \end{pmatrix}_x = \epsilon^2 \begin{pmatrix} 0 \\ \rho^{-1}\partial_x(\hat{\mu}v_x) \\ \rho^{-1}T^{-1}[\hat{\mu}v_x^2 + \partial_x(\hat{k}T_x)] \end{pmatrix}.
$$

The equation of state is normalized so that

$$
c = 1 \quad \text{and} \quad T\frac{\partial s}{\partial T}\bigg|_P = 1 \quad \text{at } \rho = 1, \quad s = 0.
$$

For an ideal gas with constant specific heats,

$$
P = \gamma^{-1}e^{\gamma s}\rho^\gamma,
$$

$$
c^2 = e^{\gamma s}\rho^{\gamma-1},
$$

$$
T = (\gamma - 1)^{-1}e^{\gamma s}\rho^{\gamma-1}.
$$

The weakly nonlinear solution for interacting waves follows by applying the expansion described in Section (11.2) to this system. The result is that

$$
(5.9) \quad \begin{pmatrix} \rho \\ v \\ s \end{pmatrix} = \begin{pmatrix} 1 \\ 0 \\ 0 \end{pmatrix} + \epsilon a_1\left[x,t,\frac{k_1(x+t)}{\epsilon}\right]\begin{pmatrix} -1 \\ 1 \\ 0 \end{pmatrix} + \epsilon a_2\left[x,t,\frac{k_2 x}{\epsilon}\right]\begin{pmatrix} G/(\gamma g - G^2) \\ 0 \\ -1 \end{pmatrix}
$$

$$
+ \epsilon a_3\left[x,t,\frac{k_3(x-t)}{\epsilon}\right]\begin{pmatrix} 1 \\ 1 \\ 0 \end{pmatrix} + O(\epsilon^2).
$$

The wave amplitudes satisfy

$$
a_{1t}(\theta) - a_{1x}(\theta) + k_1\mathcal{G}a_1(\theta)a_{1\theta}(\theta)
$$

$$
+ k_2\Gamma \lim_{T\to\infty} \frac{1}{T}\int_0^T a_2'\left[\frac{k_2\theta}{2k_1} + \frac{k_2\xi}{2k_3}\right]a_3(\xi)d\xi = \frac{1}{2}k_1^2\hat{\delta}a_{1\theta\theta}(\theta),
$$

$$
(5.10) \quad a_{2t} = k_2^2\hat{\kappa}a_{2\theta\theta},
$$

$$
a_{3t}(\theta) + a_{3x}(\theta) + k_3\mathcal{G}a_3(\theta)a_{3\theta}(\theta)
$$

$$
- k_2\Gamma \lim_{T\to\infty} \frac{1}{T}\int_0^T a_2'\left[\frac{k_2\theta}{2k_3} + \frac{k_2\xi}{2k_1}\right]a_1(\xi)d\xi = \frac{1}{2}k_3^2\hat{\delta}a_{3\theta\theta}.
$$

In (5.10), we have not shown the dependence of $a_j(x,t,\theta)$ on (x,t) explicitly, and $a_2'(\theta) = \partial_\theta a_2(\theta)$. The coefficients are

$$
\Gamma = \frac{1}{2}\left\{\frac{G\mathcal{G} - \gamma\mathcal{H}}{\gamma g - G^2}\right\},
$$

$$
\hat{\delta} = \hat{\mu}\left\{1 + \frac{1}{Pr}\left[\frac{G^2}{\gamma g - G^2}\right]\right\}.
$$

Here, the dimensionless thermodynamic parameters are defined in (5.2–3) and they are evaluated at $V = \rho_0^{-1}, s = s_0$. For an ideal gas with constant specific heats,

$$\mathcal{G} = \frac{\gamma+1}{2},$$

$$\Gamma = \frac{1}{4},$$

$$\hat{\delta} = \hat{\mu}\left[1 + \frac{\gamma-1}{Pr}\right].$$

Equation (5.10) consists of a pair of Burgers equations for the sound wave amplitudes, coupled by correlations with the entropy wave amplitude. The entropy wave is determined independently of the sound waves from the diffusion Equation (5.10) and an initial condition. This is similar to the problem of the reflection of internal waves off a steady disturbance that was discussed in Section (10.5). The entropy wave is analogous to the steady internal wave. It is not affected by the interaction, but it couples the sound waves together. This decoupling of the entropy wave from the sound waves follows from the fact that entropy is a Riemann invariant of (5.1). If no entropy wave is present ($a_2 = 0$), then each sound wave satisfies the usual Burgers equation of nonlinear acoustics (Crighton, 1986).

The coefficient for quadratically nonlinear self-interactions of the entropy wave is zero. This is because the entropy wave-field is linearly degenerate. The self-interaction coefficient of the sound waves, $\mathcal{G}$, is called the *parameter of nonlinearity* of the fluid. It is positive for any fluid with normal thermodynamic properties. It can be written as

$$\mathcal{G} = 1 + \frac{\rho}{c}c_\rho.$$

The "1" in $\mathcal{G}$ is due to convection of the wave by the velocity perturbations which it carries. The remaining part is due to the variation of sound speed with density. For air, which is a diatomic ideal gas with $\gamma = 1.4$, $\mathcal{G}$ is equal to 1.2. For water, the experimental value of $\mathcal{G}$ is about 3.5. The nonlinearity of sound waves in air is predominantly caused by convection. In water, the main source of nonlinearity is the dependence of the sound speed on density.

The dimensional diffusivity corresponding to $\hat{\delta}$ is

$$\delta = \rho_0^{-1}\bar{\mu}\left\{1 + \frac{1}{Pr}\left[\frac{G^2}{\gamma g - G^2}\right]\right\}.$$

Lighthill (1956) called δ the *diffusivity of sound*.

Next, let us consider periodic solutions of (5.10). The conservative form of the interaction term in (5.10) is

$$k_2\Gamma \lim_{T\to\infty} \frac{1}{T}\int_0^T a_2'\left[\frac{k_2\theta}{2k_1} + \frac{k_2\xi}{2k_3}\right]a_3(\xi)d\xi$$

$$= \partial_\theta\left\{2k_1\Gamma \lim_{T\to\infty} \frac{1}{T}\int_0^T a_2\left[\frac{k_2\theta}{2k_1} + \frac{k_2\xi}{2k_3}\right]a_3(\xi)d\xi\right\}.$$

Suppose that $\{a_j(\theta)\}$ are 2π-periodic functions of θ with zero mean. Then, if $k_2/2k_3$ is irrational,

$$\lim_{T \to \infty} \frac{1}{T} \int_0^T a_2 \left[\frac{k_2\theta}{2k_1} + \frac{k_2\xi}{2k_3} \right] a_3(\xi)d\xi = 0.$$

Thus, there is no reflected sound wave unless $k_2/2k_3$ is rational. This is the resonance condition given in (1.13).

The resonance condition for periodic solutions is that

$$k_2 = nk_1 + mk_3,$$
$$\omega_2 = n\omega_1 + m\omega_3,$$
$$\omega_2 = 0, \qquad \omega_1 = -k_1, \qquad \omega_3 = k_3$$

where m and n are relatively prime integers. It follows that

$$\frac{k_2}{2k_1} = n, \qquad \frac{k_2}{2k_3} = m.$$

To simplify (5.10) further, we neglect thermo-viscous effects ($\hat{\delta} = \hat{\kappa} = 0$). Then, from (5.10), $a_2 = a_2(x, \theta)$ is an arbitrary function of (x, θ). We define new independent variables

$$u(\theta, x, t) = k_3 \mathcal{G} a_3(x, t, \theta),$$
$$v(\theta, x, t) = k_1 \mathcal{G} a_1(x, t, \theta),$$

and let

(5.11) $$K(\theta, x) = 2k_2 \Gamma a_{2\theta}(x, \theta).$$

Using these equations in (5.10) gives the following pair of integro-differential equations,

(5.12)
$$u_t + u_x + uu_\theta + \frac{n}{m} \frac{1}{2\pi} \int_0^{2\pi} K(m\theta + n\xi, x)v(\xi, x, t)d\xi = 0,$$
$$v_t - v_x + vv_\theta + \frac{m}{n} \frac{1}{2\pi} \int_0^{2\pi} K(n\theta + m\xi, x)u(\xi, x, t)d\xi = 0.$$

Here, $u(\theta, x, t), v(\theta, x, t)$, and $K(\theta, x)$ are 2π-periodic functions of θ. We obtain some exact solutions of (5.12) later in this section.

Two space dimensions

In this section we consider the reflection of an oblique sound wave off a vorticity-entropy wave. The wave number vectors of the incident and reflected sound waves and the vorticity-entropy wave are coplanar. Therefore, we can restrict to two space dimensions without loss of generality. We shall begin with isentropic flow. The equations of motion are

$$
(5.13) \qquad \begin{pmatrix} \rho \\ u \\ v \end{pmatrix}_t + \begin{pmatrix} u & \rho & 0 \\ \rho^{-1}c^2 & u & 0 \\ 0 & 0 & u \end{pmatrix} \begin{pmatrix} \rho \\ u \\ v \end{pmatrix}_x + \begin{pmatrix} v & 0 & \rho \\ 0 & v & 0 \\ \rho^{-1}c^2 & 0 & v \end{pmatrix} \begin{pmatrix} \rho \\ u \\ v \end{pmatrix}_y = 0,
$$

where ρ is the mass density, u is the x-component of velocity, and v is the y-component of velocity. The sound speed is a given function of ρ, $c = c(\rho)$, $c(\rho_0) = c_0$. The linearization of (5.13) about $\rho = \rho_0$, $u = v = 0$, is

$$
(5.14) \qquad \begin{pmatrix} \rho' \\ u' \\ v' \end{pmatrix}_t + \begin{pmatrix} 0 & \rho_0 & 0 \\ \rho_0^{-1}c_0^2 & 0 & 0 \\ 0 & 0 & 0 \end{pmatrix} \begin{pmatrix} \rho' \\ u' \\ v' \end{pmatrix}_x + \begin{pmatrix} 0 & 0 & \rho_0 \\ 0 & 0 & 0 \\ \rho_0^{-1}c_0^2 & 0 & 0 \end{pmatrix} \begin{pmatrix} \rho' \\ u' \\ v' \end{pmatrix}_y = 0.
$$

The plane wave solutions of (5.14) are

$$
\begin{pmatrix} \rho' \\ u' \\ v' \end{pmatrix}_t = a(kx + \ell y - \omega t)\mathbf{R}(\omega, k, \ell),
$$

where

$$
\begin{pmatrix} -\omega & \rho_0^{\frac{k}{}} & \rho_0^{\frac{\ell}{}} \\ \rho_0^{-1}c_0^2 k & -\omega & 0 \\ \rho_0^{-1}c_0^2 \ell & 0 & -\omega \end{pmatrix} \mathbf{R}(\omega, k, \ell) = 0.
$$

The equation of the characteristic variety of (5.14) is obtained by setting the determinant of this matrix to zero, which gives

$$
(5.15) \qquad \omega(\omega^2 - c_0^2 k^2 - c_0^2 \ell^2) = 0.
$$

A null vector associated with the root $\omega^2 = c_0^2(\kappa^2 + \ell^2)$, is

$$
\mathbf{R} = \begin{bmatrix} \omega \\ c_0 k \\ c_0 \ell \end{bmatrix}.
$$

This null vector corresponds to a sound wave. The wave carries density and velocity perturbations and the velocity perturbations are parallel to the wave number vector.

The null vector corresponding to $\omega = 0$ is

$$
\mathbf{R} = \begin{bmatrix} 0 \\ \ell \\ -k \end{bmatrix}.
$$

This null vector corresponds to a vorticity wave. The associated flow is incompressible and the velocity is perpendicular to the wave number vector.

Next, we write out the asymptotic equations that describe the reflection of an oblique sound wave off a vorticity wave. We suppose that (5.13) is nondimensionalized, so that $c = 1$ when $\rho = 1$. In Section (11.3) we showed that the tri-resonance condition implies that the angle of incidence and the angle of reflection are the same. We choose coordinates with the x-axis parallel to the vorticity wave's wave number vector. The wave number vectors of the waves are given in (3.8–10). We normalize the wave numbers of the waves to one for convenience. The associated phases of the waves are then

$$\phi_0 = x, \qquad\qquad\qquad\qquad \text{vorticity wave,}$$
$$\phi_1 = x\cos\alpha + y\sin\alpha - t, \qquad \text{incident sound wave,}$$
$$\phi_2 = -x\cos\alpha + y\sin\alpha - t, \qquad \text{reflected sound wave,}$$

where $\pi/2 - \alpha$ is the angle of incidence. The weakly nonlinear solution is

$$\begin{pmatrix} \rho \\ u \\ v \end{pmatrix} = \begin{pmatrix} 1 \\ 0 \\ 0 \end{pmatrix} + \epsilon a_v\left(x, y, \frac{\phi_0}{\epsilon}\right)\begin{pmatrix} 0 \\ 0 \\ 1 \end{pmatrix} + \epsilon a_1\left(x, y, t, \frac{\phi_1}{\epsilon}\right)\begin{pmatrix} 1 \\ \cos\alpha \\ \sin\alpha \end{pmatrix}$$

$$+ \epsilon a_2\left(x, y, t, \frac{\phi_2}{\epsilon}\right)\begin{pmatrix} 1 \\ -\cos\alpha \\ \sin\alpha \end{pmatrix} + O(\epsilon^2).$$

The amplitude a_v of the vorticity wave is any function independent of t. The sound wave amplitudes satisfy the pair of equations,

(5.16)

$$(\partial_t + \cos\alpha\,\partial_x + \sin\alpha\,\partial_y)a_1(\theta) + \mathcal{G}a_1(\theta)a_{1\theta}(\theta)$$

$$- \lim_{T\to\infty}\frac{1}{T}\int_0^T K(\xi)a_2(\theta - 2\xi\cos\alpha)d\xi = 0,$$

$$(\partial_t - \cos\alpha\,\partial_x + \sin\alpha\,\partial_y)a_2(\theta) + \mathcal{G}a_2(\theta)a_{2\theta}(\theta)$$

$$+ \lim_{T\to\infty}\frac{1}{T}\int_0^T K(\xi)a_2(\theta + 2\xi\cos\alpha)d\xi = 0.$$

In (5.16) we have not shown any (x, y, t) dependence explicitly. The kernel K is defined by

$$K(x, y, \theta) = \frac{1}{2}\tan\alpha\cos(2\alpha)a_{v\theta}(\theta),$$

and $\mathcal{G}$ is given in (5.3).

The same equations also describe nonisentropic flow in two space dimensions.

The solution in that case is (Majda, Rosales and Schonbeck, 1988),

$$
\begin{pmatrix} \rho \\ u \\ v \\ s \end{pmatrix} = \begin{pmatrix} 1 \\ 0 \\ 0 \\ 0 \end{pmatrix} + \epsilon a_v\left(x,y,\frac{\phi_0}{\epsilon}\right)\begin{pmatrix} 0 \\ 0 \\ 1 \\ 0 \end{pmatrix} + \epsilon a_e\left(x,y,\frac{\phi_0}{\epsilon}\right)\begin{pmatrix} G(\gamma_g - G^2) \\ 0 \\ 0 \\ -1 \end{pmatrix}
$$

$$
+ \epsilon a_1\left(x,y,t,\frac{\phi_1}{\epsilon}\right)\begin{pmatrix} 1 \\ \cos\alpha \\ \sin\alpha \\ 0 \end{pmatrix} + \epsilon a_2\left(x,y,t,\frac{\phi_2}{\epsilon}\right)\begin{pmatrix} 1 \\ -\cos\alpha \\ \sin\alpha \\ 0 \end{pmatrix} + O(\epsilon^2)\,.
$$

The vorticity wave and entropy wave amplitudes are arbitrary functions of (x,y,θ). The sound wave amplitudes satisfy (5.16) with

$$(5.17) \qquad K(x,y,\theta) = \frac{\cos 2\alpha}{2\cos\alpha}\left\{\sin\alpha\, a_{v\theta}(x,y,\theta) + 2\Gamma a_{e\theta}(x,y,\theta)\right\}\,.$$

For solutions that are independent of y, (5.16) has exactly the same form as the one-dimensional equations (5.10).

The kernel (5.17) vanishes when $\alpha = \pi/4$. At that angle, the sound wave passes through the entropy-vortex wave without reflection. Also, it is possible for the vorticity and entropy waves to cancel, so that $K \equiv 0$. The solutions which contain vorticity waves are limited physical significance, because such waves are subject to the Kelvin-Helmholtz instability.

The initial value problem

We suppose that amplitudes in (5.12) do not depend on x i.e. there are no spatial modulations. Also, we take $m = 1$ and $n = -1$, which corresponds to a direct resonance between the fundamental harmonics of the entropy and sound waves. In that case, the wavelength of the sound waves is twice the wavelength of the entropy waves. Finally, it is convenient to use x instead of θ to stand for the phase variable. Since there is no dependence here on the original space variable this should not lead to any confusion. With these assumptions, (5.12) becomes

$$(5.18) \qquad \begin{aligned} u_t + uu_x + \frac{1}{2\pi}\int_0^{2\pi} K(x-y)v(y,t)dy &= 0\,, \\ v_t + vv_x - \frac{1}{2\pi}\int_0^{2\pi} K(-x+y)u(y,t)dy &= 0\,. \end{aligned}$$

Equation (5.18) is supplemented by initial conditions,

$$u(x,0) = u_0(x)\,, \quad v(x,0) = v_0(x)\,.$$

The system in (5.18) has been studied by Majda, Rosales, and Schonbek (1988).

One of the most interesting properties of (5.18) is that it has smooth travelling wave solutions. These solutions remain smooth because the reflection of the sound waves off the entropy wave balances their tendency to shock up at least over the time scales that the weakly nonlinear expansion is valid. For sinusoidal entropy distributions, there is an explicit solution for the travelling waves (Pego, 1988).

PROPOSITION (5.1). *Suppose that $K(x) = \sin x$. Then an exact travelling wave solution of (5.18) is*

(5.19)
$$u = c + b[1 + \alpha \cos(x - ct + \delta)]^{1/2},$$
$$v = c + b[1 + \sigma\alpha \cos(x - ct + \delta)]^{1/2},$$

where δ is an arbitrary phase shift, $0 \le \alpha \le 1$, $\sigma \in \{-1, +1\}$, and

(5.20)
$$b(\alpha, \sigma) = \frac{\sigma}{\pi\alpha} \int_0^{2\pi} \cos y \, (1 + \alpha \cos y)^{1/2} dy,$$
$$c(\alpha, \sigma) = -b(\alpha, \sigma)\frac{1}{2\pi} \int_0^{2\pi} (1 + \alpha \cos y)^{1/2} dy.$$

PROOF. We look for travelling waves,

(5.21)
$$u = c + U(x - ct + \delta),$$
$$v = c + V(x - ct + \delta).$$

Using (5.21) in (5.17), with $K(x) = \sin x$, and integrating once, implies that

$$\frac{1}{2}U^2(z) = \frac{1}{2\pi} \int_0^{2\pi} \cos(z - \varsigma)V(\varsigma)d\varsigma + K_U,$$
$$\frac{1}{2}V^2(z) = \frac{1}{2\pi} \int_0^{2\pi} \cos(z - \varsigma)U(\varsigma)d\varsigma + K_V,$$

where K_U and K_V are constants of integration. Expanding the cosines in this equation, we find that

(5.22)
$$\frac{1}{2}U^2(z) = C_V \cos z + S_V \sin z + K_U,$$
$$\frac{1}{2}V^2(z) = C_U \cos z + S_U \sin z + K_V.$$

Here, C_U, C_V, S_U and S_V are the Fourier coefficients of U and V,

(5.23)
$$C_U = \frac{1}{2\pi} \int_0^{2\pi} \cos z \, U(z)dz, \quad C_V = \frac{1}{2\pi} \int_0^{2\pi} \cos z \, V(z)dz,$$
$$S_U = \frac{1}{2\pi} \int_0^{2\pi} \sin z \, U(z)dz, \quad S_V = \frac{1}{2\pi} \int_0^{2\pi} \sin z V(z)dz.$$

By choosing δ appropriately, we can assume that $S_V = 0$. Then U is even and it follows that $S_U = 0$. Equation (5.22) shows that

(5.24)
$$U = b_U[1 + \alpha_U \cos z]^{1/2},$$
$$V = b_V[1 + \alpha_V \cos z]^{1/2},$$

where $b_U^2 = 2K_U, \alpha_U = C_V/K_U$ and $b_V^2 = 2K_V, \alpha_V = C_U/K_V$. Using (5.24) in (5.23), and eliminating C_U and C_V, implies that

$$\frac{1}{2}\alpha_V b_V^2 = b_U \frac{1}{2\pi} \int_0^{2\pi} \cos z \, (1 + \alpha_V \cos z)^{1/2} dz$$

$$\frac{1}{2}\alpha_U b_U^2 = b_V \frac{1}{2\pi} \int_0^{2\pi} \cos z \, (1 + \alpha_U \cos z)^{1/2} dz.$$

The solutions of this equation are $\alpha_U = \sigma\alpha_V = \alpha, b_U = b_V = b$, where $\sigma = \pm 1$, and b is given by (5.20). Equation (5.20) for the wave speed $c(\alpha, \sigma)$ follows from the requirement that u and v have zero mean. □

These travelling waves exist only up to a maximum amplitude, corresponding to $\alpha = 1$. The limiting wave has a corner at its crest trough. There are two one parameter families of waves, depending on the choice of $\sigma = \pm 1$.

Pego (1988) also uses bifurcation theory to prove that there are small amplitude travelling wave solutions of (5.18) for general kernels K.

Equation (5.18) has the form

(5.25)
$$u_t + \left(\frac{1}{2}u^2\right)_x + Kv = 0,$$

$$v_t + \left(\frac{1}{2}v^2\right)_x - K^*u = 0,$$

where K is an integral operator and K^* is its adjoint. Equation (5.25) therefore consists of two Burgers equation coupled by a skew-symmetric lower order term. This lower order term is dispersive. It follows that, for smooth solutions, the total acoustic wave action is conserved:

$$\int_0^{2\pi} [u^2(x,t) + v^2(x,t)]dx = \text{constant}.$$

If shocks are present, then the action decays according to

(5.26)
$$\frac{d}{dt} \int_0^{2\pi} [u^2(x,t) + v^2(x,t)]dx = -\frac{1}{6} \sum_{\text{shocks}} \{[\![u]\!]^3 + [\![v]\!]^3\} \leq 0.$$

Here, $[\![w]\!] = w_L - w_R$ is the jump in w across a shock, where w_L, w_R are the values to the left and the right of the shock, respectively. The inequality in (5.26) follows from the fact that $[\![u]\!], [\![v]\!]$ are positive for admissible shocks.

One interesting choice for K is

(5.27)
$$K(x) = \alpha[1 - 2\pi \sum_{n=-\infty}^{+\infty} \delta(x - 2\pi n)].$$

From (5.11), the entropy wave is proportional to the integral of K. For (5.27), $K(x) = \alpha S'(x - \pi)$, where S is a sawtooth wave,

(5.28) $$S(x) = x \quad |x| < \pi, \; S(x + 2\pi) = S(x).$$

Thus, the kernel (5.27) corresponds to a periodic sawtooth entropy wave. Substituting (5.27) in (5.18), and using the fact that u and v have zero mean with respect to x, we obtain a local PDE,

(5.29)
$$u_t + uu_x - v = 0,$$
$$v_t + vv_x + u = 0.$$

We have normalized α to one in (5.29) without loss of generality.

The travelling wave solutions of (5.29) are qualitatively similar to the ones for sinusoidal kernels. For given wave speed, there is a one parameter family of smooth periodic travelling waves. There is a limiting wave of maximum amplitude which has corners at its crest or trough.

Another interesting family of solutions of (5.29) are the sawtooth waves,

(5.30) $$u = a(t)S(x - \xi), \; v = b(t)S(x - \xi).$$

Here, S is the sawtooth function defined in (5.28), and ξ is an arbitrary, constant phase shift. The functions in (5.30) satisfy the jump conditions for (5.29), because the shocks are stationary and the average values of u and v across the shocks are zero. The shocks are admissible only if $a \geq 0, b \geq 0$. Use of (5.30) in (5.29) gives a pair of ODEs for the amplitudes of the sawtooth waves,

(5.31) $$\dot{a} = -a^2 + b, \; \dot{b} = -b^2 - a.$$

These have a first integral,

$$\frac{1}{2}(a^2 + b^2)(1 - a + b)^{-1} + \frac{1}{2}(a - b) + \log|1 - a + b| + \frac{1}{2}(1 - a + b) = \text{constant}.$$

A phase plane analysis of (5.31) shows that any trajectory starting in the first quadrant of the (a, b) plane crosses the a-axis after a finite time. The shocks in (5.30) are then inadmissible. It seems likely that when b changes sign, a rarefaction wave with a unusual structure will appear in v.

Majda, Rosales, and Schonbek (1988) also solved (5.25) numerically. One of their calculations corresponds to

$$K(x) = -24 \sin x, \; u_0(x) = v_0(x) = 3S(x - \pi).$$

They find that the shocks in the initial data weaken as time goes on. The shocks effectively disappear by the time $t \simeq 4$. For $4.6 < t < 7.2$, the solution looks like a smooth travelling wave of the type obtained in Proposition (4.1). Thus, the dispersive interaction terms in (5.25) slow the decay of the sound waves by weakening the shocks.

A second calculation is for the local kernel (5.27), with $\alpha = 24$, and the sinusoidal initial data,

$$u_0(x) = v_0(x) = 4\pi \sin x \,.$$

There is a complicated exchange of energy between u and v. Shocks form in u while is smooth, and then the shocks disappear from u and appear in v. The solution seems to have an approximately recurrent behaviour in time. However, the solution cannot be exactly recurrent when shocks are present because the total energy (5.26) is strictly decreasing. These numerical results suggest that solutions of (5.25) approach smooth travelling or standing waves as $t \to +\infty$.

The signaling problem

The initial value problem in the previous subsection would be difficult to set up experimentally. The following signaling problem is a more resonable experimental arrangement. Consider a long tube of gas with small periodic entropy perturbations at constant pressure. We generate a sound wave at one end of the tube. This sound wave propagates down the tube and reflects off the entropy wave. We suppose that the spatial modulations of the incident and reflected sound waves settle down to a steady state. This state is described by the time-independent version of (5.12). The strongest resonance occurs when the sound wavelength is twice the entropy wavelength ($m = -n = 1$). Also, it is convenient to make the change of variable $v \to -v$ in (5.12). With these assumptions (5.12) becomes

(5.32)
$$u_x + uu_\theta = Kv \,,$$
$$v_x + vv_\theta = K^*u \,,$$

where K, K^* are

$$Kv(\theta, x) = \frac{1}{2\pi} \int_0^{2\pi} K(\theta - \xi, x)v(\xi, x)d\xi \,,$$

$$K^*u(\theta, x) = \frac{1}{2\pi} \int_0^{2\pi} K(-\theta + \xi, x)u(\xi, x)d\xi \,.$$

Here, $K(\theta, x)$ is a given, arbitrary kernel determined by the entropy wave. The entropy wave is allowed to vary slowly in x, so that K may depend on x. To complete the specification of the problem, we add an initial condition for the incident wave amplitude $u(\theta, x)$ and a radiation condition for the reflected wave amplitude $v(\theta, x)$,

(5.33)
$$u(\theta, 0) = u_0(x) \,,$$
$$v(\theta, x) \to 0 \text{ as } x \to \infty \,.$$

For a single sound wave, the signaling problem leads to an initial value problem for Burgers equation, with space as the time-like ray variable. For interacting waves, the signaling problem differs from the initial value problem in two ways. First, the signaling problem is an initial-boundary value problem, rather than a pure initial value problem. Secondly, the interaction terms in (5.32) are symmetric, instead of skew-symmetric, as in (5.25).

We shall only consider the local equations which arise when K is given by (5.27),

(5.34)
$$u_x + uu_\theta + \alpha v = 0,$$
$$v_x + vv_\theta + \alpha u = 0.$$

Here, α can be a function of x. From (5.9) and (5.11), the corresponding entropy perturbations are

$$s = -\epsilon \frac{\alpha(x)}{2k_2\Gamma} S\left[k_2\left(\frac{x-\pi}{\epsilon}\right)\right],$$

where S is the sawtooth function defined in (5.28). The amplitude of the entropy perturbations is proportional to α. For an ideal gas with constant specific heats, the sound speed increases from left to right across the contact discontinuities in the entropy wave when $\alpha > 0$; the sound speed decreases across the contacts when $\alpha < 0$.

If $\alpha > 0$, a solution of (5.34), (5.33) is given by $v = u$ where $u(0, x)$ solves

(5.35)
$$u_x + uu_\theta + \alpha(x)u = 0, \quad u(0, x) = u_0(\theta).$$

The incident sound wave is attenuated by reflection off the entropy wave. This attenuation can prevent shock formation. The method of characteristics shows that the solution of (5.35) is smooth provided that

$$u_0'(\theta) > -\left\{\int_0^\infty \exp\left[-\int_0^y \alpha(x)dx\right]dy\right\}^{-1}.$$

In particular, if α is a positive constant, the solution is smooth provided that $u_0'(\theta) > -\alpha$.

Equation (5.34) has simple sawtooth wave solutions for the sound waves. If α is a positive constant and $u_0(\theta) = a_0 S(\theta), a_0 > 0$, then an admissible solution of (5.34) is

$$u(\theta, x) = v(\theta, x) = \frac{\alpha a_0}{(\alpha + a_0)e^{\alpha x} - a_0} S(\theta).$$

Thus, the sound waves decay exponentially as $x \to +\infty$. If α is negative, then the sawtooth wave solution of (5.34) and (5.33) is inadmissible. There do not seem to be any simple explicit solutions when $\alpha < 0$.

The difference between the cases $\alpha > 0$ and $\alpha < 0$ is explained by the following fact (Courant and Friedrich, 1948): A weak shock incident on a contact is reflected as a shock if the sound speed increases across the contact ($\alpha > 0$), and as a rarefaction if the sound speed decreases across the contact ($\alpha < 0$).

6 Elasticity

An elastic medium supports two types of body waves. One type is longitudinal, and is similar to sound waves in a fluid; the other type is transverse, and has no direct analogue in fluids. We begin by summarizing the elasticity equations. Then we shall derive the resonant interaction equations in one space dimension (Hunter, 1991). We shall also consider cubically nonlinear interactions of transverse waves.

The elasticity equations

The motion of an elastic body is described by giving the position, $\vec{x}(\vec{X}, t)$, at time t of the material points located at $\vec{X}$ in some reference configuration of the body. For a homogeneous body, the equations of motion are (Gurtin, 1981)

$$(6.1) \qquad \rho_0 \vec{x}_{tt} = \text{Div } S(\nabla \vec{x}).$$

Here, S is the (first) Piola-Kirchhoff stress tensor, which measures stress with respect to area in the reference configuration, and ρ_0 is the density of the reference configuration. Also, Div and ∇ are the divergence and gradient with respect to the material coordinates $\vec{X}$.

The Piola-Kirchhoff stress tensor is a given function of the deformation gradient, $F = \nabla \vec{x}$. The body is hyperelastic if $S(F)$ is the derivative of a scalar function $W(F)$, i.e. $S = \nabla_F W$. The scalar-valued function W is called the strain-energy density. For simplicity, we shall assume that the body is isotropic. In this case, the constitutive relation for the stress tensor is

$$(6.2) \qquad S(F) = \det(F)[\alpha_0 I + \alpha_1 B + \alpha_2 B^2] F^{-t}, \quad B = FF^t,.$$

Here, the α_j are scalar-valued functions of the eigenvalues of the left Cauchy-Green strain tensor B. The strain-energy density of an isotropic hyperelastic body depends only of the eigenvalues of B.

Equation (6.1) can be written as a first order system for the velocity $\vec{v} = \vec{x}_t$ and the deformation gradient,

$$\vec{v}_t = \text{Div } S(F)$$
$$F_t = \nabla \vec{v}.$$

For the moment, we shall work with the second order system (6.1).

First, we consider the linearised equations. We assume that the reference configuration is unstressed, so that $S(I) = 0$. We introduce the displacement, $\vec{u} = \vec{x} - \vec{X}$ and linearise (6.1). This gives

$$(6.3) \qquad \rho_0 \vec{u}_{tt} = \text{Div}(C \cdot \nabla \vec{u}),$$

Here $C = \nabla_F S(I)$ is a fourth order tensor, called the elasticity tensor. For an isotropic material, (6.2) implies that

$$(6.4) \qquad C \cdot H = 2\mu E + \lambda(tr E)I, \qquad E = \frac{1}{2}(H + H^t),$$

where the scalars λ, μ are the Lamé constants of the body.

The plane wave solutions of (6.3) are

$$\vec{u} = \exp(i\vec{k} \cdot \vec{X} - i\omega t)\vec{r},$$

where

$$\rho_0 \omega^2 \vec{r} = C \cdot (\vec{r} \otimes \vec{k})\vec{k}.$$

Using (6.4), this implies that

(6.5) $$\rho_0 \omega^2 \vec{r} = (\mu + \lambda)(\vec{k} \cdot \vec{r})\vec{k} + \mu k^2 \vec{r}.$$

There are two families of solutions of (6.5); they are longitudinal p-waves and transverse s-waves. For the p-waves, the displacement vector $\vec{r}$ is parallel to $\vec{k}$ and

(6.6) $$\omega^2 = c_p^2 k^2, \qquad c_p^2 = \frac{2\mu + \lambda}{\rho_0}.$$

For the s-waves, $\vec{r}$ is perpendicular to $\vec{k}$ and

(6.7) $$\omega^2 = c_s^2 k^2, \qquad c_s^2 = \frac{\mu}{\rho_0}.$$

Since μ and λ are positive, $c_p > c_s$.

Next, we consider one-dimensional deformations,

(6.8)
$$
\begin{aligned}
x_1 &= X + u_1(X, t), \\
x_2 &= X_2 + u_2(X, t), \\
x_3 &= X_3 + u_3(X, t).
\end{aligned}
$$

We define the displacement gradients p_j by

$$p_j = u_{j,X}.$$

The deformation gradient and the left Cauchy-Green strain tensor corresponding to (6.8) are

$$
F = \begin{pmatrix} 1 + p_1 & 0 & 0 \\ p_2 & 1 & 0 \\ p_3 & 0 & 1 \end{pmatrix}
$$

$$
B = \begin{pmatrix} (1 + p_1)^2 & (1 + p_1)p_2 & (1 + p_1)p_3 \\ (1 + p_1)p_2 & 1 + p_2^2 & p_2 p_3 \\ (1 + p_1)p_3 & p_2 p_3 & 1 + p_3^2 \end{pmatrix}
$$

The eigenvalues of B are functions of p_1 and $p_2^2 + p_3^2$. From (6.2), the first column of the Piola-Kirchhoff stress has the form

(6.9)
$$
\begin{aligned}
S_{11} &= f(p_1, p_2^2 + p_3^2), \\
S_{21} &= p_2 g(p_1, p_2^2 + p_3^2), \\
S_{31} &= p_3 g(p_1, p_2^2 + p_3^2),
\end{aligned}
$$

where $f, g \colon \mathbb{R} \times \mathbb{R} \longrightarrow \mathbb{R}$. For a hyperelastic body, $W = W(p_1, p_2^2 + p_3^2)$, and

(6.10) $$f(p, q) = \frac{\partial W}{\partial p}(p, q), \qquad g(p, q) = 2\frac{\partial W}{\partial q}(p, q).$$

Using (6.8) and (6.9) in (6.1), the equations of motion are

(6.11)
$$\rho_0 u_{1tt} = \frac{\partial}{\partial X}\left\{ f\left(u_{1X}, u_{2X}^2 + u_{3X}^2\right)\right\},$$
$$\rho_0 u_{jtt} = \frac{\partial}{\partial X}\left\{ u_{jX}\, g\left(u_{1X}, u_{2X}^2 + u_{3X}^2\right)\right\}, \quad j = 2, 3.$$

These equations can be written as a first order system for the displacement gradients, p_j, and the velocity components, $v_j = x_{jt}$,

(6.12)
$$\rho_0 v_{1t} = \frac{\partial}{\partial X}\left\{ f\left(p_1, p_2^2 + p_3^2\right)\right\},$$
$$\rho_0 v_{jt} = \frac{\partial}{\partial X}\left\{ p_j\, g\left(p_1, p_2^2 + p_3^2\right)\right\}, \quad j = 2, 3,$$
$$p_{jt} = v_{jX}, \qquad j = 1, 2, 3.$$

Interaction of elastic waves

We shall restrict ourselves to plane motions in which $u_3 = 0$. Equation (6.12) is then a 4×4 system of conservation laws,

(6.13) $$\mathbf{U}_t + \mathbf{F}(\mathbf{U})_X = 0,$$

where

(6.14) $$\mathbf{U} = \begin{bmatrix} v_1 \\ v_2 \\ p_1 \\ p_2 \end{bmatrix}, \qquad \mathbf{F} = -\begin{bmatrix} \rho_0^{-1} f(p_1, p_2^2) \\ \rho_0^{-1} p_2 g(p_1, p_2^2) \\ v_1 \\ v_2 \end{bmatrix}.$$

We assume that the constitutive functions f, g have Taylor expansions

(6.15)
$$f(p_1, p_2^2) = f_0 + a p_1 + c p_1^2 + d p_2^2 + O(p_1^3, p_1 p_2^2, p_2^4),$$
$$g(p_1, p_2^2) = b + e p_1 + O(p_1^2 + p_2^2),$$

where a–e are material constants. The constants a, b characterise the linearized response of the body. They are related to the Lamé constants by

$$a = 2\mu + \lambda = \rho_0 c_p^2, \qquad b = \mu = \rho_0 c_s^2.$$

The constants c, d, e are higher order elastic constants. For a hyperelastic body, the corresponding Taylor expansion of the strain-energy density is

(6.16) $$W(p_1, p_2^2) = f_0 p_1 + \frac{1}{2} a p_1^2 + \frac{1}{2} b p_2^2 + \frac{1}{3} c p_1^3 + d p_1 p_2^2 + O(p_1^4 + p_2^4).$$

Then, (6.10) implies that $e = 2d$ in (6.16).

The Jacobian matrix of the flux at $\mathbf{U} = 0$ is

$$\mathcal{A} = \nabla_{\mathbf{U}}\mathbf{F}(0) = -\begin{pmatrix} 0 & 0 & c_p^2 & 0 \\ 0 & 0 & 0 & c_s^2 \\ 1 & 0 & 0 & 0 \\ 0 & 1 & 0 & 0 \end{pmatrix}.$$

The eigenvalues, λ_j, and eigenvectors, $\mathbf{R}_j$ of $\mathcal{A}$ are

$$\lambda_1 = -c_p\,, \quad \lambda_2 = -c_s\,, \quad \lambda_3 = c_s\,, \quad \lambda_4 = c_p\,,$$

$$\mathbf{R}_1 = \begin{pmatrix} 1 \\ 0 \\ c_p^{-1} \\ 0 \end{pmatrix}, \quad \mathbf{R}_2 = \begin{pmatrix} 0 \\ 1 \\ 0 \\ c_s^{-1} \end{pmatrix}, \quad \mathbf{R}_3 = \begin{pmatrix} 0 \\ 1 \\ 0 \\ -c_s^{-1} \end{pmatrix}, \quad \mathbf{R}_4 = \begin{pmatrix} 1 \\ 0 \\ -c_p^{-1} \\ 0 \end{pmatrix}.$$

The 1 and 4 waves are the left and right moving p-waves; the 2 and 3 waves are the left and right moving s-waves.

The asymptotic solution for small-amplitude interacting waves is

$$\mathbf{U} = \epsilon \sum_{i=1}^{4} a_j \left[\epsilon^{-1}(k_j X - w_j t), X, t \right] \mathbf{R}_j + 0(\epsilon^2),$$

as $\epsilon \to 0^+$ with $(x,t) = 0(1)$. Here, w_j is the frequency of the jth wave and k_j is the wave number. They are related by $w_j = \lambda_j k_j$. The frequencies and wave numbers of triads consisting of two s-waves and a p-wave are related by

(6.17)
$$\frac{2c_s}{k_1}\left(\frac{w_1}{k_1}\right) - \frac{c_p + c_s}{k_2}\left(\frac{w_2}{k_2}\right) + \frac{c_p - c_s}{k_3}\left(\frac{w_3}{k_3}\right) = 0$$

$$\frac{c_p - c_s}{k_2}\left(\frac{w_2}{k_2}\right) - \frac{c_p + c_s}{k_3}\left(\frac{w_3}{k_3}\right) + \frac{2c_s}{k_4}\left(\frac{w_4}{k_4}\right) = 0.$$

The equations for the wave amplitudes $a_j(\theta, X, t)$ are derived by applying the general method described in Section (11.2) to the elasticity equations (6.13–14). The result is the following system of integro-differential equations, where we do not show explicitly the (X,t) dependence of the a_j:

(6.18) $a_{1t} - c_p a_{1X} + k_1 \mathcal{G} a_1 a_{1\theta}$

$$- \frac{2c_s}{c_p - c_s} k_3 \Gamma \lim_{T \to \infty} \frac{1}{T} \int_0^T a_2 \left[\frac{2k_2 c_s}{k_1(c_p + c_s)} \theta + \frac{k_2(c_p - c_s)}{k_3(c_p + c_s)} \xi \right] a_{3\xi}[\xi] d\xi = 0,$$

(6.19) $a_{2t} - c_s a_{2X}$

$$+ \frac{(c_p + c_s)}{(c_p - c_s)} k_3 \Lambda \lim_{T \to \infty} \frac{1}{T} \int_0^T a_1 \left[\frac{k_1(c_p + c_s)}{2k_2 c_s} \theta - \frac{k_1(c_p - c_s)}{2k_3 c_s} \xi \right] a_{3\xi}(\xi) d\xi$$

$$- \frac{(c_p - c_s)}{(c_p + c_s)} k_3 \Lambda \lim_{T \to \infty} \frac{1}{T} \int_0^T a_4 \left[-\frac{k_4(c_p - c_s)}{2k_2 c_s} \theta + \frac{k_4(c_p + c_s)}{2k_3 c_s} \xi \right] a_{3\xi}(\xi) d\xi = 0,$$

(6.20) $a_{3t} + c_s a_{3X}$

$$- \frac{(c_p - c_s)}{(c_p + c_s)} k_2 \Lambda \lim_{T \to \infty} \frac{1}{T} \int_0^T a_1 \left[-\frac{k_1(c_p - c_s)}{2k_3 c_s} \theta + \frac{k_1(c_p + c_s)}{2k_2 c_s} \xi \right] a_{2\xi}(\xi) d\xi$$

$$+ \frac{(c_p + c_s)}{(c_p - c_s)} k_2 \Lambda \lim_{T \to \infty} \frac{1}{T} \int_0^T a_4 \left[\frac{k_4(c_p + c_s)}{2k_3 c_s} \theta - \frac{k_4(c_p - c_s)}{2k_2 c_s} \xi \right] a_{2\xi}(\xi) d\xi = 0,$$

(6.21) $a_{4t} + c_p a_{4X} + k_4 \mathcal{G} a_4 a_{4\theta}$

$$+ \frac{2c_s}{(c_p + c_s)} k_3 \Gamma \lim_{T \to \infty} \frac{1}{T} \int_0^T a_2 \left[-\frac{2k_2 c_s}{k_4(c_p - c_s)} \theta + \frac{k_2(c_p + c_s)}{k_3(c_p - c_s)} \xi \right] a_{3\xi}(\xi) d\xi = 0.$$

The interaction coefficients $\mathcal{G}, \Gamma$ and Λ depend on the higher order elastic coefficients. They are

$$\mathcal{G} = -\frac{c}{\rho_0 c_p^2}, \quad \Gamma = \frac{d}{\rho_0 c_s^2}, \quad \Lambda = \frac{e}{2\rho_0 c_s c_p}.$$

If the body is hyperelastic, then $e = 2d$, and $c_p \Gamma = c_s \Lambda$.

Equations (6.18) are simpler than for a general 4×4 system because the interaction of two longitudinal waves does not generate any transverse waves. also, the self-interaction coefficient for the s-waves is zero. This is a consequence of the lack of genuine nonlinearity of the s-wave fields.

Resonant triads of elastic waves

Next, we consider resonant triads made up of two s-waves (the 2 and 3 waves) and a right-moving p-wave (the 4-wave). We suppose that the wave amplitudes are 2π-periodic functions of the phase θ, and that they satisfy the strongest form of the tri-resonance condition, namely

$$\omega_2 + \omega_3 + \omega_4 = 0, \quad k_2 + k_3 + k_4 = 0.$$

From (6.17), this occurs when

(6.22) $$k_2 = \frac{c_p - c_s}{2c_s} k_4, \quad k_3 = -\frac{c_p + c_s}{2c_s} k_4.$$

We shall assume that

(6.23) $$\left[\frac{c_p - c_s}{c_p + c_s} \right]^2 \quad \text{is irrational}.$$

Then, from (6.22),

$$\frac{k_2(c_p - c_s)}{k_3(c_p + c_s)} \quad \text{is irrational}.$$

It follows that the correlation in (6.18) is zero and no left-moving p-wave is produced when the wave speeds satisfy (6.23). We can therefore consider solutions of (6.18–21) with $a_1 \equiv 0$. Using (6.22), the resulting equations are

$$a_{2t} - c_s a_{2X} + k_2 \Lambda \frac{1}{2\pi} \int_0^{2\pi} a_4(-\theta - \xi) a_{3\xi}(\xi) d\xi = 0,$$

(6.24) $$a_{3t} + c_s a_{3X} - k_3 \Lambda \frac{1}{2\pi} \int_0^{2\pi} a_4(-\theta - \xi) a_{2\xi}(\xi) d\xi = 0,$$

$$a_{4t} + c_p a_{4X} + k_4 \mathcal{G} a_4 a_{4\theta} + k_4 \Gamma \frac{1}{2\pi} \int_0^{2\pi} a_2(-\theta - \xi) a_{3\xi}(\xi) d\xi = 0.$$

To write (6.24) in a normalized form, we introduce new variables

$$u = k_4 \mathcal{G} a_4,$$

$$v = k_4 |\mathcal{G}\Gamma|^{1/2} \left[\frac{c_p - c_s}{c_p + c_s}\right]^{1/4} a_3,$$

$$w = \operatorname{sgn}(\mathcal{G}\Gamma) k_4 |\mathcal{G}\Gamma|^{1/2} \left[\frac{c_p + c_s}{c_p - c_s}\right]^{1/4} a_2,$$

$$x = c_p^{-1} X.$$

We assume that $\mathcal{G}$ and Γ are nonzero. The rescaled equations are

$$u_t + u_x + u u_\theta + \frac{1}{2\pi} \int_0^{2\pi} v(-\theta - \xi, x, t) w_\xi(\xi, x, t) d\xi = 0,$$

(6.25) $$v_t + \beta v_x - \Omega \frac{1}{2\pi} \int_0^{2\pi} w(-\theta - \xi, x, t) u_\xi(\xi, x, t) d\xi = 0,$$

$$w_t - \beta w_x - \Omega \frac{1}{2\pi} \int_0^{2\pi} v(-\theta - \xi, x, t) u_\xi(\xi, x, t) d\xi = 0.$$

The dimensionless parameters β, Ω are

$$\beta = \frac{c_s}{c_p},$$

(6.26) $$\Omega = -\frac{1}{2} c_s^{-1} [c_p^2 - c_s^2]^{1/2} \operatorname{sgn}(\mathcal{G}\Gamma) \frac{\Lambda}{\mathcal{G}}.$$

In (6.25), $u(\theta, x, t)$ is the p-wave amplitude, and $v(\theta, x, t)$ and $w(\theta, x, t)$ are the s-wave amplitudes. Since the s-waves are not genuinely nonlinear, equations $(6.25)_2$ and $(6.25)_3$ are linear in v and w.

Equation (6.25) has two conservation laws. To derive the first one, we multiply $(6.25)_2$ by v and $(6.25)_3$ by w, substract the results, and integrate over a period with respect to θ. Using the fact that

$$\int_0^{2\pi} \int_0^{2\pi} v(\theta)w(-\theta - \xi, x, t)u_\xi(\xi, x, t)\,d\xi\,d\theta$$

$$= \int_0^{2\pi} \int_0^{2\pi} w(\theta)v(-\theta - \xi, x, t)u_\xi(\xi, x, t)\,d\xi\,d\theta\,,$$

we find that

(6.27)
$$\left\{\int_0^{2\pi} [v^2(\theta, x, t) - w^2(\theta, x, t)]\,d\theta\right\}_t$$

$$+ \left\{\int_0^{2\pi} \beta\left[v^2(\theta, x, t) + w^2(\theta, x, t)\right]d\theta\right\}_x = 0\,.$$

Similarly, multiplying $(6.25)_1$ by 2Ω, adding $(6.25)_2$ and $(6.25)_3$, and integrating over a period, implies that

$$\left\{\int_0^{2\pi} [2\Omega u^2(\theta, x, t) + v^2(\theta, x, t) + w^2(\theta, x, t)]\,d\theta\right\}_t$$

$$+ \left\{\int_0^{2\pi} [2\Omega u^2(\theta, x, t) + \beta v^2(\theta, x, t) - \beta w^2(\theta, x, t)]\,d\theta\right\}_x = -\frac{\Omega}{3}\sum_{\text{shocks}} [\![u]\!]^3\,.$$

Here, $[\![u]\!] = u_L - u_R$, where u_L and u_R are the values of u to the left and right of shock, respectively. The sum is taken over all shocks in a period. For admissible shocks, $[\![u]\!] > 0$. Therefore, (6.24) shows that, if $\Omega > 0$, the L^2-norm in (θ, x) of the wave amplitudes in bounded.

We assume that $\Omega > 0$; otherwise the unstressed state of the elastic body is unstable.

First, we use (6.25) to discuss the generation of s-waves by a single p-wave. To do this, we consider solutions of (6.25) with no spatial modulations i.e. solutions that are independent of x. They satisfy

(6.28)
$$u_t + uu_\theta + \frac{1}{2\pi}\int_0^{2\pi} v(-\theta - \xi, t)w_\xi(\xi, t)d\xi = 0\,,$$

$$v_t - \Omega\frac{1}{2\pi}\int_0^{2\pi} w(-\theta - \xi, t)u_\xi(\xi, t)d\xi = 0\,,$$

$$w_t - \Omega\frac{1}{2\pi}\int_0^{2\pi} v(-\theta - \xi, t)u_\xi(\xi, t)d\xi = 0\,.$$

Linearizing (6.28) about $v = w = 0$ gives

$$u_t + uu_\theta = 0\,,$$

(6.29)
$$v_t - \Omega\frac{1}{2\pi}\int_0^{2\pi} w(-\theta - \xi, t)u_\xi(\xi, t)d\xi = 0\,,$$

$$w_t - \Omega\frac{1}{2\pi}\int_0^{2\pi} v(-\theta - \xi, t)u_\xi(\xi, t)d\xi = 0\,.$$

In this approximation, the p-wave satisfies an inviscid Burgers equation. For simplicity, we suppose that the p-wave is initially sinusoidal i.e.

$$u(\theta,0) = u_0 \sin \theta .$$

Then (6.29), implies that u approaches a sawtooth wave for large times (Whitham, 1974),

(6.30) $$u(\theta,t) \sim t^{-1} S(\theta) \quad \text{as} \quad t \to +\infty ,$$

where the sawtooth function S is defined in (5.28).

Taking the distributional derivative of (6.30) gives

(6.31) $$u_\theta(\theta,t) \sim t^{-1} \left\{ 1 - 2\pi \sum_{n=-\infty}^{n=+\infty} \delta\left[\theta - (2n+1)\pi\right] \right\} .$$

Using (6.31) in $(6.29)_2$, $(6.29)_3$, together with the assumption that v and w are 2π-periodic and have zero mean with respect to θ, gives

(6.32) $$v_t(\theta,t) + t^{-1}\Omega w(-\theta - \pi,t) = 0 ,$$
$$w_t(\theta,t) + t^{-1}\Omega v(-\theta - \pi,t) = 0 .$$

To solve (6.32), we define $T = \log t$,

(6.33) $$V(\theta,T) = v(\theta,t) ,$$
$$W(\theta,T) = w(-\theta - \pi,t) .$$

Then V,W satisfy the linear, constant coefficient ODEs,

(6.34) $$V_T + \Omega W = 0 ,$$
$$W_T + \Omega V = 0 .$$

Solving (6.34), and using the result in (6.33) implies that

(6.35) $$v(\theta,t) = \alpha(\theta)t^\Omega + \beta(\theta)t^{-\Omega} ,$$
$$w(\theta,t) = \alpha(-\theta - \pi)t^\Omega + \beta(-\theta - \pi)t^{-\Omega} ,$$

where α, β are arbitrary functions of integration. Thus, the s-wave amplitudes grow algebrically in time. The growth of the s-waves is algebraic, rather than exponential, because shocks cause the large amplitude p-wave, which "pumps" the s-waves, to decay. Ultimately, the s-wave amplitudes become as large as the p-wave amplitude, and the linearization of (6.28) is invalid.

This result does not quite imply that the p-wave is unstable, as is the case for the analagous parametric instability of a dispersive wave. This is because shocks cause the p-wave to decay. Suppose that the p-wave, with amplitude of the order one in (6.28), is perturbed by s-waves, with amplitudes of the order $\delta \ll 1$. From

(6.30) and (6.35), the amplitudes are of the same order of magnitude, namely $\delta^{1/(\Omega+1)}$, after times of the order $\delta^{-1/(\Omega+1)}$. Thereafter, the "entropy" inequality in (6.28) shows that the L^2-norms of the wave amplitudes remain of this order. Thus, this small perturbation of a p-wave results in small perturbations of the solution for all times.

Next, we consider a signaling problem for elastic waves.

Suppose we generate right moving s- and p-waves at $x = 0$. These waves propagate into $x > 0$, and interact. If the resonance condition above (6.22) and the nonresonance condition (6.23) are both satisfied, the wave interaction will produce a left moving s-wave. The steady wave pattern is a solution of the time independent version of (6.25), namely

$$u_x + u u_\theta + \frac{1}{2\pi} \int_0^{2\pi} v(-\theta - \xi, x) w_\xi(\xi, x) d\xi = 0,$$

(6.36)
$$v_x - G\frac{1}{2\pi} \int_0^{2\pi} w(-\theta - \xi, x) u_\xi(\xi, x) d\xi = 0,$$

$$w_x + G\frac{1}{2\pi} \int_0^{2\pi} v(-\theta - \xi, x) u_\xi(\xi, x) d\xi = 0.$$

Here, $G = \Omega/\beta$. The interaction terms in $(6.36)_2$ and $(6.36)_3$ are skew-symmetric, rather than symmetric as in time-dependent equations (6.28). Equation (6.36) is supplemented by boundary and radiation conditions,

(6.37)
$$u(\theta, 0) = u_0(\theta), \qquad v(\theta, 0) = v_0(\theta),$$
$$w(\theta, x) \to 0 \text{ as } x \to +\infty.$$

We assume that none of the left-moving s-wave is reflected back into the region $x > 0$.

There is a simple explicit solution of (6.36) and (6.37) for sawtooth waves. The corresponding boundary conditions are

(6.38)
$$u_0(\theta) = \alpha S(\theta), \qquad v_0(\theta) = \beta S(\theta),$$

where S is the sawtooth function defined in (5.28). We look for solutions of the form

(6.39)
$$u = a(x)S(\theta),$$
$$v = b(x)S(\theta),$$
$$w = c(x)S(\theta).$$

Using (6.39) in (6.36) implies that

(6.40)
$$a' + a^2 + bc = 0,$$
$$b' - Gac = 0,$$
$$c' + Gab = 0.$$

The boundary conditions are

(6.41) $a(0) = \alpha$, $b(0) = \beta$, $c(x) \to 0$ as $x \to +\infty$.

The shocks in u are only admissible if $a \geq 0$. We assume that $\alpha > 0$. We also assume that $\beta > 0$. This does not entail any loss of generality, since (6.40) is invariant under the change of variables $b \to -b$, $c \to -c$.

From $(6.40)_2$, $(6.40)_3$, we can write a, b, c in terms of a single function $\phi(z)$,

$$a(x) = R\phi'(GRx),$$

(6.42) $$b(x) = R\sin[\phi(GRx)],$$

$$c(x) = R\cos[\phi(GRx)].$$

Here, R is an arbitrary constant, which we assume positive without loss of generality. The total energy of the s-waves is proportional to $b^2 + c^2 = R^2$. The fact that the s-wave energy is constant is also implied by the conservation law (6.27). Using (6.42) in $(6.40)_1$ implies that $\phi(z)$ satisfies

(6.43) $$G\phi'' + (\phi')^2 + \frac{1}{2}\sin(2\phi) = 0.$$

From (6.41) and (6.42), the boundary and admissibility conditions are

$$\phi' = \frac{\alpha}{\beta}\sin\phi \quad \text{at} \quad z = 0,$$

(6.44) $$\phi \to \frac{\pi}{2} \quad \text{as} \quad z \to +\infty,$$

$$\phi'(z) \geq 0 \quad \text{for} \quad z \geq 0.$$

Equation (6.44) implies that the solution of (6.43) lies on an intersection of the curve $\phi' = (\alpha/\beta)\sin\phi$ with the stable manifold of the fixed point

$$(\phi, \phi') = \left(\frac{\pi}{2}, 0\right).$$

The first integral of (6.43) is

(6.45) $$(\phi')^2 = C\exp\left(-\frac{1}{2}G\phi\right) + \frac{1}{2(G^2+1)}[G\cos 2\phi - \sin 2\phi],$$

where the constant of integration for trajectories which approach $(\pi/2, 0)$ is

$$C = \frac{G}{2(G^2+1)}\exp(\pi G/4).$$

The initial values $\phi_0 = \phi(0)$ and $\phi'_0 = \phi(0)$ for the solution of (6.43) and (6.44) are the solution of $(6.44)_1$ and (6.45) with $0 < \phi_0 < \phi/2$. The amplitude $b(x)$ of the right moving sawtooth wave increases monotonically from its initial value of β, and approaches $\beta\csc\phi_0$ as $x \to +\infty$. The wave energy required for this growth is extracted from the right-moving p-wave via reflection off the left moving s-wave. On the other hand, the amplitude $C(x)$ of the left-moving s-wave decays monotonically to zero as $x \to +\infty$ from its maximum value of $\beta\tan\phi_0$ at $x = 0$.

It would be interesting to compare these theoretical predictions with direct experimental observations of ultrasonic wave interactions in elastic solids.

Interacting shear waves

Transverse wave interactions have special features which are not described be the asymptotic equations derived above. As a simple prototype problem, we consider the interaction of shear waves in isotropic, incompressible elasticity. From (6.11), the equations of motion are of the form

(6.46)
$$\vec{u}_t + \vec{v}_x = 0$$
$$\vec{v}_t + [\alpha(|\vec{u}|^2)\vec{u}]_x = 0.$$

Here, $\vec{u} = -(p_2, p_3)^t$ is the transverse displacement gradient, $\vec{v} = (v_2, v_3)^t$ is the transverse velocity, and the scalar valued function $\alpha(r) = g(0, r)$. The equation for longitudinal motions is satisfied by introducing a pressure function, as required by the incompressiblity constraint. We will not assume that the transverse waves are plane polarized. In fact, we consider (6.46) for $\vec{u}, \vec{v} \in \mathbb{R}^m$, where $m = 2$ for transverse waves in three space dimensions.

Equation (6.46) is invariant under the transformations $\vec{u} \to R\vec{u}$, $\vec{v} \to R\vec{v}$ for any orthogonal matrix R. This rotational invariance has several interesting consequences (Freistühler, 1990). First, (6.46) is nonstrictly hyperbolic at $\vec{u} = 0$. This corresponds to the fact that small amplitude transverse waves of different polarizations propagate with the same velocity. Secondly, the transverse wave velocity is an even function of $\vec{u}$, so that the waves fail to be genuinely nonlinear at $\vec{u} = 0$. The dominant nonlinear effect on a single weakly nonlinear transverse wave is cubic. Finally, since the only wave velocities at $\vec{u} = 0$ are $\pm\alpha_0$, $\alpha_0 = \alpha(0)$, there are no resonant three wave interactions between different waves. Consequently, the asymptotic equations describing the interaction between left and right moving transverse waves are cubically nonlinear. This is an exceptional situation for hyperbolic waves (unlike dispersive waves) since it requires all relevant interaction coefficients for the quadratically nonlinear resonant self-interactions to vanish. The time-scale for cubically nonlinear interactions is of the order period/amplitude^2. We therefore look for an asymptotic expansion of the form

(6.47)
$$\vec{u} = \epsilon\vec{u}_1(x, t, \epsilon^2 t) + \epsilon^3\vec{u}_2(x, t, \epsilon^2 t) + 0(\epsilon^5),$$
$$\vec{v} = \epsilon\vec{v}_1(x, t, \epsilon^2 t) + \epsilon^3\vec{v}_2(x, t, \epsilon^2 t) + 0(\epsilon^5),$$

as $\epsilon \to 0$ with $\tau = \epsilon^2 t = 0(1)$. Using (6.47) in (6.46), Taylor expanding, and equating coefficients of ϵ to zero shows that

(6.48)
$$\vec{u}_{1t} + \vec{v}_{1x} = 0,$$
$$\vec{v}_{1t} + \vec{u}_{1x} = 0.$$

We suppose that $\alpha_0 = 1$, without loss of generality. Equating coefficients of ϵ^3 gives

(6.49)
$$\vec{u}_{2t} + \vec{v}_{2x} + \vec{u}_{1\tau} = 0,$$
$$\vec{v}_{2t} + \vec{u}_{2x} + \vec{v}_{1\tau} + (\alpha_0'|\vec{u}_1|^2\vec{u}_1)_x = 0.$$

The solution of (6.48) is

(6.50) $\qquad \vec{u}_1 = \vec{a}(t - x, \tau) + \vec{b}(t + x, \tau), \quad \vec{v}_1 = \vec{a}(t - x, \tau) - \vec{b}(t + x, \tau).$

Here, $\vec{a}$ is the vector-valued amplitude of the right-moving wave, and $\vec{b}$ is the amplitude of the left moving wave. We use (6.50) in (6.49), and replace (t, x) by characteristic coordinates (ξ, η), where $\xi = t - x$ and $\eta = t + x$. Then, after adding and subtracting the resulting equations, we obtain that

(6.51)
$$\partial_\eta [\vec{u}_2 + \vec{v}_2] + \vec{a}_\tau + \frac{1}{2}\alpha_0'[\vec{q}_\eta - \vec{q}_\xi] = 0$$
$$\partial_\xi [\vec{v}_2 - \vec{u}_2] + \vec{b}_\tau + \frac{1}{2}\alpha_0'[\vec{q}_\eta - \vec{q}_\xi] = 0$$

where
$$\vec{q} = |\vec{a} + \vec{b}|^2 (\vec{a} + \vec{b}).$$

To obtain a uniformly valid solution, we require that $\vec{u}_2$ and $\vec{v}_2$ are bounded, or at least sublinear, functions of ξ and η. Averaging $(6.51)_1$ with respect to η and $(6.51)_2$ with respect to ξ then gives the following system of equations for $\vec{a}(\xi, \tau)$ and $\vec{b}(\eta, \tau)$,

(6.52)
$$\vec{a}_\tau + \frac{1}{2}\alpha_0'[(\vec{a} \cdot \vec{a} + trB)\vec{a} + 2B\vec{a}]_\xi = 0,$$
$$\vec{b}_\tau + \frac{1}{2}\alpha_0'[(\vec{b} \cdot \vec{b} + trA)\vec{b} + 2A\vec{b}]_\eta = 0.$$

Here, the $m \times m$ matrices $A(\tau)$ and $B(\tau)$ are defined by

$$A = <\vec{a} \otimes \vec{a}> = \lim_{L \to \infty} \frac{1}{L}\int_0^L \vec{a}(\xi, \tau) \otimes \vec{a}(\xi, \tau)d\xi,$$
$$A_{ij} = <a_i a_j>,$$
$$B = <\vec{b} \otimes \vec{b}>,$$

where we assume that these averages exist. For example, they exist if $\vec{a}$ and b are periodic functions of the phase variables ξ and η. If $\vec{a}$ and $\vec{b}$ are compactly supported (corresponding to pulses) then the averages are zero, and the waves do not interact, to leading order in the wave amplitude. In that case, (6.52) reduces to two decoupled cubically nonlinear equations of the type derived in Brio and Hunter (1990).

According to (6.52) the left-moving wave influences the right-moving wave only through an average of its wave amplitude. The term proportional to $tr B$ represents a change in the wave speed due to the presence of the other wave. The term proportional to Ba tends to align the $\vec{a}$-wave along a mean direction of the $\vec{b}$-wave. For plane polarized waves, with $\vec{a} = a(\xi, \tau)\vec{e}$ and $\vec{b} = b(\xi, \tau)\vec{e}$ where $\vec{e}$ is a constant unit vector, this latter effect does not occur. In that case (6.52) reduces to

$$a_\tau + \frac{1}{2}\alpha_0'[(a^2 + 3 < b^2 >)a]_\xi = 0\,,$$

$$b_\tau + \frac{1}{2}\alpha_0'[(b^2 + 3 < a^2 >)b]_\eta = 0\,.$$

The evolution of $< a^2 >$ and $< b^2 >$ can be computed a priori form these equations. For smooth solutions, these averages are constant. Therefore the equations decouple, and the only effect of the interaction is to change the wave speed.

REFERENCES

- R. F. Almgren, Modulated high-frequency waves. Stud. Appl. Math. LXXXIII, 1659 (1990).
- M. Artola, A.J. Majda, *Nonlinear development of instabilities in supersonic vortex sheets II: resonant interaction among kink modes*, SIAM J. Appl. Math. **49**, 1310 (1989).
- M. Brio, J.K. Hunter, *Rotationally invariant hyperbolic waves*, Comm. Pure Appl. Math., XLIII, 1037 (1990).
- P. Ceheleskly, R.R. Rosales, *Resonantly interacting weakly nonlinear waves in the presence of shocks: a single space variable in a homogeneous time independent medium*, Stud. Appl. Math. **74**, 117 (1986).
- R. Courant, K.O. Friedrichs, *Supersonic Flow and Shock Waves*, Springer-Verlag, New York (1948).
- D.G. Crighton, *Basic theoretical nonlinear acoustic*, in Frontiers in Physical Acoustic, Proc. Int. School of Physics "Enrico Fermi", Course 93, North-Holland, Amsterdam (1986).
- H. Freistühler, *Rotational degeneracy of hyperbolic systems of conservation laws*, Arch. Rat. Mech. Anal. **113**, 39 (1990).
- M.E. Gurtin, *An Introduction to Continuum Mechanics*, Academic Press, New York (1981).
- J. K. Hunter, *Interaction of elastic waves*, to appear in Stud. Appl. Math. (1991).
- J.K. Hunter, A. Majda and R.R. Rosales, *Resonantly interacting weakly nonlinear hyperbolic waves, II: several space variables*, Stud. Appl. Math. **75**, 187 (1986).
- J. L. Joly, G. Metivier, J. Rauch, *Resonant one dimensional nonlinear geometrical optics*, Preprint no. 9007, Centre de Recherche en Mathématiques de Bordeaux (1990).
- J.L. Joly, J. Rauch, *Nonlinear resonance can create dense oscillations*, to appear in *Nonlinear Waves and Microlocal Analysis*, IMA Volumes in Mathematics and its Applications, Springer-Verlang (1991).
- P. Lax, *Hyperbolic system of conservation laws and the mathematical theory of shock waves*, Conf. Board Math. Sci. 11. SIAM, Philadelphia (1973).
- M.J. Lighthill, *Viscosity effects in sound waves of finite amplitude*, in Surveys in Mechanics, Editors G.K. Batchelor and R.M. Davis, 255, Cambridge University Press (1956).
- A. Majda, R.R. Rosales, *Resonantly interacting weakly nonlinear hyperbolic waves, I: a single space variable*, Stud. Appl. Math. **71**, 149 (1984).
- A. Majda, R.R. Rosales and M. Schonbeck, *A canonical system of integro-differential equations arising in nonlinear acoustics*, Stud. Appl. Math. **79**, 205 (1988).
- V.P. Maslov, *Mathematical Aspects of Integral Optics*, Moscow Institute of Electronic Machine-building, Moscow (1983).
- V.P. Maslov, *Resonance Processes in the Wave Theory and Self-focalization*, Moscow Institute of Electronic Machine-biulding, Moscow (1983).

- V.P. Maslov, *Coherent structures, resonances and asymptotical non-unique-ness for Navier-Stokes equations under large Reynolds numbers*, Russ. Math. Surveys **41** (6), 19 (1986).
- R. Menikoff, B.J. Plohr, *The Riemann problem for fluid flow of real mate-rials*, Rev. Mod. Phy. **61**, 75 (1989).
- R. Pego, *Some explicit resonating waves in weakly nonlinear gas dynamics*, Stud. Appl. Math. **79**, 263 (1988).
- O. V. Rudenko and S. I. Soluyan, *Theoretical Foundations of Nonlinear Acoustics*, Consultants Bureau, Plenum, New York (1977).
- G.B. Whitham, *Linear and Nonlinear Waves*, Wiley, New York (1974).

PROOF of Equation (3.29) of Chapter 2.

Let us expand matrix A as follows:

$$A = A_0 + (\nabla A)_0 \left\{ \sum_{\alpha=1}^{+\infty} \varepsilon^\alpha \sum_{l=-\infty}^{+\infty} \mathbf{U}_l^{(\alpha)} e^{il(kx-\omega t)} \right\}$$

$$+ \frac{1}{2}(\nabla\nabla A)_0 \left\{ \sum_{\alpha=1}^{\infty} \varepsilon^\alpha \sum_{r=-\infty}^{+\infty} \mathbf{U}_r^{(\alpha)} e^{ir(kx-\omega t)} \right\}$$

$$\cdot \left\{ \sum_{\beta=1}^{\infty} \varepsilon^\beta \sum_{s=-\infty}^{+\infty} \mathbf{U}_s^{(\beta)} e^{is(kx-\omega t)} \right\} + \cdots$$

Whence

$$A = A_0 + \epsilon(\nabla A)_0 \sum_{l=\infty}^{+\infty} \mathbf{U}_l^{(1)} e^{il(kx-\omega t)} + \epsilon^2 (\nabla A)_0 \sum_{l=-\infty}^{+\infty} \mathbf{U}_l^{(2)} e^{il(kx-\omega t)}$$

$$+ \epsilon^2 \frac{1}{2}(\nabla\nabla A_0) \left\{ \sum_{r,s=-\infty}^{+\infty} \mathbf{U}_r^{(1)} \mathbf{U}_s^{(1)} e^{i(r+s)(kx-\omega t)} \right\} + O(\epsilon^3)$$

and by writing $l = r + s$, it follows

$$A = A_0 + \epsilon(\nabla A)_0 \sum_{l=\infty}^{+\infty} \mathbf{U}_l^{(1)} e^{il(kx-\omega t)}$$

(1)

$$+ \epsilon^2 \sum_{l=-\infty}^{+\infty} \left\{ (\nabla A)_0 \mathbf{U}_l^{(2)} + \frac{1}{2}(\nabla\nabla A)_0 \sum_{r=-\infty}^{+\infty} \mathbf{U}_r^{(1)} \mathbf{U}_{l-r}^{(1)} \right\} \cdot$$

$$e^{il(kx-\omega t)} + O(\epsilon^3).$$

Similarly we expand the vector $\mathbf{B}$:

$$\mathbf{B} = \mathbf{B}_0 + \epsilon(\nabla\mathbf{B})_0 \sum_{l=-\infty}^{+\infty} \mathbf{U}_l^{(1)} e^{il(kx-\omega t)}$$

$$+ \epsilon^2 \sum_{l=-\infty}^{+\infty} \left\{ (\nabla\mathbf{B})_0 \mathbf{U}_l^{(2)} + \frac{1}{2}(\nabla\nabla\mathbf{B})_0 \sum_{r=-\infty}^{+\infty} \mathbf{U}_r^{(1)}\mathbf{U}_{l-r}^{(1)} \right\} e^{il(kx-\omega t)}$$

(2)

$$+ \epsilon^3 \sum_{l=-\infty}^{+\infty} \left\{ (\nabla\mathbf{B})_0 \mathbf{U}_l^{(3)} + \frac{1}{6}(\nabla\nabla\nabla\mathbf{B})_0 \sum_{r,s=-\infty}^{+\infty} \mathbf{U}_{l-r-s}^{(1)}\mathbf{U}_r^{(1)}\mathbf{U}_s^{(1)} \right.$$

$$\left. + (\nabla\nabla\mathbf{B})_0 \sum_r \mathbf{U}_r^{(1)}\mathbf{U}_{l-r}^{(2)} \right\} + O(\epsilon^4).$$

Substituting into (3.19) and using (3.22), (3.23), yields, to order ϵ^2,

$$- \sum_{l=-\infty}^{+\infty} \frac{\partial \mathbf{U}_l^{(1)}}{\partial \xi} e^{il(kx-\omega t)} + (-i\omega l) \sum_{l=-\infty}^{+\infty} \mathbf{U}_l^{(2)} e^{il(kx-\omega t)}$$

$$+ A_0 \frac{d\mathbf{U}^{(0)}}{d\eta} + \frac{A_0}{\lambda_0} \left\{ \sum_{l=-\infty}^{+\infty} \frac{\partial \mathbf{U}_l^{(1)}}{\partial \xi} e^{il(kx-\omega t)} \right\} + A_0 \left\{ \sum_{l=-\infty}^{+\infty} il \frac{d(k\eta)}{d\eta} \mathbf{U}_l^{(2)} e^{il(kx-\omega t)} \right\}$$

$$+ \sum_{l=-\infty}^{+\infty} \sum_{r=-\infty}^{+\infty} ir \frac{d(k\eta)}{d\eta} (\nabla A)_0 \mathbf{U}_{l-r}^{(1)} \mathbf{U}_r^{(1)} e^{il(kx-\omega t)} + \sum_{l=-\infty}^{+\infty} (\nabla\mathbf{B})_0 \mathbf{U}_l^{(2)} e^{il(kx-\omega t)}$$

$$+ \frac{1}{2} \sum_{l=-\infty}^{+\infty} \sum_{r=-\infty}^{+\infty} (\nabla\nabla\mathbf{B})_0 \mathbf{U}_r^{(1)} \mathbf{U}_{l-r}^{(1)} e^{il(kx-\omega t)} = 0.$$

By taking the Fourier component of wavenumber l we obtain

$$\mathcal{W}_l \mathbf{U}_l^{(2)} + \left(-I + \frac{A_0}{\lambda_0} \right) \frac{\partial \mathbf{U}_l^{(1)}}{\partial \xi} + A_0 \frac{d\mathbf{U}^{(0)}}{d\eta} \delta_{0l}$$

(3)

$$+ \sum_{r=-\infty}^{+\infty} \left[ir \frac{d(k\eta)}{d\eta} (\nabla A)_0 \mathbf{U}_{l-r}^{(1)} \mathbf{U}_r^{(1)} \right] + \frac{1}{2} \sum_{r=-\infty}^{+\infty} (\nabla\nabla\mathbf{B})_0 \mathbf{U}_r^{(1)} \mathbf{U}_{l-r}^{(1)} = 0.$$

Making use of Eq. (3.28) of Chapter 2, the above equation, for $l = 1$, gives

(4)
$$\mathcal{W}_1 \mathbf{U}_1^{(2)} + \left(-I + \frac{1}{\lambda_0} A_0 \right) \frac{\partial \mathbf{U}_1^{(1)}}{\partial \xi} = 0.$$

The group velocity λ_0 is given in this case by

(5)
$$\frac{1}{\lambda_0} = \frac{\partial k}{\partial \omega} + \frac{\partial^2 k}{\partial \omega \partial \eta} \eta.$$

This formula is obtained from the expression of the phase variable $\theta = kx - \omega t$. By definition the local wave number is

$$\theta_x = \frac{\partial}{\partial \eta}(k\eta)$$

and the local frequency $\theta_t = -\omega$. Therefore, for the group velocity

$$\lambda_0 = -\frac{\partial \theta_t}{\partial \theta_x},$$

we have

$$\frac{1}{\lambda_0} = -\frac{\partial \theta_x}{\partial \theta_t} = \frac{\partial}{\partial \omega}\left(\frac{\partial(k\eta)}{\partial \eta}\right) = \frac{\partial k}{\partial \omega} + \frac{\partial^2 k}{\partial \omega \partial \eta}\eta.$$

In order to prove (3.29) we need some intermediate results.

PROPOSITION (1). *One has*

(6)
$$-i\frac{\partial \mathcal{W}_1}{\partial \omega} = -I + \frac{\mathcal{A}_0}{\lambda_0}.$$

PROOF. From

$$\mathcal{W}_1 = -i\omega I + i\frac{d(k\eta)}{d\eta}\mathcal{A}_0 + \nabla \mathbf{B}_0$$

by differentiating with respect to ω gives

$$\frac{\partial \mathcal{W}_1}{\partial \omega} = -iI + i\frac{\partial k}{\partial \omega}\mathcal{A}_0 + i\frac{\partial^2 k}{\partial \omega \partial \eta}\eta.$$

☐

PROPOSITION (2).

(7)
$$\frac{\partial \mathcal{W}_1}{\partial \omega}R = -\mathcal{W}_1\frac{\partial \mathbf{R}}{\partial \omega}.$$

PROOF. From $\mathcal{W}_1 \mathbf{R} = 0$ by differentiating $\mathcal{W}_1$ with respect to ω. ☐

Now Eqs. (6), (7) imply

(8)
$$\left(-I + \frac{1}{\lambda_0}\mathcal{A}_0\right)\mathbf{R} = i\mathcal{W}_1\frac{\partial \mathbf{R}}{\partial \omega}.$$

Hence, the general solution of (3) is, for $l = 1$,

(9)
$$\mathbf{U}_1^{(2)} = \mathbf{R}\hat{\psi}(\xi, \eta) - i\frac{\partial \mathbf{R}}{\partial \omega}\frac{\partial \varphi}{\partial \xi},$$

where $\hat{\psi}$ is an arbitrary function and relations (3.28) of Chapter 2 have been used.

From (3), for $l = 0$ we have

$$\mathcal{W}_0 \mathbf{U}_0^{(2)} + \mathcal{A}_0 \frac{d\mathbf{U}^{(0)}}{d\eta} + i\frac{d(k\eta)}{d\eta}(\nabla \mathcal{A})_0 \mathbf{U}_{-1}^{(1)} \mathbf{U}_1^{(1)} - i\frac{d(k\eta)}{d\eta}(\nabla \mathcal{A}_0)\mathbf{U}_1^{(1)} \mathbf{U}_{-1}^{(1)} +$$
$$+ \frac{1}{2}(\nabla\nabla\mathbf{B})_0(\mathbf{U}_1^{(1)}\mathbf{U}_{-1}^{(1)} + \mathbf{U}_{-1}^{(1)}\mathbf{U}_1^{(1)}) = 0 .$$

Now

$$\mathbf{U}_1^{(1)} = \mathbf{R}(\eta)\varphi(\xi,\eta) \text{ hence } \mathbf{U}_{-1}^{1)} = \mathbf{R}^*\varphi * .$$

Therefore the above equation gives

$$\mathcal{W}_0 \mathbf{U}_0^{(2)} + \mathcal{A}_0 \frac{d\mathbf{U}^{(0)}}{d\eta} + \left\{ \left(i\frac{d(k\eta)}{d\eta}\nabla \mathcal{A}_0 \mathbf{R}^*\mathbf{R} - c.c. \right) \right.$$
$$\left. + \frac{1}{2}((\nabla\nabla\mathbf{B})_0\mathbf{R}\mathbf{R}^* + c.c.) \right\} |\varphi|^2 = 0$$

whence

(10) $$\mathbf{U}_0^{(2)} = \mathbf{R}_0^{(2)}|\varphi|^2 + \mathbf{V}(\eta),$$

with

(11) $$\begin{cases} \mathbf{R}_0^{(2)} = -\mathcal{W}_0^{-1} \left\{ \left(i\frac{d(k\eta)}{d\eta}\nabla \mathcal{A}_0 \mathbf{R}^*\mathbf{R} - c.c. \right) + \frac{1}{2}(\nabla\nabla\mathbf{B}_0\mathbf{R}\mathbf{R}^* + c.c.) \right\} \\ \mathbf{V}(\eta) = -\mathcal{W}_0^{-1}\mathcal{A}_0 \dfrac{d\mathbf{U}^{(0)}}{d\eta} \end{cases} .$$

From (3), for $l = 2$ we have

$$\mathcal{W}_2 \mathbf{U}_2^{(2)} + i\frac{d(k\eta)}{d\eta}\nabla \mathcal{A}_0 \mathbf{U}_1^{(1)}\mathbf{U}_1^{(1)} + \frac{1}{2}\nabla\nabla\mathbf{B}_0\mathbf{U}_1^{(1)}\mathbf{U}_1^{(1)} = 0$$

hence

(12) $$\mathbf{U}_2^{(2)} = \mathbf{R}_2^{(2)}\varphi^2 ,$$

with

(13) $$\mathbf{R}_2^{(2)} = -\mathcal{W}_2^{-1} \left\{ i\frac{d(k\eta)}{d\eta}\nabla \mathcal{A}_0 \mathbf{R}\mathbf{R} + \frac{1}{2}\nabla\nabla\mathbf{B}_0\mathbf{R}\mathbf{R} \right\} .$$

From (3), for $|l| \geq 3$ we have

(14) $$\mathbf{U}_l^{(2)} = 0, \quad |l| \geq 3 .$$

To the order ϵ^3 we obtain

$$
-i\omega l\, U_l^{(3)} \frac{\partial U_i^{(2)}}{\partial \xi} + il\frac{d}{d\eta}(k\eta)\mathcal{A}_0 U_i^{(3)} + \frac{\mathcal{A}_0}{\lambda_0}\frac{\partial U_i^{(2)}}{\partial \xi} + \mathcal{A}_0\frac{\partial U_i^{(1)}}{\partial \eta}
$$

$$
+\frac{1}{\lambda_0}(\nabla \mathcal{A}_0)\sum_{r=-\infty}^{+\infty} U_{l-r}^{(1)}\frac{\partial U_r^{(1)}}{\partial \xi} + \nabla B_0 U_l^{(3)}
$$

(15)
$$
+i\frac{d(k\eta)}{d\eta}\left\{\sum_{r=-\infty}^{+\infty} r\nabla \mathcal{A}_0(U_{l-r}^{(1)}U_r^{(2)} + U_{l-r}^{(2)}U_r^{(1)})\right.
$$

$$
+\left(\nabla \mathcal{A}_0 U_l^{(1)}\frac{dU^{(0)}}{d\eta} + \frac{1}{2}\sum_{r,s}\nabla\nabla \mathcal{A}_0 U_{l-r-s}^{(1)}U_r^{(1)}U_s^{(1)}\right)
$$

$$
+\frac{1}{2}\sum_r \nabla\nabla B_0 U_{l-r}^{(1)}U_r^{(2)} + \frac{1}{6}\sum_{r,s}(\nabla\nabla\nabla B)_0 U_{l-r-s}^{(1)}U_r^{(1)}U_s^{(1)}\right\} = 0.
$$

For $l = 1$ Eq. (15) reads

$$
\mathcal{W}_1 U_1^{(3)} + \left(\frac{\mathcal{A}_0}{\lambda_0} - I\right)\frac{\partial U_1^{(2)}}{\partial \xi} + \mathcal{A}_0\frac{\partial U_1^{(1)}}{\partial \eta} + \nabla \mathcal{A}_0 U_1^{(1)}\frac{dU^{(0)}}{d\eta}
$$

$$
+i\frac{d(k\eta)}{d\eta}\left[2\nabla \mathcal{A}_0 U_{-1}^{(1)}U_2^{(2)} + \nabla \mathcal{A}_0 U_0^{(2)}U_1^{(1)} - \nabla \mathcal{A}_0 U_{(2)}^{(2)}U_{-1}^{(1)}\right]
$$

$$
+\frac{1}{2}\nabla\nabla \mathcal{A}_0\left[U_{-1}^{(1)}U_1^{(1)}U_1^{(1)} + U_1^{(1)}U_{-1}^{(1)}U_1^{(1)} + U_1^{(1)}U_1^{(1)}U_{-1}^{(1)}\right]
$$

$$
+\nabla\nabla B_0 U_1^{(1)}U_0^{(2)} + \nabla\nabla B_0 U_{-1}^{(1)}U_2^{(2)}
$$

$$
+\frac{1}{6}(\nabla\nabla\nabla B_0)\left[U_{-1}^{(1)}U_1^{(1)}U_1^{(1)} + U_1^{(1)}U_{-1}^{(1)}U_1^{(1)} + U_1^{(1)}U_1^{(1)}U_{-1}^{(1)}\right] = 0.
$$

Let L be the left eigenvector of $\mathcal{W}_1$, $L\mathcal{W}_1 = 0$, then, multiplying the above equation by L gives

$$
L\left(\frac{\mathcal{A}_0}{\lambda_0} - I\right)\left[R\frac{\partial \hat{\Psi}}{\partial \xi} - i\frac{\partial R}{\partial \omega}\frac{\partial^2 \varphi}{\partial \xi^2}\right] + L\mathcal{A}_0\left[R\frac{\partial \varphi}{\partial \eta} + \varphi\frac{\partial R}{\partial \eta}\right]
$$

$$
+ L\nabla \mathcal{A}_0\varphi R\frac{dU^{(0)}}{d\eta} + L_i\frac{d(k\eta)}{d\eta}\left\{2\nabla \mathcal{A}_0\varphi^* R\varphi^2 R_2^{(2)} + \nabla \mathcal{A}_0 R_0^{(2)}|\varphi|^2 R\varphi\right.
$$

(16)
$$
+ \nabla \mathcal{A}_0 V R\varphi - \nabla \mathcal{A}_0 R_2^{(2)}\varphi^2 R^*\varphi^* + \nabla\nabla \mathcal{A}_0 R^* R R\varphi|\varphi|^2
$$

$$
+\frac{1}{2}\nabla\nabla \mathcal{A}_0 R R^* R\varphi|\varphi|^2\right\} + L\left[\nabla\nabla B_0 R R_0^{(2)}\varphi|\varphi|^2 + \nabla\nabla B_0 R\varphi V\right.
$$

$$
+\nabla\nabla B_0^* R^*\varphi R_2^{(2)}\varphi^2 + \frac{1}{6}\nabla\nabla\nabla B_0(2R^* R R\varphi|\varphi|^2 + R R^* R\varphi|\varphi|^2)\right] = 0.
$$

From (8), we have

$$\mathbf{L}\left(-I + \frac{A_0}{\lambda_0}\right)\mathbf{R} = i\mathbf{L}W_1\frac{\partial\mathbf{R}}{\partial\omega} = 0$$

$$\mathbf{L}\left(\frac{A_0}{\lambda_0} - I\right)\frac{\partial\mathbf{R}}{\partial\omega} = -i\mathbf{L}\frac{\partial W_1}{\partial\omega}\frac{\partial\mathbf{R}}{\partial\omega}$$

Substituting these expressions into (16) we obtain Eq. (3.29) of Chapter 2. The coefficient α in Eq. (3.31) is given by

$$\alpha = -i\mathbf{L}\frac{\partial W_1}{\partial\omega}\frac{\partial\mathbf{R}}{\partial\omega}\Big/\mathbf{L}A_0\mathbf{R}.$$

It is possible to express α in a more convenient form. It is

$$\frac{\partial W_1}{\partial\omega}\mathbf{R} = -W_1\frac{\partial\mathbf{R}}{\partial\omega}$$

hence

$$\frac{\partial^2 W_1}{\partial\omega^2}\mathbf{R} + \frac{\partial W_1}{\partial\omega}\frac{\partial\mathbf{R}}{\partial\omega} = -\frac{\partial W_1}{\partial\omega}\frac{\partial\mathbf{R}}{\partial\omega} - W_1\frac{\partial^2\mathbf{R}}{\partial\omega^2}\mathbf{L}\frac{\partial W_1}{\partial\omega}\frac{\partial\mathbf{R}}{\partial\omega} = -\frac{1}{2}\mathbf{L}\frac{\partial^2 W_1}{\partial\omega^2}\mathbf{R}$$

and therefore

$$-i\mathbf{L}\frac{\partial W_1}{\partial\omega}\frac{\partial\mathbf{R}}{\partial\omega} = \frac{1}{2}i\mathbf{L}\frac{\partial^2 W_1}{\partial\omega^2}\mathbf{R}.$$

Moreover

$$\frac{\partial W_1}{\partial\omega} = -iI + i\frac{\partial}{\partial\omega}\frac{\partial(k\eta)}{\partial\eta}A_0,$$

and

$$\frac{\partial^2 W_1}{\partial\omega^2} = i\frac{\partial^2}{\partial\omega^2}\frac{\partial(k\eta)}{\partial\eta}A_0$$

hence

$$\alpha = -\frac{1}{2}\frac{\partial^2}{\partial\omega^2}\frac{\partial(k\eta)}{\partial\eta}.$$

PROOF of the relations (2.4–5) of Chapter 5.

Let us consider the identities:

(1)
$$(\mathcal{M}^j n_j - \lambda I)\mathbf{R}^I = 0,$$

(2)
$$\mathbf{L}_J(\mathcal{M}^j n_j - \lambda I) = 0.$$

Let us take the derivative of (1) with respect to n^i, left multiply by $\mathbf{L}_J$ and make use of (2). We obtain

(3)
$$\mathbf{L}_J \mathcal{M}^i \mathbf{R}^I = \frac{\partial \lambda}{\partial n_i} \mathbf{L}_J \mathbf{R}^I.$$

Let us take the derivative of (1) with respect to U^a, left multiply by $\mathbf{L}_J$ and make use of (2). We obtain

(4)
$$\mathbf{L}_J \mathcal{N}_a^i n_i \mathbf{R}^I = \nabla_a \lambda \mathbf{L}_J \mathbf{R}^I$$

where $\mathcal{N}_a^i = \partial \mathcal{M}^i / \partial U^a$. □

PROOF of relations (2.11–13) of Chapter 5.

(i) $\chi_{ij} = \chi_{ji}$

It is:

$$h^s_j \partial_k n_s \equiv \partial_k n_j - n_j n^s \partial_k n_s = \partial_k n_j,$$

because $\partial_k(n^s n_s) = 0$. From the definition of n^i:

$$n^i = a\phi_i,$$

with $\phi_i \equiv \partial\phi/\partial x^i$, $a \equiv |\nabla\phi|^{-1}$. Then:

$$\chi_{ij} = \tilde{\partial}_i n_j = h^k_i \partial_k n_j = h^k_i h^s_j \partial_k(a\phi_s) = h^k_i h^s_j(\partial_k a)\phi_s + ah^k_i h^s_j \phi_{ks}.$$

Note that the first term in the last expression is zero and the second is symmetric in i and j, thus $\chi_{ij} = \chi_{ji}$

(ii) $\dfrac{\delta\tilde{\partial}_j}{\delta t} - \tilde{\partial}_j \dfrac{\delta}{\delta t} = (\tilde{\partial}^k V_\Sigma n_j - V_\Sigma \chi^k_j)\tilde{\partial}_k$

where V_Σ is the normal speed of propagation of the singular surface. For an arbitrary generic function $\xi(x^i, t)$ we have:

$$\frac{\delta}{\delta t}\tilde{\partial}_j\xi = h^k_j(\partial_t + V_\Sigma n^s \partial_s)\partial_k\xi + \frac{\delta h^k_j}{\delta t}\partial_k\xi$$

$$= h^k_j\partial_k(\partial_t + V_\Sigma n^s \partial_s)\xi - h^k_j\partial_k(V_\Sigma n^s)\partial_s\xi - \left(n^k\frac{\delta n_j}{\delta t} + n_j\frac{\delta n^k}{\delta t}\right)\partial_k\xi$$

$$= \tilde{\partial}_j\frac{\delta\xi}{\delta t} - \tilde{\partial}_j V_\Sigma n^s \partial_s\xi - V_\Sigma\tilde{\partial}_j n^s \partial_s\xi - \left(n^k\frac{\delta n_j}{\delta t} + n_j\frac{\delta n^k}{\delta t}\right)\partial_k\xi.$$

Making use of the transport equation (1.15) for n^i this expression becomes:

$$\frac{\delta}{\delta t}\tilde{\partial}_j\xi = \tilde{\partial}_j\frac{\delta\xi}{\delta t} - V_\Sigma\chi^s_j\partial_s\xi + n_j\tilde{\partial}^k V_\Sigma\partial_k\xi = \tilde{\partial}_j\frac{\delta\xi}{\delta t} + (\tilde{\partial}^k V_\Sigma n_j - V_\Sigma\chi^k_j)\tilde{\partial}_k\xi.$$

If $V_\Sigma = \lambda(x^i, n^j)$ then

$$\frac{\delta}{\delta t}\tilde{\partial}_j\xi - \tilde{\partial}_j\frac{\delta\xi}{\delta t} = \left[\left(\frac{\partial\lambda}{\partial x^k} + \frac{\partial\lambda}{\partial n^s}\chi^k_s\right)n_j - \lambda\chi^k_j\right]\tilde{\partial}_k\xi.$$

□

(iii) $\tilde{\partial}_j\tilde{\partial}_i - \tilde{\partial}_i\tilde{\partial}_j = (n_j\chi^s_i - n_i\chi^s_j)\tilde{\partial}_s$.

PROOF. For an arbitrary function $\xi(x^i)$ one has:

$$\tilde{\partial}_j\tilde{\partial}_i\xi = h^s_i h^k_j \partial^2_{ks}\xi + (\tilde{\partial}_j h^s_i)\partial_s\xi$$

$$= h^s_i\partial_s(h^k_j\partial_k\xi) - h^s_i(\partial_k\xi)(\partial_s h^k_j) + (\tilde{\partial}_j h^s_i)\partial_s\xi$$

$$= \tilde{\partial}_i\tilde{\partial}_j\xi + (\tilde{\partial}_j h^s_i - \tilde{\partial}_i h^s_j)\partial_s\xi$$

$$= \tilde{\partial}_i\tilde{\partial}_j\xi + (\chi_{ij}n^s + n_j\chi^s_i)\partial_s\xi - (\chi_{ij}n^s + n_i\chi^s_j)\partial_s\xi$$

$$= \tilde{\partial}_i\tilde{\partial}_j\xi + (n_j\chi^s_i - n_i\chi^s_j)\partial_s\xi = \tilde{\partial}_i\tilde{\partial}_j\xi + (n_j\chi^s_i - n_i\chi^s_j)\tilde{\partial}_s\xi.$$

□

PROOF of relation (2.3) of Chapter 6.

Let $A(t) = [a_{ij}(t)]$ be a $N \times N$ matrix, whose elements are differentiable functions of t, and let

$$J = \det(A) \neq 0.$$

Then

(1)
$$\frac{dJ}{dt} = \sum_{i,j} \frac{\partial J}{\partial a_{ij}} \frac{da_{ij}}{dt}.$$

Expanding the determinant along the elements of the j-th row:

(2)
$$J = \sum_{i=1}^{N} a_{ij} A_{ij}^*$$

where A_{ij}^* is the cofactor of a_{ij} and does not depend on a_{ij}, therefore

(3)
$$\frac{\partial J}{\partial a_{ij}} = A_{ij}^* = Jb_{ji}$$

where $B = [b_{ij}] = A^{-1}$.

Let us consider now a regular mapping

$$x^i = x^i(t, \vec{X})$$

which satisfies the equations:

$$\begin{cases} \dfrac{\partial x^i}{\partial t} = \Lambda^i(\vec{x}, t) \\[2mm] x^i(0, \vec{X}) = x^i \end{cases}$$

and let $A = [\partial x^i / \partial X^j]$. Then, making use of (1) and (2), the Jacobian $J = \det(A)$ satisfies the equation

$$\frac{dJ}{dt} = \left(\frac{\partial J}{\partial t}\right)_{\vec{X}=\text{const}} = J \sum_{i,j} \frac{\partial X^j}{\partial x^i} \frac{\partial}{\partial t} \frac{\partial x^i}{\partial X^j} = J \sum_{i,j} \frac{\partial X^j}{\partial x^i} \frac{\partial}{\partial X^j} \frac{\partial x^i}{\partial t}$$

$$= J \sum_{i,j} \frac{\partial X^j}{\partial x^i} \frac{\partial}{\partial X^j} \Lambda^i = J \sum_{i,j,k} \frac{\partial X^j}{\partial x^i} \frac{\partial x^k}{\partial X^j} \frac{\partial \Lambda^i}{\partial x^k} = J \sum_{i,k} \delta_i^k \frac{\partial \Lambda^i}{\partial x^k} = J \sum_{i} \frac{\partial \Lambda^i}{\partial x^i}.$$

$\square$

PROOF of relations (2.5–6) of Chapter 6.

We start from the identities:

(1) $$(\mathcal{M}^i n_i - \lambda I)\mathbf{R} = 0,$$

(2) $$\mathbf{L}(\mathcal{M}^i n_i - \lambda I) = 0.$$

Eq. (2.5) follows from the identity (4) of Appendix A2, in case of single eigenvalue and normalized eigenvectors. In the same conditions the identity (3) of Appendix A2 reads:

(3) $$\mathbf{L}\mathcal{M}^j\mathbf{R} = \frac{\partial\lambda}{\partial n_j}.$$

Let us take the derivative of (3) with respect to n_k:

(4) $$\frac{\partial\mathbf{L}}{\partial n_k}\mathcal{M}^j\mathbf{R} + \mathbf{L}\mathcal{M}^j\frac{\partial\mathbf{R}}{\partial n_k} = \frac{\partial^2\lambda}{\partial n_j\partial n_k}.$$

Let us take the derivative of (1) with respect to n_j and left multiply it by $(\partial\mathbf{L}/\partial n_k)$:

(5) $$\frac{\partial\mathbf{L}}{\partial n_k}\mathcal{M}^j\mathbf{R} + \frac{\partial\mathbf{L}}{\partial n_k}\mathcal{M}^s n_s\frac{\partial\mathbf{R}}{\partial n_j} - \frac{\partial\lambda}{\partial n_j}\frac{\partial\mathbf{L}}{\partial n_k}\mathbf{R} - \lambda\frac{\partial\mathbf{L}}{\partial n_k}\frac{\partial\mathbf{R}}{\partial n_j} = 0.$$

Let us take now the derivative of (2) with respect to n_j and right multiply it by $\partial\mathbf{R}/\partial n_k$:

(6) $$\mathbf{L}\mathcal{M}^j\frac{\partial\mathbf{R}}{\partial n_k} + \frac{\partial\mathbf{L}}{\partial n_j}\mathcal{M}^s n_s\frac{\partial\mathbf{R}}{\partial n_k} - \frac{\partial\lambda}{\partial n_j}\mathbf{L}\frac{\partial\mathbf{R}}{\partial n_k} - \lambda\frac{\partial\mathbf{L}}{\partial n_j}\frac{\partial\mathbf{R}}{\partial n_k} = 0.$$

Multiplying (5) and (6) by χ and subtracting we get

(7) $$\left(\frac{\partial\mathbf{L}}{\partial n_k}\mathcal{M}^i\mathbf{R} - \mathbf{L}\mathcal{M}^i\frac{\partial\mathbf{R}}{\partial n_k} + 2\frac{\partial\lambda}{\partial n_i}\mathbf{L}\frac{\partial\mathbf{R}}{\partial n_k}\right)\chi_{ik} = 0$$

where the identity

(8) $$0 = \frac{\partial}{\partial n_k}(\mathbf{LR}) = \frac{\partial\mathbf{L}}{\partial n_k}\mathbf{R} + \mathbf{L}\frac{\partial\mathbf{R}}{\partial n_k}$$

has been used.

Making use of (7), Eq. (4) gives

(9) $$\mathbf{L}\mathcal{M}^i\frac{\partial\mathbf{R}}{\partial n_k}\chi_{ik} = \frac{1}{2}\frac{\partial^2\lambda}{\partial n_i\partial n_k}\chi_{ik} + \frac{\partial\lambda}{\partial n_i}\mathbf{L}\frac{\partial\mathbf{R}}{\partial n_k}\chi_{ik}.$$

□

PROOF of relation (2.7) of Chapter 6.

From (2.3) and (2.4) we have:

(10)
$$\frac{d\log J}{dt} = \partial_i \Lambda^i = \frac{\partial \Lambda^i}{\partial x^i} + \frac{\partial \Lambda^i}{\partial n^k} \frac{\partial n^k}{\partial x^i}.$$

We remind that

$$\Lambda^i = \lambda n^i + \mathbf{L} M^j \mathbf{R} h^i_j = \frac{\partial \lambda}{\partial n_i}.$$

It is easy to show that

(11)
$$\frac{\partial \lambda^i}{\partial n^k} n_i n^j \frac{\partial n_k}{\partial x^j} = 0$$

and therefore

$$\frac{\partial \Lambda^i}{\partial n^k} \frac{\partial n^k}{\partial x^i} = \frac{\partial \Lambda^i}{\partial n^k} \frac{\partial \tilde{n}^k}{\partial x^i}.$$

From this relation and from (10) it follows

$$\frac{d}{dt} \log J - \frac{\partial \Lambda^i}{\partial x^i} = \frac{\partial^2 \lambda}{\partial n_k \partial n_s} \chi_{ks}.$$ □

PROOF of relation (2.7) of Chapter 7.

From the definition of $P \equiv M^i n_i - V_\Sigma I$ it follows:

(1)
$$\mathbf{L}P = (\lambda - V_\Sigma)\mathbf{L}$$

and, after right multiplication by $P^{-1}/(\lambda - V_\Sigma)$

(2)
$$\frac{\mathbf{L}}{\lambda - V_\Sigma} = \mathbf{L}P^{-1}.$$

In order to prove the second relation let us differentiate the identity

$$\mathbf{L}(M^i n_i - \lambda I) = 0$$

with respect to n_i:

$$\mathbf{L}M^i = \frac{\partial \lambda}{\partial n_i}\mathbf{L} - \frac{\partial \mathbf{L}}{\partial n_i}(M^j n_j - \lambda I).$$

Multiplication by P^{-1} yields:

$$
\begin{aligned}
\mathbf{L}M^i P^{-1} &= \frac{\partial \lambda}{\partial n_i}\mathbf{L}P^{-1} - \frac{\partial \mathbf{L}}{\partial n_i}[P + (V_\Sigma - \lambda)I]P^{-1} \\[2mm]
&= \frac{\partial \lambda}{\partial n_i}\frac{\mathbf{L}}{\lambda - V_\Sigma} - \frac{\partial \mathbf{L}}{\partial n_i} + (\lambda - V_\Sigma)\frac{\partial \mathbf{L}}{\partial n_i}P^{-1} \\[2mm]
&= \frac{\partial \lambda}{\partial n_i}\frac{\mathbf{L}}{\lambda - V_\Sigma} - \frac{\partial \mathbf{L}}{\partial n_i} + (\lambda - V_\Sigma)\frac{\partial}{\partial n_i}(\mathbf{L}P^{-1}) - (\lambda - V_\Sigma)\mathbf{L}\frac{\partial P^{-1}}{\partial n_i} \\[2mm]
&= \frac{\partial(\lambda - V_\Sigma)}{\partial n_i}\frac{\mathbf{L}}{\lambda - V_\Sigma} + (\lambda - V_\Sigma)\frac{\partial}{\partial n_i}\left(\frac{\mathbf{L}}{\lambda - V_\Sigma}\right) + \frac{\partial V_\Sigma}{\partial n_i}\frac{\mathbf{L}}{\lambda - V_\Sigma} \\[2mm]
&\quad - \frac{\partial \mathbf{L}}{\partial n_i} + (\lambda - V_\Sigma)\mathbf{L}P^{-2}\frac{\partial P}{\partial n_i} \\[2mm]
&= \frac{\partial V_\Sigma}{\partial n_i}\frac{\mathbf{L}}{\lambda - V_\Sigma} + \frac{\mathbf{L}}{\lambda - V_\Sigma}\left(M^i - \frac{\partial V_\Sigma}{\partial n_i}\right) = \mathbf{L}P^{-1}M^i .
\end{aligned}
$$

□

PROOF of relation (4.15) of Chapter 7.

In the rest frame of the fluid ahead the jump conditions can be solved in the form

(1)
$$\rho = \rho_0 \frac{V_\Sigma}{V_\Sigma - u}$$

(2)
$$P = P_0 + \rho_0 u V_\Sigma$$

(3)
$$2(E - E_0) = (P + P_0)(\tau_0 - \tau)$$

where $\tau \equiv 1/\rho$ is the specific volume.

Differentiating (3) we obtain

(4)
$$(2E'_\tau + P + P_0)\hat{\tau} = (\tau_0 - \tau - 2E'_P)\hat{P}$$

where, for convenience, we suppose that the equation of state is given in the form

$$E = E(\tau, P)$$

and

$$\hat{\cdot} \equiv \frac{d\cdot}{dV_\Sigma} \equiv \left(\frac{\partial \cdot}{\partial V_\Sigma}\right)_{U_+, \hat{n}}.$$

From (1)–(2) we have:

(5)
$$\hat{\tau} = -\tau \frac{\hat{u} - u/V_\Sigma}{V_\Sigma - u} = -\tau \frac{u}{V_\Sigma} \frac{\psi - 1}{V_\Sigma - u}$$

(6)
$$\hat{P} = -\rho_0 (V_\Sigma \hat{u} + u) = \rho \frac{u}{V_\Sigma}(V_\Sigma - u)(\psi + 1)$$

where

$$\psi \equiv \frac{V_\Sigma}{u} \hat{u}.$$

From the first law of thermodynamics:

(7)
$$E'_\tau = \rho^2 c_s^2 E'_P - P.$$

Substituting in (4) and making use of (1)–(2):

$$(8) \quad (-2\rho^2 c_s^2 E_p' + \rho_0 u V_\Sigma)\tau \frac{\psi - 1}{V_\Sigma - u} = (k - 1 - 2\rho E_P')(V_\Sigma - u)(\psi + 1)$$

where $k \equiv \tau_0/\tau$ is the compression ratio.

This expression can be rewritten in the form:

$$(9) \qquad \left(k - 1 - \frac{2\rho E_P'}{M^2}\right)(\psi - 1) = (k - 1 - 2\rho E_P')(\psi + 1)$$

where $M \equiv (V_\Sigma - u)/c_s$ is the Mach number.

Solving for ψ we obtain:

$$(10) \qquad\qquad \psi = \frac{1 + M^2 - M^2(k - 1)/\rho E_P'}{1 - M^2}$$

which is equivalent to (4.15). $\qquad\qquad\qquad\qquad\qquad\qquad\qquad\qquad$ □

SUBJECT INDEX

Milton Keynes UK
Ingram Content Group UK Ltd.
UKHW040105071024
449327UK00019B/838